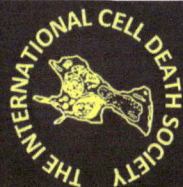

20 YEARS OF CELL DEATH

The International Cell Death Society

Edited by

Richard A Lockshin and Zahra Zakeri

Table of Contents

20 Years of Cell Death

Richard A. Lockshin and Zahra Zakeri, Editors

For the

INTERNATIONAL CELL DEATH SOCIETY

Copyright 2015 Richard A. Lockshin for ICDS

Cover Photo courtesy of Benjamin Loos

ISBN (ELECTRONIC EDITION): 978-0-9894674-4-5

ISBN (HARDCOPY EDITION): 978-0-9894674-5-2

Dedication and Acknowledgment

We would like to dedicate this book to all the individuals who have been contributing so much to the field of cell death who are no longer among us to fully see where the field have moved to and the impact that it has had on all the different fields of investigation.

We dedicate this book to all the young researchers who are getting started in the field and will no doubt move the field beyond our imagination in the future. The aim of ICDS has always been to push young investigators to display their work. The society has been a forum for many now well-known investigators to get their first exposure to the cell death community.

This book is also dedicated to all the women in the field who have been instrumental to make sure that ICDS moves forward. ICDS was the first to establish workshops for women to discuss how to excel in this field of science. These workshops are still a main part of our yearly meetings and have been including both men and women; they aim to enlighten the junior investigator on how to survive and excel in research.

We dedicate this book to all our selected honorees for their contributiona to the field. This was also a first for ICDS as we established a forum for recognition of contributors to the field.

We thank all of our colleagues in the field who have directly or indirectly contributed to the success of the International Cell Death Society. Among them are all the silent partners who worked in the back offices to make sure we succeeded such as the ICDS secretaries i.e. Ms Victoria Matssov, Fiorella Tapia Penaloza, and Lynnmarie Alafnourian. Also the administrators of Queens College who have allowed the ICDS to establish its home there and have supported it by taking care of all the financial administration of it.

Zahra Zakeri and Richard Lockshin

Table of Contents

Chapter 1: Introduction (Lockshin and Zakeri)

Richard A. Lockshin[1,2], Ph.D. Zahra Zakeri[2], Ph.D.

[1]Dept. of Biological Sciences, St. John's University, Jamaica, NY USA (Emeritus)

[2]Dept. of Biology, Queens College of CUNY, Flushing, NY USA

rlockshin@gmail.com

The year 2014 marks several anniversaries. This book was compiled with the intent of commemorating the 20th anniversary of the International Cell Death Society, but it also marks the 50th anniversary of the appearance of the term "programmed cell death" in the visible [1] scientific literature. Although there were many valuable experiments and observations that preceded that point, especially some of the experiments by John Saunders (Saunders and Gasseling, 1962; Saunders, 1966; Fallon and Saunders, 1968), in a sense it marks the beginning of a new era. However, interest built only slowly. 1972 marked the culmination of a thoughtful approach initiated by the pathologist John Kerr, with the publication of the recognition by Kerr, Wyllie, and Currie that the, at that time inexplicable, mode of cell death was quite common if not universal. Kerr had previously pointed out that cell shrinkage and blebbing, combined with condensation and margination of chromatin, was very common but not not interpretable by any obvious osmotic mechanism. Intending to highlight that generality, they christened the mode "apoptosis"(Kerr, 2002; Kerr et al., 1972). Even so, interest continued to grow slowly. It was not until the 1990's that the manner in which cells die

---

[1]"Programmed cell death" was the title of a doctoral thesis accepted in 1963; publications from that thesis appeared as journal articles in 1964 and 1965.

began to attract attention. What happened was a series of discoveries: The genetics of a major path to cell death was worked out for *Caenorhabditis*, leading to the recognition that the primary effector of cell death was a protease that was evolutionarily conserved, with an apparently similar role in humans(Horvitz, 2003); recognition that at least two types of cancer could be attributed to mutations of genes that affected cell death, the genes ultimately proving to be members of larger families now known as the bcl-2 and tumor necrosis factor families (see Vaux, 2002), with the gene most commonly mutated in cancer (p53) also being shown to affect the ability of cells to undergo apoptosis (Yonish-Rouach et al., 1993); and, finally, the description of a simple and cheap technique by which the existence of apoptosis could be documented in many types of cells.

The impact of these discoveries was immediately obvious. Within a few years of each other, conference series on cell death were founded at the Gordon Conferences, the Keystone Symposia, Cold Spring Harbor Laboratories Conferences, the International Cell Death Society, and the European Cell Death Organization. From that point the field grew exponentially, to the extent that by 2013, 75 papers on various aspects of cell death were appearing every day.

In 1994 Raymond Birge, Michael Hengartner, Richard Lockshin, and Zahra Zakeri founded The International Cell Death Society (ICDS), nicknamed also "The Death Poet's Society" as small group in New York at Rockefeller University. The society promulgates a better understanding of the mechanisms of cell death, establishing communication among the various branches of the research and communicating and coordinating the application of research findings. Soon after we got started, others learned of our activities, come to our meetings, and asked if we had considered the possibility of taking the meetings to other parts of the world. The society first had biannual meetings, which were later changed, due to demand, to annual meetings. So far the society has directed 20 meetings.

In conjunction with the meetings, the ICDS has established specific workshops on the topic of cell death as well as advice for young scientists, women, and scientists from the third world. We have also honored distinguished contributions to the field. as subsequently has also been done by other organizations. We celebrated our 20 years of establishment at a meeting in South Africa in 2014.

Like all topics in the sciences, excitement grew to certainty; troubling observations, some previously well-known but ignored, began to gnaw at that certainty; alternative hypotheses were erected, vigorously contested, and modified; a new level of complexity was admitted; and that complexity has led us to recognize that there is another layer to the onion, and that we need to know much more about the prodromal as opposed to what are essentially the final phases of cell death. We can summarize what has happened since 1994 as a series of steps, each of which is more specifically addressed by the authors of the several chapters, many of whom develop their subject in a very personal style, with anecdotal descriptions of the birth of the field. Altogether they provide not only an accurate and highly readable statement of the state of the art, but also a personalized sense of history as it is lived.

Thanks to the contributors, this overview of the last 20 years addresses most of the issues that have appeared and are now under consideration. In Part 1, "Components and Pathways of Apoptosis," Patrizia Agostinis describes how we came to understand how endoplasmic reticulum stress and the unfolded protein response influence the fate of cells. Bodvael Pennarun, Octavian Bucur, and Roya Khosravi-Far describe an important negative regulator for programmed cell death, and go on to point out how this knowledge is being exploited for cancer therapy. Mauro Piacentini continues, telling the story of how transglutaminase came to be recognized as a marker for apoptosis, how it works, and how it affects the fate of apoptotic cells. Finally in this section, Boris Zhivotovsky addresses the issue of the other functions of caspases, including those that function in the differentiation of lymphoid tissues and functions of the apoptotic caspases that have nothing to do with the death of the cell. This theme is expanded by J. Marie Hardwick, who has used highly innovative approaches to question the alternative roles of many proteins that are presumed to function primarily for apoptosis. In Part 2, "Processes of Apoptosis," we turn to mechanisms by which cells are killed. Zakeri and Lockshin, reviewing the modern history of the field, emphasize how understanding has gone from a rather narrow certainty to a broader recognition of the importance of many overlapping and sometimes competing processes, such as autophagy and necroptosis. Birge describes a currently very active field, the role of the dying and fragmenting cell in invoking (or not invoking) an immune response. How the immune system responds to a dying cell can determine the difference between autoimmune disease and health, or between health and cancer. Domagoj Vucic tells the story of the discovery of inhibitor of apoptosis proteins and the progress of the exploration to its current status as a target for therapy. Doug Green looks at the centrality of mitochondrial metabolism in determining cell fate, and asks why very similar cells differ in terms of timing or extent of their response to perturbation.

In Part 3, "Autophagy and Necroptosis," we explore alternatives to apoptosis. De Zio and Cecconi lead this section by asking what the genes Apaf1 and Ambra do; they find that both genes are very important for both apoptosis and autophagy, and that their activities determine the extent to which either process proceeds. Gozuacik and Kig follow with a review of how autophagy came to be closely analyzed, and how we now understand its role in cell fate and its interaction with apoptosis. Next, Ben Loos reflects, in a manner similar to Green, on how the energy is handled and how availability of energy affects apoptosis, autophagy, and cell fate. He emphasizes the newest, extremely high resolution microscopy tools. Junying Yuan recounts the discovery of necroptosis and evaluates its meaning and importance for today's research.

In Part 4, "Viruses and Cancer," Zakeri et al describe the battles waged between cells and viruses for control of their destinies. Viruses often protect cells against other stresses, most commonly by activating autophagy, thereby preserving the cells to ensure the reproduction of the virus. Ultimately however the virus overwhelms the cell, allowing an unregulated autophagy that eventually kills the cell. Next, Simone Fulda looks at the issue of pediatric cancers and considers how understanding of apoptosis has determined modern approaches and indicated which targets are most appropriate to pursue. Finally, van der Walt and Cronjé take a different perspective, discussing what is known about medicinal plants, which ones induce apoptosis, and suggesting some that might prove to be important for the future.

We are pleased to offer this work. We feel that it, particularly thanks to the thoughtful contributions of our colleagues, very well serves the purposes of the society. It provides a detailed, interesting, and reflective look at the past; an up-to-date view of our understanding of our chosen field of research; and thoughtful estimations of the near future and potential of this field. In all these aspects it addresses the goals of the society. We hope that you agree.

References

Fallon, J.F., and Saunders, J.W., Jr. (1968). In vitro analysis of the control of cell death in a zone of prospective necrosis from the chick wing bud. Developmental biology *18*, 553-570.

Horvitz, H.R. (2003). Nobel lecture. Worms, life and death. Bioscience reports *23*, 239-303.

Kerr, J.F. (2002). History of the events leading to the formulation of the apoptosis concept. Toxicology *181-182*, 471-474.

Kerr, J.F., Wyllie, A.H., and Currie, A.R. (1972). Apoptosis: a basic biological phenomenon with wide-ranging implications in tissue kinetics. British journal of cancer *26*, 239-257.

Saunders, J.W., Jr. (1966). Death in embryonic systems. Science (New York, NY) *154*, 604-612.

Saunders, J.W., Jr., and Gasseling, M.T. (1962). Cellular death in morphogenesis of the avian wing. Developmental biology *5*, 147-178.

Vaux, D.L. (2002). Apoptosis timeline. Cell death and differentiation *9*, 349-354.

Yonish-Rouach, E., Grunwald, D., Wilder, S., Kimchi, A., May, E., Lawrence, J.J., May, P., and Oren, M. (1993). p53-mediated cell death: relationship to cell cycle control. Molecular and cellular biology *13*, 1415-1423.

History of the Society in Slides

20 YEARS OF CELL DEATH
Richard A Lockshin and Zahra
Zakeri

History of ICDS

MEETINGS

- Started as a club in 1995
- 1996: sponsored a successful one-day symposium
- 1998: Second New York meeting

INTERNATIONAL MEETINGS

- 2000: El Escorial, Spain
- 2002: Noosa, Australia
- 2004: Maynooth, Ireland
- 2005: Tehran, Iran

History of ICDS continued

- 2006: Angra dos Reis, Brazil
- 2007: (a) Nice, France; (b) Tehran, Iran; (c) Special Symposium, New York, USA
- 2008: Shanghai, China

- 2009: (a) near Johannesburg, South Africa; (b) Tehran, Iran
- 2010: Side, Turkey
- 2011: (a) Cape Town, South Africa; (b) Cheiro de Mato, Brazil

Satellite meeting   Special meeting

History of ICDS continued

- 2012: Singapore
- 2013: Fuengirola, Spain
- 2014: Stellenbosch, South Africa

PUBLICATIONS
- Annal of the New York Academy of Sciences (release date, fall 1999)

PUBLICATIONS
- Annal of the New York Academy of Sciences (released December 2000).
- When Cells Die II (edited by Lockshin and Zakeri, Wiley Press, 2003)

Birth of the Cell Death Society
The Death Poet's Society

Established 1994 by

Richard Lockshin

Zahra Zakeri

Ray Birge

Michael Hengartner

Birth of the Cell Death Society
First Meeting,
1994 Rockefeller University New York

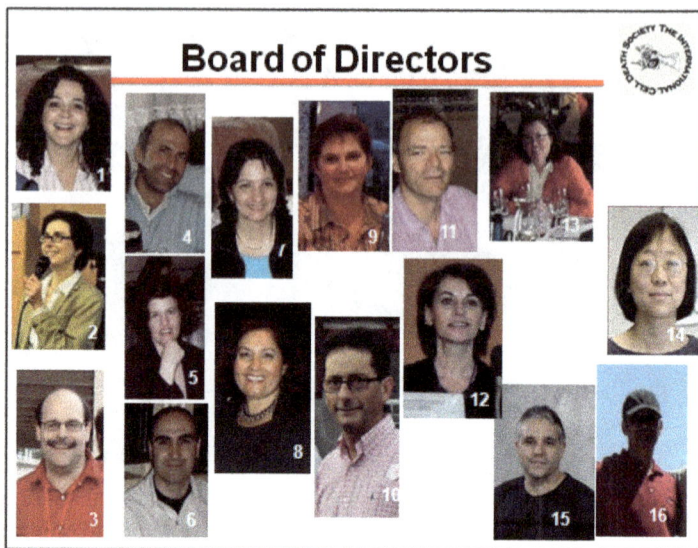

Board of Directors

1. Soraya Smaili
2. Simone Fulda
3. Raymond Birge
4. Francesco Cecconi
5. Katharina D'Herde
6. Devrim Gozuacik
7. Roya Khosravi-Far
8. Zahra Zakeri
9. Marianne Cronje
10. Tom Cotter
11. Christoph Borner
12. Patrizia Agostinis
13. Eileen White
14. Junying Yuan
15. Eli Arama
16. Nader Maghsoudi

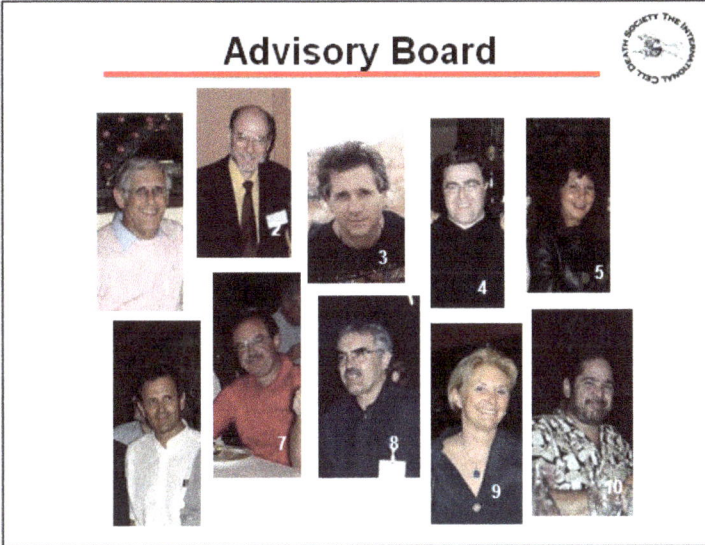

Advisory Board Members

1. Richard Lockshin
2. Robert Horvitz
3. Mauro Piacentini
4. Seamus Martin
5. Adi Kimchi
6. Carlos Martinez-A
7. Boris Zhivotovsky
8. Laszlo Fesus
9. Marie-Lise Gougeon
10. Douglas Green

Honorees

2002 John Kerr   2004 Peter Krammer   2004 Shigekazu Nagata 2006 Junying Yuan 2006 Richard Lockshin

2008 Zahra Zakeri   2008 H R Horvitz   2009 Marie-Lise Gougeon   2009 Douglas Green

Honorees

2010 Eileen White   2011 Guido Kroemer   2012 Adi Kimchi   2013 Mauro Piacentini

2013 Sten Orrenius   2014 Vishva Dixit   2015 Scott Lowe

Intro-(Lockshin and Zakeri)

Cell Death Society Meeting, 1996 and 1998 Queens College New York

Cell Death Meeting El Escorial, Spain 2000

Cell Death Meeting
Noosa, Australia
2002

Cell Death Meeting
Maynooth, Ireland
2004

Cell Death Meeting Angra dos Reis, Brazil 2006

Cell Death Meeting Nice, France 2007

Cell Death
Meeting
Shanghai, China
2008

Cell Death Meeting
Jo'Burg, 2009;
Cape Town, 2012
South Africa

Cell Death
Meeting
Tehran, Iran
2007, 2009, 2011,
2014

Cell Death
Meeting
Side, Turkey
2010

Cell Death Meeting Cheiro de Mato Brazil 2011

Cell Death Meeting Singapore 2012

Cell Death Meeting
Fuengirola, Spain
2013

Cell Death Meeting
Stellenbosch,
South Africa
2014

20 YEARS OF ICDS!!!!

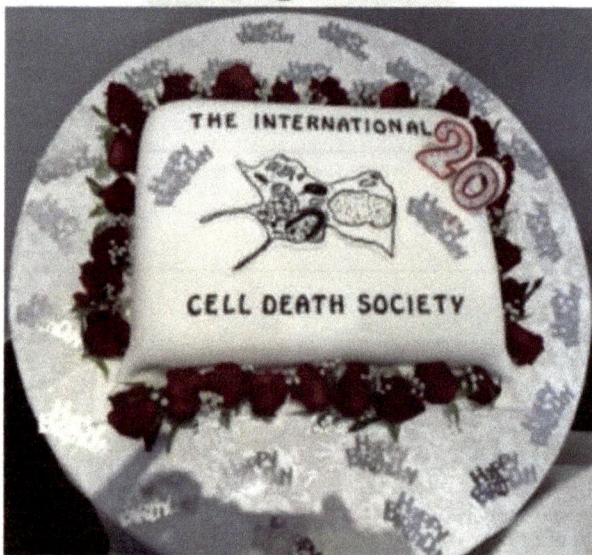

PART 1. COMPONENTS AND PATHWAYS OF *APOPTOSIS*

Chapter 2: Endoplasmic Reticulum stress in Cancer and Cell Death: a time lapse (Agostinis)

Patrizia Agostinis

Cell Death Research and Therapy Unit, Dept. Cellular and Molecular Medicine, Campus Gasthuisberg, O&N1, Herestraat 49, Box 802, 3000 Leuven, Belgium

Abstract

The endoplasmic Reticulum (ER) is a central organelle for a number of vital cell-intrinsic as well as cell-extrinsic processes, principally regulated by the ability of the ER to maintain cellular protestasis and secretion. Research over the last couple of decades has largely contributed to our understanding of how loss of proteostasis, leading to a condition known as ER stress, and the signal transduction pathway that is engaged to re-establish ER homeostasis, known as the unfolded protein response or UPR, is coupled to the induction of cell death pathways. We also learned how defects exacerbating or suppressing the diverse functions of the UPR underlie severe pathologies, like cancer. On the other hand, increased knowledge on the effectors and ER stress responses they modulate will also offer us novel therapeutic opportunities to exploit the beneficial facet of ER stress in fighting cancer.

From the ER to the unfolded protein response: a bird's eye view

At the beginning of the 20th century, with the advent of microscopy and optimized staining methods, scientists started to identify vital organelles within the cell. In spite of being one of the largest structures in a cell, the ER was the last organelle to be recognized in 1902, by Emilio Verratti, a student of Camillo Golgi (Schuldiner and Schwappach, 2013). However, the existence of the ER as bona fide organelle was completely revealed only later in the mid 50, when thin-sectioning electron microscopy techniques became available along with optimization of centrifugation techniques allowing the fractionation of subcellular components. In the period 1953-56, Keith Porter and George Palade provided the first high-resolution images of the ER and a new era in ER biology research was launched (for recent reviews on the history of ER and ER stress research see Garg et al, 2014b; Schuldiner and Schwappach, 2013). Subsequent research unraveled the major functional roles of the ER and/or sarcoplasmic reticulum, in Ca^{2+} signaling during muscle contraction and lipid biosynthesis, and placed the ER at the core of vital signaling and cellular functions. Later on, in the early 1970s, seminal studies from George Palade, Günter Blobel and David Sabatini identified the crucial role of ER in governing the first step of the secretory pathway, by deciphering how newly synthesized proteins enter the ER as unfolded polypeptides and traffic through the Golgi on their way to the plasma membrane (Blobel and Sabatini, 1971; Blobel et al 1979). These crucial findings shed light on the crucial role of the ER in secretion and trafficking. Soon after, it also became clear that these crucial processes are subjected to stringent ER quality control mechanisms enabling only correctly folded and post-translationally modified proteins to leave the ER and flow through the Golgi in order to reach their final destination (Schuldiner and Schwappach, 2013; Garg et al, 2014b). Considering that approximately one-third of the polypeptides synthesized by a cell enter the ER, and upon folding are trafficked across the cell partly through the secretory pathway, this imposes on the ER a demanding task.

In the period between mid-70s to mid-80s, the main mechanisms regulating oxidative folding, disulfide bridge formation and glycosylation as signals of a protein's folding state in the ER were revealed. These studies led to the identification of several crucial ER molecular chaperones like calreticulin (CRT), calnexin and the glucose-sensitive GRP78 (glucose regulated protein 78, also known as immunoglobulin binding protein or BiP), and their role in assisting key folding processes and preventing aberrant interactions and aggregation of protein-folding intermediates (Schuldiner and Schwappach, 2013; Garg et al, 2014b). In 1987, Munro and Pelham (1987) posited the concept of ER protein retrieval (i.e. avoidance of ER escape of ER lumen proteins), by demonstrating that ER luminal proteins contain a C-terminal KDEL sequence that is required for their retention in the ER. With the discoveries elucidating key roles of the ER in protein folding, Ca^{2+} handling and secretion and the molecular mechanisms governing these processes, scientists also realized that under conditions causing disturbances of ER quality control, adaptive mechanisms are instigated in order to re-establish ER proteostasis. Around the end of the 80's- beginning of the 90's two exciting discoveries moved the ER field forward. The first was the discovery of a dedicated machinery for the recognition and retrotranslocation of misfolded proteins from the ER to the cytosol for degradation, the ER associated degradation (ERAD) (Lippincott-Schwartz et al, 1988; Vembar and Brodsky, 2008). At approximately the same time, the expression of GRP78/BiP, a chaperone encoded by the *KAR2* gene in yeast was found to be transcriptionally induced by the accumulation of unfolded proteins in the ER. This signaling pathway was then baptized 'the unfolded protein response' or UPR (Mori, et al 1992). Following this finding, using a genetic screen for mutations that would block the activation of an UPR-inducible reporter in yeast, the groups of Walter (Cox et al, 1993) and Sambrook (Mori et al, 1993) independently identified a gene encoding an ER transmembrane protein *i.e.* the Inositol Requiring-1 (IRE1), which harbored a Ser/Thr kinase activity. These studies identified IRE1 as an evolutionary conserved proximal ER stress sensor in the UPR pathway. The IRE1 site-specific endoribonuclease (RNase) activity was further recognized to be vital for the transcriptional induction of pro-survival UPR genes, a process conserved from yeast to mammals. Soon after the discovery of IRE1, the two other UPR effectors *i.e.* the Ser/Thr kinase PERK (EIF2AK3) and the transcription factor ATF6, were identified in higher eukaryotes (Shi et al, 1998; Harding et al, 1999, 2000; Haze et al, 1999).

These discoveries incited a massive interest in the role of the UPR in cellular homeostasis and contributed to delineate how these UPR effectors sense and decode disturbances in the protein folding status of the ER, and transmit *via* the activation of three key transcription factors (i.e. XBP1, ATF4 and ATF6), a signal to the nucleus to 'switch on' the complex UPR gene expression program that we know now (readers are referred to recent extensive reviews in this subject: Xu et al, 2005; Woehlbier and Hetz, 2011; Hetz et al, 2013; Clarke et al, 2014).

ER stress and the regulation of cell death; when, where, how and why?

The cell death field in the 1990s, when the UPR came of age, was dominated by a mitochondrion-centric view, and although ER stress and the UPR were recognized to be essential for recovery of cellular homeostasis, they were not considered to actively take part of cell death decisions. On the other hand, the last two decades have witnessed an avalanche of studies using different model systems, which demonstrated how the intrinsic pro-survival and adaptive role of the UPR, which is primarily engaged to rescue proteostasis, redox balance and ER secretory capacity, can be turned into a cell death mechanism, which prevalently –albeit not uniquely- occurs through mitochondrial apoptosis (Xu et al, 2005).

Although our understanding of the molecular mechanisms linking the UPR to apoptosis remains currently still partial, on the bases of recent studies and identification of crucial molecular mediators, some models can be postulated. ER stress mediated cell death is governed by the spatiotemporal coordination of the three different signaling branches of the UPR. A sustained PERK signaling has been found to be crucial to mount an apoptotic threshold level of CHOP expression, leading to transcriptional up-regulation or down regulation of certain Bcl-2 family members, thus directly linking the UPR to mitochondrial apoptosis (Xu et al, 2005). Although this is considered a major pro-apoptotic mechanism under conditions of persistent ER stress, CHOP-mediated exacerbation of protein translation and protein oxidation in the ER, thus shifting the UPR into a lethal pathway is an additional documented mechanism (Marciniak et al, 2004). Also the regulation of the amplitude and duration of IRE1 signaling, along with the scaffolding function of IRE1 allowing the recruitment of a pro-apoptotic signaling platform through the recently defined UPRosome (Woehlbier and Hetz, 2011), appears to be a determining factor.

Accumulating evidence also indicates that when adaptive mechanisms, including the induction of autophagy as a downstream cytoprotective response inherently associated with ER stress (Bernales et al, 2006; Ogata et al, 2006), have failed to rescue ER homeostasis and ER stress persists, both caspase-dependent and caspase-dispensable mechanisms of cell death are put in place to ensure cell death (reviewed in Xu C et al, 2005). However, the precise mechanisms linking ER stress, or the UPRosome to other forms of non-apoptotic cell death, such as necroptosis, are still unknown and will require further studies in the near future.

Last but not least, in the last decade several elegant reports have documented the relevance of the ER-mitochondria interface and the relevance of this inter-organellar communication for cell death initiation and propagation (Grimm, 2012; Naon and Scorrano, 2014). It became clear that these key organelles are dynamically maintained in contact through proteinaceous ER subdomains juxtaposed to mitochondria, called mitochondria-associated membranes (MAMs). Through these subdomains, which were first isolated as specific structural entities by Jean Vance in 1990, the ER and mitochondria exchange lipids and second messengers like Ca^{2+} or reactive oxygen species (ROS) and by so doing shape and regulate the (mutual) spatiotemporal propagation of cell death signals during ER stress (reviewed in Van Vliet et al, 2014). The recent finding that PERK, one of the major ER stress sensors, is located at the ER-mitochondria contact site along with other known ER resident proteins, and has scaffolding functions allowing the fast transfer of cell death mediators from the stressed ER to the mitochondria (Verfaillie et al, 2012), reveals new facets of ER functions in cell death which will be the focus of future studies (Figure 2.1). For instance, it will be crucial in future studies to reveal which PERK interacting partners are located at the MAMs and the impact this interactome has on vital cellular functions that are more and more recognized to be modulated at the ER-mitochondria contact site, such as energy metabolism, autophagy, inflammation (Garg AD, et al 2012a; van Vliet et al, 2014). The concept that ER stress sensors may have evolved to control other homeostatic functions that reach beyond the UPR-mediated transcriptional machinery is perhaps not surprising given the unique ability of these transmembrane proteins to sense stress signals in the ER lumen and decode them into cytoplasmic effects through their cytoplasmic domains. With the development of novel and more sophisticated proteomics approaches the molecular identity and full repertoire of the PERK, IRE1 and ATF6 interacting partners will be soon revealed.

Finally, the continuous effort by many laboratories during the last two decades has delineated not only how deregulation of UPR signaling and/or chronic ER stress underlies a variety of pathologies, including cancer, but also how this pathway can be explored therapeutically. Considering cancer as a disease hallmarked by loss of proteostasis, a number of important studies coupled evidence of altered ER morphology accompanied by biochemical signature of chronically activated ER stress signaling in solid tumors (Garg et al, 2014b). Given that signals that alter ER functions and disturb the folding environment, such as glucose or oxygen deprivation and oxidative stress, are predominantly found in the tumor microenvironment, it is perhaps not surprising that the UPR is found activated in a variety of tumors. In line with this, the heightened metabolic and proliferation rates of cancer cells along with an increased demand to secrete factors assisting pro-tumorigenic processes like angiogenesis and invasion, impose an increased folding and secretory load on the ER that results in a constitutive activation of the UPR (Dejeans et al, 2014). Thus at the beginning of the new millennium the UPR was theorized to serve both a pro-tumorigenic role, through its cytoprotective function, or an anti-tumorigenic function, mainly by inducing dormancy or increasing cancer cell's vulnerability to cell death

(Ma Y, Hendersho LM, 2004). Recent advances in the ER stress field have underscored that constitutive activation of cytoprotective UPR, indeed facilitates angiogenesis, invasion and dissemination and assist cancer cells survival by increasing resistance to oxidative stress, especially in hypoxic tumors (Bi et al, 2005; Blais et al, 2006; Drogat et al, 2007; Auf et al, 2010; Bobrovnikova-Marjon et al, 2010; Feng et al, 2014). These important findings proposed that targeting the UPR effectors and forcing proteotoxicity could be a novel strategy with potential clinical implications. In line with this Bortezomib, a selective proteasome inhibitor, has shown significant success in the treatment of patients with advanced, refractory myeloma (Crawford et al, 2011; Garg et al, 2014b). The recent exciting development of novel and specific small molecule inhibitors of IRE1 and PERK (Papandreou I et al, 2011; Axten et al, 2012; Mimura et al, 2012; Atkins et al, 2013) and the promising results obtained might help in revealing new facets of proximal UPR function or UPR-independent functions of these ER stress sensors in cancer and cell death, and facilitate their clinical application in cancer therapy and other diseases (Garg et al, 2014b). However, the challenge in the future will be we to reach a more complete understanding of how blockage of these ER stress effectors, which by causing suppression of the cell-autonomous cytoprotective functions of the UPR should tip the balance towards cancer cell death, impact cancer cell-tumor stroma interactions.

Indeed another exciting direction in the field is the recognition of the crucial extracellular role of ER-resident proteins such as CRT, GRP78, GRP94 and PDIs, usually functioning as pro-survival and pro-tumorigenic signals in proliferating cancer cells (Garg et al, 2014b; Xu et al, 2014). In the context of cell death though, a remarkable exception is the mobilization of CRT at the plasma membrane (ecto-CRT) during immunogenic cell death (Zitvogel L et al, 2010). Gardai and Henson (see Gardai et al, 2005) provided the first evidence indicating that ecto-CRT in response to apoptotic stimuli acts as an 'eat-me signal' inciting the removal of the dying cells by phagocytes. Following this discovery a series of seminal studies from Guido Kroemer's laboratory and others, including ours, further underscored the ability of certain anticancer agents to induce danger signaling causing the pre-apoptotic mobilization of ecto-CRT on the surface of cancer cells (Obeid et al, 2007; Panaretakis et al, 2009; Garg et al, 2012b; reviewed in Krysko et al, 2012). This danger pathway was shown to be reliant on the induction ER stress and ROS and the presence of a functional PERK-modulated secretory pathway or UPR module (Figure 2.1). At the surface of dying cancer cells exposure this specific ER-resident protein acts as a danger-associated molecular pattern (DAMP) and has been recognized to have a vital role in the cascade of events driving anti-tumor immunity, an essential component of successful anticancer therapy (Apetoh et al, 2008; Zitvogel et al, 2010; Krysko et al, 2012).

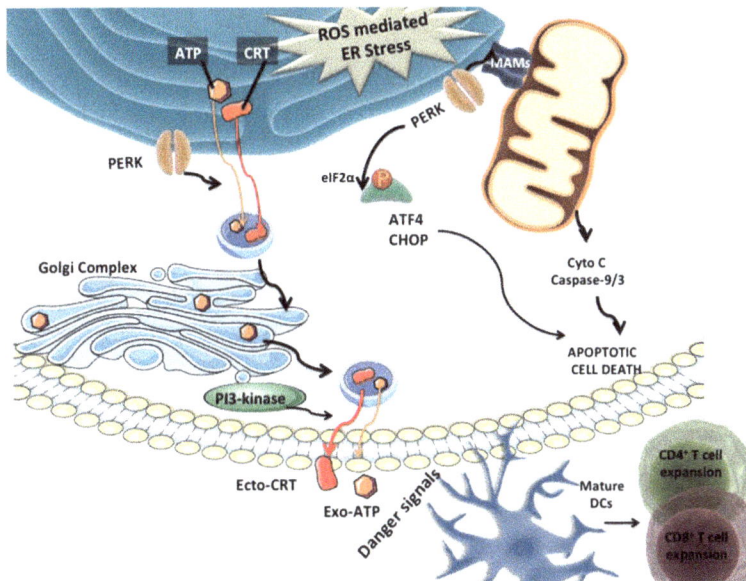

Fig. 2.1 ER stress and the danger pathway.

In conclusion, these novel findings underscore that the regulated and targeted induction of lethal ER stress could be a promising therapeutic strategy to target simultaneously multiple cancer cell autonomous and non-autonomous functions; future studies will tell us whether this prediction holds true.

References

Apetoh, L., Mignot, G., Panaretakis, T., Kroemer, G., and Zitvogel, L. (2008). Immunogenicity of anthracyclines: moving towards more personalized medicine. Trends Mol Med *14*, 141-151.

Atkins, C., Liu, Q., Minthorn, E., Zhang, S.Y., Figueroa, D.J., Moss, K., Stanley, T.B., Sanders, B., Goetz, A., Gaul, N., *et al.* (2013). Characterization of a novel PERK kinase inhibitor with antitumor and antiangiogenic activity. Cancer Res *73*, 1993-2002.

Auf, G., Jabouille, A., Guerit, S., Pineau, R., Delugin, M., Bouchecareilh, M., Magnin, N., Favereaux, A., Maitre, M., Gaiser, T., *et al.* (2010). Inositol-requiring enzyme 1alpha is a key regulator of angiogenesis and invasion in malignant glioma. Proc Natl Acad Sci U S A *107*, 15553-15558.

Axten, J.M., Medina, J.R., Feng, Y., Shu, A., Romeril, S.P., Grant, S.W., Li, W.H., Heerding, D.A., Minthorn, E., Mencken, T., *et al.* (2012). Discovery of 7-methyl-5-(1-{[3-(trifluoromethyl)phenyl]acetyl}-2,3-dihydro-1H-indol-5-yl)-7H-p yrrolo[2,3-d]pyrimidin-4-amine (GSK2606414), a potent and selective first-in-class inhibitor of protein kinase R (PKR)-like endoplasmic reticulum kinase (PERK). Journal of medicinal chemistry *55*, 7193-7207.

Banerjee, A., Lang, J.Y., Hung, M.C., Sengupta, K., Banerjee, S.K., Baksi, K., and Banerjee, D.K. (2011). Unfolded protein response is required in nu/nu mice microvasculature for treating breast tumor with tunicamycin. J Biol Chem *286*, 29127-29138.

Bernales, S., McDonald, K.L., and Walter, P. (2006). Autophagy counterbalances endoplasmic reticulum expansion during the unfolded protein response. PLoS Biol *4*, e423.

Bi, M., Naczki, C., Koritzinsky, M., Fels, D., Blais, J., Hu, N., Harding, H., Novoa, I., Varia, M., Raleigh, J., *et al.* (2005). ER stress-regulated translation increases tolerance to extreme hypoxia and promotes tumor growth. EMBO J *24*, 3470-3481.

Blais, J.D., Addison, C.L., Edge, R., Falls, T., Zhao, H., Wary, K., Koumenis, C., Harding, H.P., Ron, D., Holcik, M., *et al.* (2006). Perk-dependent translational regulation promotes tumor cell adaptation and angiogenesis in response to hypoxic stress. Mol Cell Biol *26*, 9517-9532.

Blobel, G., and Sabatini, D.D. (1971). Ribosome/membrane interaction in eukaryotic cells. Biomembranes *2*, 193-195.

Blobel, G., Walter, P., Chang, C.N., Goldman, B.M., Erickson, A.H., and Lingappa, V.R. (1979). Translocation of proteins across membranes: the signal hypothesis and beyond. Symp Soc Exp Biol *33*, 9-36.

Bobrovnikova-Marjon, E., Grigoriadou, C., Pytel, D., Zhang, F., Ye, J., Koumenis, C., Cavener, D., and Diehl, J.A. (2010). PERK promotes cancer cell proliferation and tumor growth by limiting oxidative DNA damage. Oncogene *29*, 3881-3895.

Clarke, H.J., Chambers, J.E., Liniker, E., and Marciniak, S.J. (2014). Endoplasmic reticulum stress in malignancy. Cancer cell *25*, 563-573.

Cox, J.S., Shamu, C.E., and Walter, P. (1993). Transcriptional induction of genes encoding endoplasmic reticulum resident proteins requires a transmembrane protein kinase. Cell *73*, 1197-1206.

Crawford, L.J., Walker, B., and Irvine, A.E. (2011). Proteasome inhibitors in cancer therapy. J Cell Commun Signal *5*, 101-110.

Dejeans, N., Manie, S., Hetz, C., Bard, F., Hupp, T., Agostinis, P., Samali, A., and Chevet, E. (2014). Addicted to secrete - novel concepts and targets in cancer therapy. Trends Mol Med *20*, 242-250.

Ding, W.X., Ni, H.M., Gao, W., Hou, Y.F., Melan, M.A., Chen, X., Stolz, D.B., Shao, Z.M., and Yin, X.M. (2007). Differential effects of endoplasmic reticulum stress-induced autophagy on cell survival. J Biol Chem *282*, 4702-4710.

Drogat, B., Auguste, P., Nguyen, D.T., Bouchecareilh, M., Pineau, R., Nalbantoglu, J., Kaufman, R.J., Chevet, E., Bikfalvi, A., and Moenner, M. (2007). IRE1 signaling is essential for ischemia-induced vascular endothelial growth factor-A expression and contributes to angiogenesis and tumor growth in vivo. Cancer Res *67*, 6700-6707.

Feng, Y.X., Sokol, E.S., Del Vecchio, C.A., Sanduja, S., Claessen, J.H., Proia, T.A., Jin, D.X., Reinhardt, F., Ploegh, H.L., Wang, Q., *et al.* (2014). Epithelial-to-mesenchymal transition activates PERK-eIF2alpha and sensitizes cells to endoplasmic reticulum stress. Cancer discovery *4*, 702-715.

Gardai, S.J., McPhillips, K.A., Frasch, S.C., Janssen, W.J., Starefeldt, A., Murphy-Ullrich, J.E., Bratton, D.L., Oldenborg, P.A., Michalak, M., and Henson, P.M. (2005). Cell-surface calreticulin initiates clearance of viable or apoptotic cells through trans-activation of LRP on the phagocyte. Cell *123*, 321-334.

Garg, A.D., Kaczmarek, A., Krysko, O., Vandenabeele, P., Krysko, D.V., and Agostinis, P. (2012a). ER stress-induced inflammation: does it aid or impede disease progression? Trends Mol Med*18*, 589-598.

Garg, A.D., Krysko, D.V., Verfaillie, T., Kaczmarek, A., Ferreira, G.B., Marysael, T., Rubio, N., Firczuk, M., Mathieu, C., Roebroek, A.J., *et al.* (2012b). A novel pathway combining calreticulin exposure and ATP secretion in immunogenic cancer cell death. EMBO J *31*, 1062-1079.

Garg, A.D., Martin, S., Golab, J., and Agostinis, P. (2014a). Danger signalling during cancer cell death: origins, plasticity and regulation. Cell Death Differ *21*, 26-38.

Garg, A.D., van Vliet, A., Maes, H., and Agostinis, P. (2014b). Targeting the Hallmarks of Cancer with Therapy-induced Endoplasmic Reticulum (ER) stress. Molecular and Cellular Oncology, *in press*

Grimm, S. (2012). The ER-mitochondria interface: the social network of cell death. Biochim Biophys Acta *1823*, 327-334.

Harding, H.P., Zhang, Y., Bertolotti, A., Zeng, H., and Ron, D. (2000). Perk is essential for translational regulation and cell survival during the unfolded protein response. Mol Cell *5*, 897-904.

Harding, H.P., Zhang, Y., and Ron, D. (1999). Protein translation and folding are coupled by an endoplasmic-reticulum-resident kinase. Nature *397*, 271-274.

Haze, K., Yoshida, H., Yanagi, H., Yura, T., and Mori, K. (1999). Mammalian transcription factor ATF6 is synthesized as a transmembrane protein and activated by proteolysis in response to endoplasmic reticulum stress. Mol Biol Cell *10*, 3787-3799.

Hetz, C., Chevet, E., and Harding, H.P. (2013). Targeting the unfolded protein response in disease. Nat Rev Drug Discov *12*, 703-719.

Krysko, D.V., Garg, A.D., Kaczmarek, A., Krysko, O., Agostinis, P., and Vandenabeele, P. (2012). Immunogenic cell death and DAMPs in cancer therapy. Nat Rev Cancer *12*, 860-875.

Lippincott-Schwartz, J., Bonifacino, J.S., Yuan, L.C., and Klausner, R.D. (1988). Degradation from the endoplasmic reticulum: disposing of newly synthesized proteins. Cell *54*, 209-220.

Ma, Y., Hendershot ,L.M. (2004) The role of the unfolded protein response in tumour development: friend or foe? Nat Rev Cancer *4*, 966-977.

Marciniak, S.J., Yun, C.Y., Oyadomari, S., Novoa, I., Zhang, Y., Jungreis, R., Nagata, K., Harding, H.P., and Ron, D. (2004). CHOP induces death by promoting protein synthesis and oxidation in the stressed endoplasmic reticulum. Genes Dev *18*, 3066-3077.

Mimura, N., Fulciniti, M., Gorgun, G., Tai, Y.T., Cirstea, D., Santo, L., Hu, Y., Fabre, C., Minami, J., Ohguchi, H., *et al.* (2012). Blockade of XBP1 splicing by inhibition of IRE1alpha is a promising therapeutic option in multiple myeloma. Blood *119*, 5772-5781.

Mori, K., Ma, W., Gething, M.J., and Sambrook, J. (1993). A transmembrane protein with a cdc2+/CDC28-related kinase activity is required for signaling from the ER to the nucleus. Cell *74*, 743-756.

Mori, K., Sant, A., Kohno, K., Normington, K., Gething, M.J., and Sambrook, J.F. (1992). A 22 bp cis-acting element is necessary and sufficient for the induction of the yeast KAR2 (BiP) gene by unfolded proteins. EMBO J *11*, 2583-2593.

Munro, S., and Pelham, H.R. (1987). A C-terminal signal prevents secretion of luminal ER proteins. Cell *48*, 899-907.

Naon, D., and Scorrano, L. (2014). At the right distance: ER-mitochondria juxtaposition in cell life and death. Biochim Biophys Acta *1843*, 2184-2194.

Obeid, M., Tesniere, A., Ghiringhelli, F., Fimia, G.M., Apetoh, L., Perfettini, J.L., Castedo, M., Mignot, G., Panaretakis, T., Casares, N., *et al.* (2007). Calreticulin exposure dictates the immunogenicity of cancer cell death. Nat Med *13*, 54-61.

Ogata, M., Hino, S., Saito, A., Morikawa, K., Kondo, S., Kanemoto, S., Murakami, T., Taniguchi, M., Tanii, I., Yoshinaga, K., *et al.* (2006). Autophagy is activated for cell survival after endoplasmic reticulum stress. Mol Cell Biol *26*, 9220-9231.

Panaretakis, T., Kepp, O., Brockmeier, U., Tesniere, A., Bjorklund, A.C., Chapman, D.C., Durchschlag, M., Joza, N., Pierron, G., van Endert, P., *et al.* (2009). Mechanisms of pre-apoptotic calreticulin exposure in immunogenic cell death. EMBO J *28*, 578-590.

Papandreou I, Goliasova T, Denko NC. (2011) Anticancer drugs that target metabolism: Is dichloroacetate the new paradigm? Int J Cancer 128, 1001-1008.

Schuldiner, M., and Schwappach, B. (2013). From rags to riches - the history of the endoplasmic reticulum. Biochim Biophys Acta *1833*, 2389-2391.

Shi, Y., Vattem, K.M., Sood, R., An, J., Liang, J., Stramm, L., and Wek, R.C. (1998). Identification and characterization of pancreatic eukaryotic initiation factor 2 alpha-subunit kinase, PEK, involved in translational control. Mol Cell Biol *18*, 7499-7509.

van Vliet, A.R., Verfaillie T., Agostinis, P. New functions of mitochondria associated membranes in cellular signaling. (2014) Biochim Biophys Acta. *1843*,2253-2262.

Vembar, S.S., and Brodsky, J.L. (2008). One step at a time: endoplasmic reticulum-associated degradation. Nat Rev Mol Cell Biol *9*, 944-957.

Verfaillie, T., Rubio, N., Garg, A.D., Bultynck, G., Rizzuto, R., Decuypere, J.P., Piette, J., Linehan, C., Gupta, S., Samali, A., *et al.* (2012). PERK is required at the ER-mitochondrial contact sites to convey apoptosis after ROS-based ER stress. Cell Death Differ *19*, 1880-1891.

Woehlbier, U., and Hetz, C. (2011). Modulating stress responses by the UPRosome: a matter of life and death. Trends in biochemical sciences *36*, 329-337.

Xu, C., Bailly-Maitre, B., and Reed, J.C. (2005). Endoplasmic reticulum stress: cell life and death decisions. The Journal of clinical investigation *115*, 2656-2664.

Xu, S., Sankar, S., and Neamati, N. (2014). Protein disulfide isomerase: a promising target for cancer therapy. Drug discovery today *19*, 222-240.

Zitvogel, L., Kepp, O., Senovilla, L., Menger, L., Chaput, N., and Kroemer, G. (2010). Immunogenic tumor cell death for optimal anticancer therapy: the calreticulin exposure pathway. Clin Cancer Res *16*, 3100-3104.

Chapter 3: c-FLIP: A negative regulator of programmed cell death and a target for cancer therapy (Pennarun, Bucor, and Khosravi-Far)

Bodvael Pennarun#, Octavian Bucur# and Roya Khosravi-Far*

Department of Pathology, Beth Israel Deaconess Medical Center and Harvard Medical School, Boston, MA 02215, USA

ABSTRACT

c-FLIP is an endogenous inhibitor of programmed cell death and is commonly overexpressed in many cancers, leading to acquired or innate resistance of cancer cells to cytokines and chemotherapeutic agents. Functional studies have shown that c-FLIP protein levels play a role in resistance of cancer cells to therapy, suggesting c-FLIP as a promising anti-cancer drug target. However, direct inhibitors of c-FLIP have yet to be identified. This review discusses the role of c-FLIP as an inhibitor of cell death and the potential of c-FLIP as a drug target in cancer therapy.

Apoptotic pathways

Apoptosis or programmed cell death is an orderly and tightly regulated signal transduction pathway. One of the two main branches of apoptotic program, the so-called extrinsic pathway, is mediated by death receptors, such as CD95/Fas, TNFR1, DR4/TRAIL-R1, and DR5/TRAIL-R2 (Figure 1). Ligation of these receptors by their cognate ligand at the cell surface initiates apoptotic response, a sequence of molecular events proceeding via the oligomerization of death receptors. This oligomerization is driven by extracellular ligands and homotypic interactions between intracellular death domains of ligated receptors, and the assembly of the death-inducing signaling complex (DISC) at the intracellular domains of aggregated receptors. The DISC assembly proceeds via the recruitment of the adaptor protein FADD and procaspase-8 and/or -10 (Plati et al., 2011; Fulda, 2012; Safa, 2012). According to the accepted views, when local concentration of recruited procaspases exceeds a threshold value, concentrated caspases, which possess residual catalytic activity, are activated by autoprocessing (Maelfait and Beyaert,

2008). Released into the cytosol as fully active tetramers, apical caspases activate downstream caspases in a cascade fashion, leading to cleavage of specific intracellular targets and, as a consequence, organized cell demise (Nagata, 1997; Guicciardi and Gores, 2009; Plati et al., 2011).

The mitochondria-dependent intrinsic pathway on the other hand can be triggered by a variety of stress signals such as DNA damage and UV that are sensed by mitochondria. Activation of the mitochondrial pathway causes changes in the mitochondrial membrane and its permeabilization leading to the release of cytochrome c. The released cytochrome c will then associate with procaspase-9, ATP and apoptotic protease activating factor 1 (APAF-1) to form a complex named the apoptosome. Caspase-9 is activated at the apoptosome leading to cleavage and activation of other effector caspases, including caspase-3, -6, -7, and -10 (Figure 1).

It is now well-established that there is also cross-talk between the intrinsic and extrinsic pathways of apoptosis. An example of this is when the activated caspase-8 cleaves the pro-apoptotic protein Bid during the extrinsic pathway of apoptosis. Following the cleavage, truncated Bid (tBid) translocates to the mitochondria and through its interactions with the pro-apoptotic Bcl-2 proteins Bax and Bak, initiates the intrinsic pathway of apoptosis (Fig. 1) (Strasser et al., 2000).

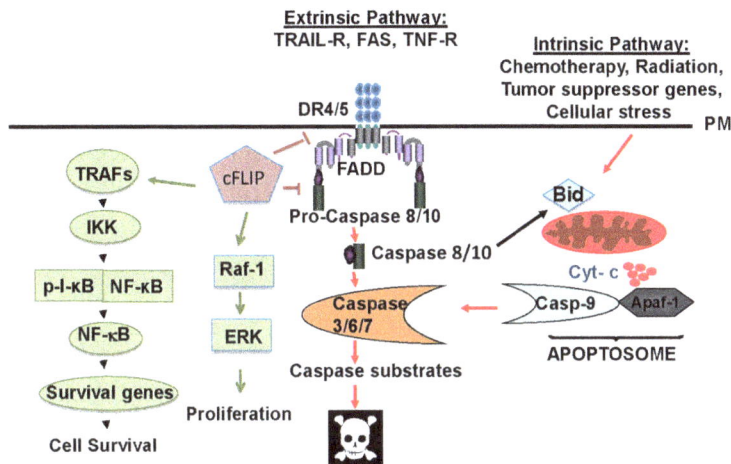

Figure 1: c-FLIP regulation of apoptotic machinery, the cell survival and proliferative pathways.

Inhibitors of apoptosis

There are at least three major families of proteins that act as endogenous inhibitors of apoptosis. These proteins include the mitochondrial proteins B-cell leukemia/lymphoma protein 2 (Bcl-2) family, the inhibitor of apoptosis proteins (IAPs), and c-FLIP. These inhibitors of apoptosis are often upregulated in transformed cells, and block the apoptotic program. The Bcl-2 family of mitochondrial protein prevents mitochondrial permeability transition and the release of pro-apoptotic molecules (Scarfo and Ghia, 2013). On the other hand, IAPs, such as c-IAP1, c-IAP2, and XIAP inhibit apoptosis as direct binders and inhibitors of caspases (Fulda, 2014). IAPs bind and inhibit caspases directly, and can also target them to the ubiquitin-proteasomal pathway for degradation. On the other hand c-FLIP acts as a major inhibitor of death receptor signaling through inhibition of death receptors directly or via inhibition of FADD or caspase-8 (Fig. 1) (Jin et al., 2004; Yang, 2008; Plati et al., 2011; Fulda, 2012; Safa, 2012). Modulators of Bcl-2 family of proteins and IAPs are currently in clinical trials, while a potent inhibitor of c-FLIP has yet to be developed (Scarfo and Ghia, 2013; Fulda, 2014).

c-FLIP in programmed cell death

c-FLIP (Irmler et al., 1997) protein is an inactive homolog of procaspase-8 that was originally discovered in homology searches (Shu et al., 1997) . C-FLIP is a death effector domain (DED)-containing protein (Safa et al., 2008) that also consists of an inactive caspase domain lacking the amino acid residues required for the catalytic function (Safa et al., 2008). Thirteen splice variants of the c-FLIP gene and only three protein isoforms of c-FLIP, the long form c-FLIP$_L$ (55 kDa), and two truncated forms, c-FLIP$_S$ (26 kDa) and c-FLIP$_R$ (24 kDa), have been identified (Fig. 2) (Safa et al., 2008; Bagnoli et al, 2010; Shirley and Micheau, 2010). Additionally, c-FLIP$_L$ has a caspase-8 cleavage site at position Asp-376. Cleavage of c-FLIP$_L$ at this site produces two other variants, c-FLIP$_{p43}$ and c-FLIP$_{p12}$.

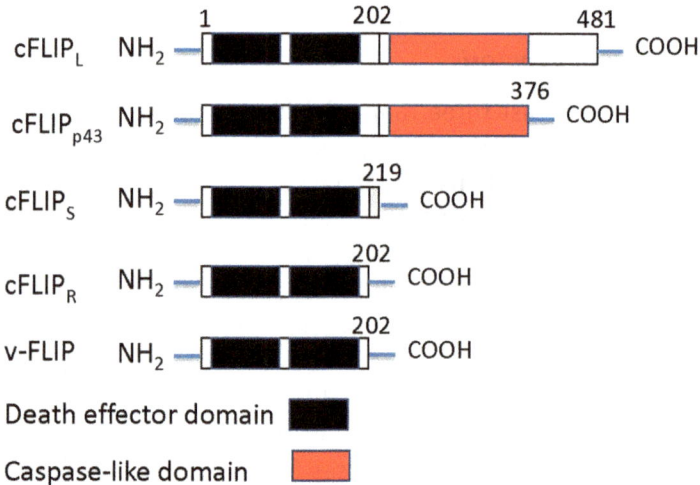

Figure 2: c-FLIP isoforms and variants.

c-FLIP is recruited to the DISC through DED-DED domain interactions with FADD. At the DISC, c-FLIP heterodimerizes with procaspase-8 and/or -10, acting as a dominant-negative form of procaspase-8, inhibiting its autoprocessing (Irmler et al., 1997; Rasper et al., 1998). On the other hand, depending on the ratio between c-FLIP$_L$ and caspase-8, c-FLIP$_L$ has also been reported to act as an activator of procaspase-8 at the DISC

(Dohrman et al., 2005; Micheau et al., 2002). Thus, the long isoform c-FLIP$_L$ exhibits both pro- and anti-apoptotic properties, while the short c-FLIP isoforms, similar to v-FLIP appear to strictly function as anti-apoptotic molecules.

In addition to modulating the activity of apical caspases at the DISC, c-FLIP is also involved in the regulation of the ripoptosome (Dickens et al., 2012). The ripoptosome is a large multiprotein cytosolic complex that contains FADD, caspase-8, RIP1 and RIP3. c-FLIP at the ripoptosome can activate several survival pathways to inhibit cell death. c-FLIP$_L$ activates ERK and NF-kappaB signaling by binding to the adaptor proteins in each pathway (Figure 1) (Dickens et al., 2012; Safa, 2012). c- FLIP$_L$ by interacting with Raf-1 activates the ERK pathway while, via its interaction with TRAF1, TRAF2 and RIP, activates the NF-KappaB pathway (Shirley and Micheau, 2010; Safa, 2012). The activation of ERK and NF-kappaB inhibits apoptosis via a mechanism involving upregulation of various proliferative and cell survival genes, including c-FLIP, Akt, Bcl-xL, and IAPs (Wang et al., 1998; Kreuz et al., 2001; Micheau et al., 2001; Safa, 2012). Other pathways that are involved in the regulation of cell survival, such as PI3K, AKT, JNK and Wnt signaling pathway have also been reported to be activated in response to c-FLIP (Safa, 2012).

c-FLIP is also thought to have a more diverse function and to regulate necroptosis and autophagy. Necroptosis is a form of programmed necrosis that is mediated through the necrosome, a RIP1/RIP3 containing cytoplasmic complex (Dickens et al., 2012). Both Caspase-8 and c-FLIP have been reported to play a role in suppressing formation of the necrosome (Vanlangenakker et al., 2011; He and He, 2013). Moreover, c-FLIP also regulates autophagy by interfering with autophagosome formation through inhibition of Atg3-mediated conjugation of LC3 (Lee et al., 2009).

Regulation of c-FLIP
The expression of c-FLIP is tightly regulated at transcriptional and post-translational levels. A number of different

transcription factors have been implicated in the regulation of c-FLIP expression. c-FLIP has a large promoter with binding sites for a variety of transcription factors including NFκB, FOXO, p53, p63, AP1, CREB, E2F1, STAT3, NFAT, c-myc, and SP1 (Fulda, 2012; Safa, 2012). Some of these transcription factors such as FOXO and E2F1 negatively regulate the expression of c-FLIP, while others such as NFκB positively regulate FLIP expression (Kreuz et al., 2001; Salon et al., 2006; Park et al., 2009).

At the posttranslational level, c-FLIP is regulated through the ubiquitin-mediated proteasomal degradation pathway. E3 ubiquitin ligase Itch leads to ubiquitination and proteasomal degradation of c-FLIP$_L$ (Chang et al., 2006). Furthermore, c-Cbl has been associated with proteasomal degradation of c-FLIP$_S$ (Kundu et al., 2009). These posttranslational modifications of c-FLIP are partly responsible for short half-life of c-FLIP proteins.

c-FLIP is a Promising Target in Cancer Therapy

c-FLIP is frequently overexpressed in various types of cancer, including breast, prostate, colorectal, ovarian, glioblastoma, pancreatic, endometrial, mesothelial, and lung cancer (Safa et al., 2008; Yang, 2008; Mahalingam et al., 2009; Shirley and Micheau, 2010; Safa, 2012). The expression of c-FLIP in these cancers correlates with the resistance of cancer cells to chemotherapy and also with poor prognosis (Safa et al., 2008; Yang, 2008; Mahalingam et al., 2009). Additionally, knockdown of c-FLIP sensitizes resistant tumor cells to cell death (Day et al., 2008; Day and Safa, 2009). Moreover, a variety of anti-cancer agents, including proteasome inhibitors, topoisomerase inhibitors, histone-deacetylase inhibitors, and alkylating agents, has been found to induce apoptosis in cancer cells, at least in part, by activating the death receptor signaling pathway, where c-FLIP can act as a major inhibitor (Mahalingam et al., 2009). These findings together have led to the evaluation of c-FLIP as a promising drug target in cancer treatment (Safa et al., 2008; Yang, 2008; Mahalingam et al., 2009; Plati et al., 2011; Fulda, 2012; Ozturk et al., 2012; Safa, 2012).

The majority of c-FLIP inhibitors that have so far been reported affect c-FLIP transcription, decrease translation, or trigger c-FLIP degradation through the ubiquitin-proteasome pathway (Safa et al., 2008; Plati et al., 2011; Fulda, 2012; Safa, 2012). One of the promising classes of inhibitors of c-FLIP is the histone deacetylase inhibitors (HDACi) (Fulda, 2012; Safa, 2012). These inhibitors have been shown to significantly inhibit c-FLIP at the level of transcriptional, translational and proteasome degradation. SAHA (suberoylanilide hydroxamic acid) is the most potent inhibitor of c-FLIP (Yerbes et al., 2012). Moreover, Safa and Colleagues have also identified a new HDACi, 4-(4-chloro-2-methylphenoxy)-N-hydroxybutanaimde (CMH) that inhibits c-FLIP transcription and translation leading to apoptosis of breast cancer cell line MCF-7 (Bijangi-Vishehsaraei et al., 2010).

Targeting of c-FLIP by metabolic inhibitors, such as cycloheximide or actinomycin D, has also been shown to sensitize cancer cells to ligand-induced CD95-mediated apoptosis. These inhibitors act, at least in part, by inducing the downregulation of c-FLIP expression (Plati et al., 2011).

The significant homology of c-FLIP with caspase-8 makes it difficult, but not impossible, to generate small molecule inhibitors that act just on c-FLIP. Recently, through an *in silico* chemical screen for compounds with affinity for the caspase 8 homodimer's interface, we have identified and experimentally validated a small molecule named CaspPro, that directly binds caspase 8, and promotes a caspase 8-dependent cell death in both single living cells or populations of cells, after TRAIL stimulation (Bucur et al, manuscript submitted for publication). Our approach is a proof-of-concept strategy leading to the discovery of a novel small molecule that can target the caspase 8-caspase 8 interface. This strategy can also be utilized for the identification of novel small molecules that can target the caspase 8-cFLIP heterodimer's interface. Directly targeting the c-FLIP interaction with caspase 8 or other DISC members by using small molecules is a promising strategy with great potential in inhibiting the recruitment of c-FLIP at the DISC and overcoming the suppression of apoptosis by c-FLIP.

A small peptide, killerFLIP, which contains an amino acid sequence derived from the C-terminal domain of c-FLIP$_L$ has also been identified. When this peptide is introduced into cells, it shows a significant cytotoxicity in a variety of cancer cell lines and inhibits tumor growth *in vivo* while sparing normal cells and tissues (Pennarun et al., 2013). Using electron microscopy, we have observed that the bioactive core of killerFLIP triggers cell death via rapid plasma membrane permeabilization (Pennarun et al., 2013). This active core possesses amphiphilic properties and self-assembles into micellar structures in aqueous solution. While this peptide sequence is derived from c-FLIP and induces cell death, its mechanism of action does not appear to be through the regulation of c-FLIP at the DISC but it is through a lytic action. Furthermore, this sequence is present only in c-FLIP$_L$ and the p12 variant of c-FLIP suggesting that this site within these c-FLIP variants may act on apoptosis through a unique mechanism.

Conclusions

In more than a decade from the discovery of c-FLIP, the mechanisms of action and the complex roles of c-FLIP in inhibiting cell death and inducing cell survival and proliferation have become clear. It is also clear that c-FLIP variants are overexpressed in many forms of cancer and their expression is accompanied by a poor prognosis. Furthermore, knock-down of c-FLIP in a variety of cancer cell lines induces tumor cell death. Together, these studies indicate that c-FLIP is a promising clinical target for cancer therapy. However, thus far the best inhibitors of c-FLIP act indirectly through regulation of its transcription, translation or proteasomal degradation. Additionally, a c-FLIP-derived peptide can induce cell death in vitro and in vivo through a lytic mechanism. The difficulty in generating an inhibitor of c-FLIP arises from its high homology and structural similarity with caspase-8. In silico screening strategies that would consider the differences in the structure of c-FLIP and caspase-8 could lead to small molecules that are selective for c-FLIP.

References:

Bagnoli, M., S. Canevari, and D. Mezzanzanica. 2010. Cellular FLICE-inhibitory protein (c-FLIP) signalling: a key regulator of receptor-mediated apoptosis in physiologic context and in cancer. *Int J Biochem Cell Biol*. 42:210-213.

Bijangi-Vishehsaraei, K., M.R. Saadatzadeh, S. Huang, M.P. Murphy, and A.R. Safa. 2010. 4-(4-Chloro-2-methylphenoxy)-N-hydroxybutanamide (CMH) targets mRNA of the c-FLIP variants and induces apoptosis in MCF-7 human breast cancer cells. *Mol Cell Biochem*. 342:133-142.

Chang, L., H. Kamata, G. Solinas, J.L. Luo, S. Maeda, K. Venuprasad, Y.C. Liu, and M. Karin. 2006. The E3 ubiquitin ligase itch couples JNK activation to TNFalpha-induced cell death by inducing c-FLIP(L) turnover. *Cell*. 124:601-613.

Day, T.W., S. Huang, and A.R. Safa. 2008. c-FLIP knockdown induces ligand-independent DR5-, FADD-, caspase-8-, and caspase-9-dependent apoptosis in breast cancer cells. *Biochem Pharmacol*. 76:1694-1704.

Day, T.W., and A.R. Safa. 2009. RNA interference in cancer: targeting the anti-apoptotic protein c-FLIP for drug discovery. *Mini Rev Med Chem*. 9:741-748.

Dickens, L.S., I.R. Powley, M.A. Hughes, and M. MacFarlane. 2012. The 'complexities' of life and death: death receptor signalling platforms. *Exp Cell Res*. 318:1269-1277.

Dohrman, A., J.Q. Russell, S. Cuenin, K. Fortner, J. Tschopp, and R.C. Budd. 2005. Cellular FLIP long form augments caspase activity and death of T cells through heterodimerization with and activation of caspase-8. *J Immunol*. 175:311-318.

Fulda, S. 2012. Targeting c-FLICE-like inhibitory protein (CFLAR) in cancer. *Expert Opin Ther Targets*. 17:195-201.

Fulda, S. 2014. Inhibitor of Apoptosis (IAP) proteins in hematological malignancies: molecular mechanisms and therapeutic opportunities. *Leukemia*. 28:1414-1422.

Guicciardi, M.E., and G.J. Gores. 2009. Life and death by death receptors. *Faseb J*. 23:1625-1637.

He, M.X., and Y.W. He. 2013. A role for c-FLIP(L) in the regulation of apoptosis, autophagy, and necroptosis in T lymphocytes. *Cell death and differentiation*. 20:188-197.

Irmler, M., M. Thome, M. Hahne, P. Schneider, K. Hofmann, V. Steiner, J.L. Bodmer, M. Schroter, K. Burns, C. Mattmann, D. Rimoldi, L.E. French, and J.

Tschopp. 1997. Inhibition of death receptor signals by cellular FLIP. *Nature*. 388:190-195.

Jin, T.G., A. Kurakin, N. Benhaga, K. Abe, M. Mohseni, F. Sandra, K. Song, B.K. Kay, and R. Khosravi-Far. 2004. Fas-associated protein with death domain (FADD)-independent recruitment of c-FLIPL to death receptor 5. *J Biol Chem*. 279:55594-55601.

Kreuz, S., D. Siegmund, P. Scheurich, and H. Wajant. 2001. NF-kappaB inducers upregulate cFLIP, a cycloheximide-sensitive inhibitor of death receptor signaling. *Molecular and cellular biology*. 21:3964-3973.

Kundu, M., S.K. Pathak, K. Kumawat, S. Basu, G. Chatterjee, S. Pathak, T. Noguchi, K. Takeda, H. Ichijo, C.B. Thien, W.Y. Langdon, and J. Basu. 2009. A TNF- and c-Cbl-dependent FLIP(S)-degradation pathway and its function in Mycobacterium tuberculosis-induced macrophage apoptosis. *Nat Immunol*. 10:918-926.

Lee, J.S., Q. Li, J.Y. Lee, S.H. Lee, J.H. Jeong, H.R. Lee, H. Chang, F.C. Zhou, S.J. Gao, C. Liang, and J.U. Jung. 2009. FLIP-mediated autophagy regulation in cell death control. *Nat Cell Biol*. 11:1355-1362.

Maelfait, J., and R. Beyaert. 2008. Non-apoptotic functions of caspase-8. *Biochem Pharmacol*. 76:1365-1373.

Mahalingam, D., E. Szegezdi, M. Keane, S. Jong, and A. Samali. 2009. TRAIL receptor signalling and modulation: Are we on the right TRAIL? *Cancer Treat Rev*. 35:280-288.

Micheau, O., S. Lens, O. Gaide, K. Alevizopoulos, and J. Tschopp. 2001. NF-kappaB signals induce the expression of c-FLIP. *Molecular and cellular biology*. 21:5299-5305.

Micheau, O., M. Thome, P. Schneider, N. Holler, J. Tschopp, D.W. Nicholson, C. Briand, and M.G. Grutter. 2002. The long form of FLIP is an activator of caspase-8 at the Fas death-inducing signaling complex. *J Biol Chem*. 277:45162-45171.

Nagata, S. 1997. Apoptosis by death factor. *Cell*. 88:355-365.

Ozturk, S., K. Schleich, and I. Lavrik. 2012. Cellular Flice-like inhibitory proteins (c-FLIPs): Fine-tuners of life and death decisions. *Exp Cell Res*. Epub ehead of print.

Park, S.J., H.Y. Sohn, J. Yoon, and S.I. Park. 2009. Down-regulation of FoxO-dependent c-FLIP expression mediates TRAIL-induced apoptosis in activated hepatic stellate cells. *Cellular signalling*. 21:1495-1503.

Pennarun, B., G. Gaidos, O. Bucur, A. Tinari, C. Rupasinghe, T. Jin, R. Dewar, K. Song, M.T. Santos, W. Malorni, D. Mierke, and R. Khosravi-Far, 2013. killerFLIP:

a novel lytic peptide specifically inducing cancer cell death. *Cell Death Dis.* 4:e894.

Plati, J., O. Bucur, and R. Khosravi-Far. 2011. Apoptotic cell signaling in cancer progression and therapy. *Integr Biol (Camb)*.

Rasper, D.M., J.P. Vaillancourt, S. Hadano, V.M. Houtzager, I. Seiden, S.L. Keen, P. Tawa, S. Xanthoudakis, J. Nasir, D. Martindale, B.F. Koop, E.P. Peterson, N.A. Thornberry, J. Huang, D.P. MacPherson, S.C. Black, F. Hornung, M.J. Lenardo, M.R. Hayden, S. Roy, and D.W. Nicholson. 1998. Cell death attenuation by 'Usurpin', a mammalian DED-caspase homologue that precludes caspase-8 recruitment and activation by the CD-95 (Fas, APO-1) receptor complex. *Cell death and differentiation*. 5:271-288.

Safa, A.R. 2012. c-FLIP, a master anti-apoptotic regulator. *Exp Oncol*. 34:176-184.

Safa, A.R., T.W. Day, and C.H. Wu. 2008. Cellular FLICE-like inhibitory protein (C-FLIP): a novel target for cancer therapy. *Curr Cancer Drug Targets*. 8:37-46.

Salon, C., B. Eymin, O. Micheau, L. Chaperot, J. Plumas, C. Brambilla, E. Brambilla, and S. Gazzeri. 2006. E2F1 induces apoptosis and sensitizes human lung adenocarcinoma cells to death-receptor-mediated apoptosis through specific downregulation of c-FLIP(short). *Cell death and differentiation*. 13:260-272.

Scarfo, L., and P. Ghia. 2013. Reprogramming cell death: BCL2 family inhibition in hematological malignancies. *Immunol Lett*. 155:36-39.

Shirley, S., and O. Micheau. 2010. Targeting c-FLIP in cancer. *Cancer Lett*. 332:141-150.

Shu, H.B., D.R. Halpin, and D.V. Goeddel. 1997. Casper is a FADD- and caspase-related inducer of apoptosis. *Immunity*. 6:751-763.

Strasser, A., L. O'Connor, and V.M. Dixit. 2000. Apoptosis signaling. *Annu Rev Biochem*. 69:217-245.

Vanlangenakker, N., M.J. Bertrand, P. Bogaert, P. Vandenabeele, and T. Vanden Berghe. 2011. TNF-induced necroptosis in L929 cells is tightly regulated by multiple TNFR1 complex I and II members. *Cell death & disease*. 2:e230.

Wang, C.Y., M.W. Mayo, R.G. Korneluk, D.V. Goeddel, and A.S. Baldwin, Jr. 1998. NF-kappaB antiapoptosis: induction of TRAF1 and TRAF2 and c-IAP1 and c-IAP2 to suppress caspase-8 activation. *Science*. 281:1680-1683.

Yang, J.K. 2008. FLIP as an anti-cancer therapeutic target. *Yonsei Med J*. 49:19-27.

Yerbes, R., A. Lopez-Rivas, M.J. Reginato, and C. Palacios. 2012. Control of FLIP(L) expression and TRAIL resistance by the extracellular signal-regulated kinase1/2 pathway in breast epithelial cells. *Cell death and differentiation*. 19:1908-1916.

Chapter 4: Transglutaminase Type2 and Cell Death: an historical overview. (Piacentini)

Mauro Piacentini

Department of Biology, University of Rome 'Tor Vergata', 00133 Rome, Italy.

and

National Institute for Infectious Diseases IRCCS 'L. Spallanzani', 00149 Rome, Italy.

Abstract:

There are several types of transglutaminases, of which Type 2 (TG2), the most ubiquitous, is multifunctional and participates in many cellular functions. Over 30 years ago we realized that TG2 was activated during apoptosis. By crosslinking, TG2 limits leakage of intracellular components from dying cells, thus limiting inflammation. Today we know that it is an early marker of apoptosis and that in some circumstances, in a different conformation, it can also protect cells. It may function as part of the permeability transition pore complex, and may play a role in mitophagy or other forms of autophagy. TG2 transamidation is important to assembly of protein aggregates, and polyamination may stabilize microtubules. Thus the field has come a long way and remains exciting. Younger scientists now coming into the field should find plenty of ideas, discoveries, and people to stimulate them.

Background

The Transglutaminase protein family

Transglutaminases are a class of enzymes that catalyze thiol- and calcium-dependent transamidation reactions. In mammals, nine distinct TGase iso-enzymes have been identified and partially characterized (see Table 1): (a) the blood coagulation zymogen Factor XIII, a key enzyme involved in stabilization of fibrin clots and in wound healing, which is converted by a thrombin-dependent proteolysis into the active TGase Factor XIIIa (plasma TGase); (b) the keratinocyte TGase (type 1 TGase), which exists in membrane-bound and soluble forms, is activated by proteolysis and is involved in the terminal differentiation of keratinocytes; (c) the ubiquitous type 2 tissue TGase (TG2)

(Fesus and Piacentini, 2002); (d) the epidermal/hair follicle TGase (type 3 TGase), which also requires proteolysis to become active, involved in the terminal differentiation of the keratinocyte (Candi et al. 2005); (e) the prostatic secretory TGase (type 4 TGase), essential for fertility in rodents (Dubbink et al., 1998); (f) the last characterized type 5 and type 6 TGases respectively involved in epidermal differentiation and central nervous system development (Grenard et al., 2001; Thomas et al., 2011); and (g) type 7 TGase with unknown function (Grenard et al. 2001). All mammalian forms have considerable structural homology and are members of the papain-like superfamily of cysteine proteases (Grenard et al. 2001). The primary structures of human TGs show identities in few regions, such as the active site region; in fact all members of this superfamily possess a catalytic triad of Cys-His-Asp or Cys-His-Asn. However, high sequence conservation and, therefore, a high degree of preservation of residue secondary structure among TG2, TG3 and FXIIIa indicates that these TGs all share four-domain tertiary structures that could be similar to those of other TGs (N-terminal b-sandwich, core domain, containing the catalytic and regulatory sites, and C-terminal b-barrels 1 and 2) (Fesus and Piacentini 2002).

Table 1. The transglutaminase protein family.

Name	Gene	Function
Factor XIII (fibrin-stabilizing factor)	F13A1, F13B	blood coagulation
Keratinocyte transglutaminase	TGM1	skin terminal differentiation
Tissue transglutaminase	TGM2	cell death/autophagy
Epidermal transglutaminase	TGM3	skin terminal differentiation
Prostate transglutaminase	TGM4	fertilization
TGM X	TGM5	skin differentiation
TGM Y	TGM6	unclear
TGM Z	TGM7	testis, lung
Band 4.2	Band 4.2	red blood cell, unclear

Type2 transglutaminase

Tissue or type 2 transglutaminase (TG2) is a 78 kDa, calcium dependent enzyme and it represents the most ubiquitous isoform of the TGs family. In humans, TG2 is encoded by a single

TGM2 gene located on chromosome 20q11–12 and it has been reported to be 32.5 kb in size and contains 13 exons and 12 introns (Gentile et al., 1991; Fraij and Gonzales, 1997). Since its discovery in 1957, a large number of its enzymatic substrates have been identified in various intracellular and extracellular compartments, including the cytosol, nucleus, and mitochondria and cell surface (Csosz et al., 2008; Facchiano and Facchiano, 2009).

The multi-functionality of TG2 is dependent on the structural features. The structure of the enzyme, is composed of 4 domains: an N-terminal β-sandwich domain with fibronectin and integrin binding sites (aa 1–140), the catalytic core containing the catalytic triad for the acyl-transfer reaction (aa 141–460) and two C-terminal b-barrel domains (aa 461–586 and 587–687). In cells TG2 is regulated by reversible conformational changes that include Ca^{2+}-dependent activation, which shifts TG2 to the "open" (extended) conformation thereby unmasking the enzyme's active center, and inhibition by GTP, GDP, and ATP, which constrains it in the "closed" (compact) conformation (Pinkas et al., 2007; Gundemir et al. 2012). In the absence of Ca^{2+}, TG2 assumes the basic latent conformation and the reactivity of Cys277 is decreased either by hydrogen-binding with the phenolic hydroxy group of Tyr516 or by formation of a disulphide with a neighboring cysteine residue, namely Cys336 (Noguchi et al., 2001). Although it was initially identified and studied as a typical cytoplasmic protein, TG2 was later described to localize in other compartments, including the nucleus, mitochondria, endolysosomes and the extracellular space (Malorni et al., 2008; Park et al., 2010; Zemskov et al., 2006, Gundemir and Johnson, 2009).

The transamidase activity of TG2 consists in the incorporation/deamidation of primary amines and in protein cross-linking. In particular, TG2 can use a primary amine as acyl-acceptor to modify the glutamine residue of a protein substrate resulting in a post-translational modification; on the other hand it can use a water molecule as an acyl-acceptor to deaminate a peptide-bound glutamine residue. This reaction is very important in the pathogenesis of Celiac Disease (Molberg et al. 1998).

In addition to its transamidase activity, TG2 has a GTPase activity involved in intracellular G protein signaling. In 1994 it was discovered that the GTP binding protein termed Ghα, co-isolated with the α_{1B} adrenergic receptor, was identical to TG2 (Nakaoka et al., 1994). By analogy, TG2 was also shown to associate with: TPα thromboxane, A2 receptor (Vezza et al., 1999) and oxytocin receptor (Park et al., 1998) by linking them to activation of PLCδ1, thereby increasing inositol-1,4,5-trisphosphate (IP3) levels upon stimulation of these receptors with agonists. TG2, due to its high affinity for GDP as well as GTP, is likely to spend significant time in the GDP bound form after the hydrolysis of GTP; however is very clear that the intracellular average GTP/calcium ratio (~150 μM/~100 nM) is more than sufficient to keep TG2 in a relatively latent state as a transamidase.

TG2 has also been reported to possess intracellular serine/threonine kinase activity, with insulin-like growth factor binding protein 3 (IGFBP), p53 tumor suppressor protein and histones (Mishra and Murphy, 2004) as substrates. Moreover a protein disulphide isomerase activity for TG2 has been established and several mitochondrial substrates have been identified (Piacentini et al., 2002; Malorni et al., 2009).

The beginning: 30 years ago

In 1984 I started my postdoctoral experience at National Institute of Dental Research at National Institutes of Health in Bethesda in the laboratory directed by Jack Folk, a pioneer of the transglutaminase field whose seminal work defined the biochemistry of these enzymes. At that time was already clear that the epidermal transglutaminases are Ca^{2+}-dependent enzymes that catalyze the formation of Nε-(γ-glutamyl)lysine bonds between the cornified envelope (CE) structural proteins thus conferring the characteristic resistance and insolubility to the skin. In fact, the transglutaminase enzymes, in this case type 1, 3 and 5, catalyze the formation of crosslinks at the level of glutamines and lysines of structural proteins of epidermis, such involucrin and loricrin, to form the protein scaffold of the cornified envelope. This process called "Cornification" is a coordinated process in space and time, which allows the formation of a layer of dead cells (corneocytes) to create the physico/chemical barrier of the skin. In my research stages in Jack Folk's lab I demonstrated for the first time that di- and polyamines could be covalently incorporated into liver proteins as well as into the CE proteins during the death/terminal differentiation of mouse primary keratinocytes (Piacentini et al. 1988). During my stay (1984-1986) at NIH I had the opportunity to meet many important scientists working in the transglutaminase field, in particular, I had the pleasure to meet Laszlo Fesus who was also spending a research experience in the same lab. Together with Laszlo we started to discuss the possible physiological role of the most ubiquitous transglutaminase, the type 2, since at that time there were no ideas about its possible functions. Before joining the lab in Bethesda, in my first Italian laboratory at the University of Rome "La Sapienza" I was working on liver and in particular on the turnover of endoplasmic reticulum intrinsic proteins (Piacentini et al. 1983), although I never published these observations I noticed that when you treat rats with phenobarbital to induce the ER biosynthesis there was an increase of transglutaminase activity during the catabolic phase leading to re-establishment of the ER homeostasis. I guess these discussions inspired Laszlo to use the liver as a model system to study the role of TG2 in rats treated with lead nitrate, which induces hyperplasia in the liver followed by a compensatory wave of apoptosis. Indeed, during the involuting phase leading to the homeostatic reduction of the hyperplastic liver the removal of dying hepatocytes by apoptosis coincides with the induction of TG2 expression.

Following my stage in Bethesda I started a very productive collaboration with Laszlo Fesus and in 1988 I visited his lab in

Debrecen for the first time. During that visit we performed some very exciting experiments in the attempt to clarify the role of TG2 in apoptosis and we demonstrated that during hepatocyte death, the enzyme crosslinking activity leads to the formation of protein polymers insoluble in detergents, urea, guanidine hydrochloride, reducing agents (Fesus et al. 1989). In a series of subsequent publications we could demonstrate that this proteic scaffold reduces the leakage of intracellular components from the dying cells and thus limiting inflammation (Fesus and Piacentini 2002).

25 years of TG2 (1989-2014) research on cell death/survival.

The discovery of the involvement of TG2 in cell death and the formation of intracellular protein crosslinks in dying cells represented, in addition to the DNA fragmentation, the only specific biochemical marker available during the first years of apoptosis research. In fact, the detection of TG2 induction was used as a hallmark of apoptosis *in vivo* and *in vitro* in hundreds of scientific publications in the early 90s. Now, in PubMed there are more than 600 publications dealing with the role of TG2 in cell death. In the last three decades it has become clear that TG2 plays a complex role in cell death/autophagy, although its definitive role in these processes has to be fully elucidated. An involvement of TG2 in the apoptotic pathway is well established, as is the observation that the enzyme can play both pro- and anti- cell death functions depending on the cell type and apoptotic stimulus and the conformational status of the enzyme (Piacentini et al. 2011). TG2 plays a pro-apoptotic function under physiological conditions acting as Ca^{++}-dependent transamidating enzyme in its "open conformation", while in highly transformed cells it can also switch its activity toward a cell protective function acting as g-protein in the "closed conformation" (Piacentini et al. 2011, Gundemir et al. 2012). The observed induction of TG2 gene expression during apoptosis onset *in vivo* was coupled with *in vitro* observations about an increased sensitivity as well as a protection against apoptosis upon induction or down-regulation of TG2 expression, respectively (Oliverio et al., 1999; Piacentini et al., 2011). TG2 shares 70% identity with the BH3 domain of Bcl-2 family proteins, suggesting that TG2 represents a BH3-only protein that regulates apoptosis. In this regard, it has been demonstrated that through this domain the enzyme is able to interact with the pro-apoptotic proteins Bax and Bak, which might be favored in their mitochondrial localization and pro-apoptotic function (Rodolfo et al., 2004). It is now accepted that the transamidation activity of TG2, which is mostly silent under physiological conditions, is necessary for its pro-death role. Furthermore, it has been shown that the transamidase inactive TG2 localized in the nucleus generally displays a protective function against apoptosis (Gundemir et al, 2012).

Interestingly, we revealed that TG2 plays a major role in mitochondrial physiology and energy metabolism acting as a protein disulfide isomerase (PDI) (Mastroberardino et al., 2006; Malorni et al., 2009). TG2 might also localize inside the mitochondrial inter-membrane space, where its PDI activity modifies and stabilizes some members of the respiratory chain as well as the activity of the ANT1, a protein involved in ADP/ATP exchange that has been proposed to act as an important component of the permeability transition pore complex (Mastroberardino et al., 2006; Malorni et al., 2009). These findings have been recently supported by showing that TG2 plays an important role in mitophagy (Rossin et al., 2014). Interestingly, previous studies carried out in TG2$^{-/-}$ knockout mice expressing mutated huntingtin suggested a possible role for TG2 in autophagy during the neurodegenerative processes occurring in Huntington Disease (HD) (Mastroberardino et al., 2002). TG2$^{-/-}$ HD transgenic mice showed that the HD onset is associated with a large reduction in non-apoptotic cell death and with an increased number of nuclear protein inclusions, clearly suggesting an impairment in the autophagic pathway (Mastroberardino et al. 2002). Indeed, the first report claiming an involvement of TG2 in autophagy was published describing highly metastatic pancreatic carcinoma cells where the knockdown of TG2 protein by siRNA in pancreatic cancer cells leads to the accumulation of autophagic vacuoles in the cytoplasm and a marked induction of LC3 II (Akar et al., 2007). In 2009, we established the involvement of TG2 in autophagy under physiological conditions. In fact, we have shown that the ablation of TG2 protein, both *in vivo* and in primary mouse embryonic fibroblasts, resulted in the accumulation of the cleaved isoform of LC3 (LC3 II) on pre-autophagic vesicles (D'Eletto et al. 2009). Moreover, the formation of the acidic vesicular organelles in the same cells was very limited, indicating an impairment of the maturation of autophagosomes (D'Eletto et al. 2009). We have also demonstrated that TG2 regulation of autophagy occurs via its transamidating activity (Rossin et al., 2011). Recently, we showed that TG2 knockout mice display impaired autophagy and accumulate ubiquitinated protein aggregates upon starvation (D'Eletto et al. 2012). Furthermore, the p62-dependent peroxisome degradation is also impaired in the absence of TG2. Under stressful cellular conditions, TG2 physically interacts with p62 and they co-localize in the cytosolic protein aggregates, which are then recruited into autophagosomes, where TG2 is degraded (D'Eletto et al. 2012). Interestingly, the enzyme's crosslinking activity is activated during autophagy and its inhibition leads to the accumulation of ubiquitinated proteins. These data indicate that the TG2 transamidating activity plays an important role in the assembly of protein aggregates, as well as in the clearance of damaged organelles by macroautophagy. Another interesting effect caused by the TG2 ablation is the accumulation of a large number of fragmented mitochondria that display decreased membrane potential along with increased levels of Drp1 and PINK1, two key proteins regulating the mitochondrial fission (Rossin et al. 2014). As a consequence of accumulation of damaged mitochondria, cells lacking TG2 increased their aerobic glycolysis and became sensitive to the glycolytic inhibitor 2-deoxy-D-glucose (2-DG). These data indicate that TG2 plays a key role in cellular dynamics and consequently influences the energetic metabolism (Rossin et al. 2014).

Beside this evident progress in the understanding of the TG2 functions there are still many important aspects that need to be addressed. In particular, I think is essential to definitively clarify the enzyme's role under physiological conditions. A key question about the TG2 physiological role is related to which one of its reported biochemical activities is essential under physiological conditions. To this aim, the identification of the TG2 protein partners is crucial as well as how the enzyme's interactome changes in various cell types and under stressful cellular conditions that generally lead to the activation of its transamidase activity. In keeping with this assumption, it is interesting to mention the recent identification of the transglutaminase-catalyzed posttranslational incorporation of polyamines in neuronal tubulin. Interestingly, the inhibition of polyamine synthesis or transglutaminase activity significantly decreases microtubule stability in vitro and in vivo (Song et al. 2013). These data suggest that transglutaminase-catalyzed polyamination of tubulins stabilizes microtubules thus playing an important function in the nervous system structure and functions.

Concluding remarks

In conclusion, I admit that I was particularly lucky to be part of this research adventure that has allowed me to learn a lot having the privilege to meet many important scientists worldwide who become very good friends of mine. The fact that TG2 was reported to be involved in cell death early in its history has allowed me to be involved since the 80s in this research field that exploded in the following decade. I think an important contribution to the development of the cell death field was the launch of the first scientific journal specialized in cell death "Cell Death and Differentiation" that I made together with Gennaro "Gerry" Melino, Alessandro Finazzi-Agrò and Richard "Dicko" Knight in 1994. This year we are happy to celebrate the 20th anniversary of its appearance. The story started in 1992 after the first Summer Course on "Cell Differentiation and Death" organized at International Center "Ettore Majorana", Erice

(Sicily, Italy), in which participated many of the pioneers of this field including the Nobel Laureate (in 2002) Robert Horvitz. I was also particularly lucky in participating in the very first exciting meetings in cell death as well as to organize several of them both in Italy and abroad. The early 90s were years of great enthusiasm in the field and the people started to organize themselves in scientific societies such as the International Cell Death Society (ICDS) and the European Cell Death Organization (ECDO) which are still very active and have been very important for the promotion and development of this scientific community. I contributed to start the ECDO back in 1992 and I am also very close to the ICDS being member of its Board of Directors for many years. I had the privilege to be invited as a speaker to one of initial ICDS Conference held in October 12, 1996 at the Queens College in New York and recently to the awarded by the same Society (2013) in Spain for my contribution to the cell death field. I would like to conclude wishing to the younger generations of scientists the fortune to live a similar experience that I believe was almost unique both as human and scientific experience.

References

Akar, U., Ozpolat, B., Mehta, K., Fok, J., Kondo, Y., and Lopez-Berestein, G. (2007). Tissue transglutaminase inhibits autophagy in pancreatic cancer cells. Mol. Cancer Res. *3*, 241-249.

Beninati S., Piacentini M., Argento-Cerù M.P., Russo-Caia S., and Autuori F. Presence of di- and polyamines covalently bound to protein in rat liver. (http://www.ncbi.nlm.nih.gov/pubmed/2861856) Biochim Biophys Acta. 1985 Jul 26;841(1):120-6.

Candi E., Schmidt R., Melino G. The cornified envelope: a model of cell death in the skin. (http://www.ncbi.nlm.nih.gov/pubmed/15803139) Nat Rev Mol Cell Biol. 2005 Apr;6(4):328-40

Csosz, E., Bagossi, P., Nagy, Z., Dosztanyi, Z., Simon, I., and Fesus, L. (2008). Substrate preference of transglutaminase 2 revealed by logistic regression analysis and intrinsic disorder examination. J. Mol. Biol. *383*, 390-402.

D'Eletto, M., Farrace, M.G., Falasca, L., Reali, V., Oliverio, S., Melino, G., Griffin, M., Fimia, G.M., and Piacentini, M. (2009). Transglutaminase 2 is involved in autophagosome maturation. Autophagy *5*, 1145-1154.

D'Eletto M., Farrace M.G., Rossin F., Strappazzon F., Giacomo G.D., Cecconi F., Melino G., Sepe S., Moreno S., Fimia G.M., Falasca L., Nardacci, R., and Piacentini M. (2012) Type 2 transglutaminase is involved in the autophagy-dependent clearance of ubiquitinated proteins. (http://www.ncbi.nlm.nih.gov/pubmed/22322858) Cell Death Differ. 19(7):1228-38.

Dubbink, H.J., de Waal, L., van Haperen, R., Verkaik, N.S., Trapman, J., and Romijn, J.C. (1998). The human prostate-specific transglutaminase gene (TGM4): genomic organization, tissue-specific expression and promoter characterization. Genomics *51*, 434-444.

Facchiano, A., and Facchiano, F. (2009). Transglutaminases and their substrates in biology and human diseases: 50 years of growing. Amino Acids *36*, 599-614.

Fesus L, Thomazy V, Falus A. (1987) Induction and activation of tissue transglutaminase during programmed cell death. (http://www.ncbi.nlm.nih.gov/pubmed/2890537) FEBS Lett. 224(1):104-8.

Fesus, L., Thomazy, V., Autuori, F., Ceru, M.P., Tarcsa, E., and Piacentini, M. (1989). Apoptotic hepatocytes become insoluble in detergents and chaotropic agents as a result of transglutaminase action. FEBS Lett. *245*, 150-154.

Fesus, L., and Piacentini, M. (2002). Transglutaminase 2: an enigmatic enzyme with diverse functions. Trends Biochem. Sci. *27*, 534-539.

Fraij, B.M., and Gonzales, R.A. (1997) Organization and structure of the human tissue transglutaminase gene. Biochim. Biophys. Acta *1354*, 65-71.

Folk, J.E. (1983). Mechanism and basis for specificity of transglutaminase-catalyzed epsilon-(gamma-glutamyl) lysine bond formation. Adv. Enzymol. Relat. Areas Mol. Biol. *54*, 1-56.

Gentile, V., Saydak, M., Chiocca, E.A., Akande, O., Birckbichler, P.J., Lee, K.N., Stein, J.P., and Davies, P.J. (1991) Isolation and characterization of cDNA clones to mouse macrophage and human endothelial cell tissue transglutaminases. J. Biol. Chem. *266*, 478-483.

Grenard, P., Bates, M.K., and Aeschlimann, D. (2001). Evolution of transglutaminase genes: identification of a transglutaminases gene cluster on human chromosome 15q. Structure of the gene encoding transglutaminase X and a novel gene family member, transglutaminase Z. J. Biol. Chem. *276*, 33066-33078.

Gundemir, S., Colak, G., Tucholski, J., and Johnson, G.V. (2012). Transglutaminase 2: a molecular Swiss army knife. Biochim. Biophys. Acta 1823, 406-419.

Molberg O, Mcadam SN, Körner R, Quarsten H, Kristiansen C, Madsen L, Fugger L, Scott H, Norén O, Roepstorff P, Lundin KE, Sjöström H, Sollid LM. Tissue transglutaminase selectively modifies gliadin peptides that are recognized by gut-derived T cells in celiac disease. (http://www.ncbi.nlm.nih.gov/pubmed/9623982) Nat Med. 1998 Jun;4(6):713-7.

Malorni, W., Farrace, M.G., Rodolfo, C., and Piacentini, M. (2008). Type 2 transglutaminase in neurodegenerative diseases: the mitochondrial connection. Curr. Pharm. Des. 14, 278–288.

Malorni, W., Farrace, M.G., Matarrese, P., Tinari, A., Ciarlo, L., Mousavi-Shafaei, P., D'Eletto, M., Di Giacomo, G., Melino, G., Palmieri, L., Rodolfo, C., and Piacentini, M. (2009). The adenine nucleotide translocator 1 acts as a type 2 transglutaminase substrate: implications for mitochondrial-dependent apoptosis. Cell Death Differ. 16, 1480-1492.

Mastroberardino, P.G., Iannicola, C., Nardacci, R., Bernassola, F., De Laurenzi, V., and Melino, G., Moreno, S., Pavone, F., Oliverio, S., Fesus, L., and Piacentini, M. (2002). 'Tissue' transglutaminase ablation reduces neuronal death and prolongs survival in a mouse model of Huntington's disease. Cell Death Differ. 9, 873–880.

Mastroberardino, P.G., Farrace, M.G., Viti, I., Pavone, F., Fimia, G.M., Melino, G., Rodolfo, C., and Piacentini, M. (2006). "Tissue" transglutaminase contributes to the formation of disulphide bridges in proteins of mitochondrial respiratory complexes. Biochim. Biophys. Acta, 1757, 1357-1365.

Mishra, S., and Murphy, L.J. (2004). Tissue transglutaminase has intrinsic kinase activity: identification of transglutaminase 2 as an insulin-like growth factor-binding protein-3 kinase. J. Biol. Chem. 279, 23863-23868.

Nakaoka, H., Perez, D.M., Baek, K.J., Das, T., Husain, A., Misono, K., Im, M.J., and Graham, R.M. (1994). Gh: a GTP-binding protein with transglutaminase activity and receptor signaling function. Science 264, 1593-1596.

Noguchi, K., Ishikawa, K., Yokoyama, K.I., Ohtsuka, T., Nio, N. and Suzuki, E. (2001). Crystal structure of red sea bream transglutaminase. J. Biol. Chem. 276, 12055-12059.

Oliverio, S., Amendola, A., Rodolfo, C., Spinedi, A., and Piacentini, M. (1999). Inhibition of "tissue" transglutaminase increases cell survival by preventing apoptosis. J. Biol. Chem. 274, 34123-34128.

Park, E.S., Won, J.H., Han, K.J., Suh, P.G., Ryu, S.H., Lee, H.S., Yun, H.Y., Kwon, N.S., and Baek, K.J. (1998). Phospholipase C-delta1 and oxytocin receptor signalling: evidence of its role as an effector. Biochem. J. 331, 283-289.

Park, D., Choi, S.S., and Ha, K.S. (2010). Transglutaminase 2: a multi-functional protein in multiple subcellular compartments. Amino Acids 39, 619-631.

Piacentini M, Spinedi A, Beninati S, Autuori F. (1983) Mechanism of release of integral proteins from rat liver microsomal membranes. (http://www.ncbi.nlm.nih.gov/pubmed/6303417) Biochim Biophys Acta. 731(2):151-60.

Piacentini M, Martinet N, Beninati S, Folk JE. (1988) Free and protein-conjugated polyamines in mouse epidermal cells. Effect of high calcium and retinoic acid. (http://www.ncbi.nlm.nih.gov/pubmed/3346223) J Biol Chem. 263(8):3790-4.

Piacentini, M., Farrace, M.G., Piredda, L., Matarrese, P., Ciccosanti, F., Falasca, L., Rodolfo, C., Giammarioli, A.M., Verderio, E., Griffin, M., Malorni, W. (2002). Transglutaminase overexpression sensitizes neuronal cell lines to apoptosis by increasing mitochondrial membrane potential and cellular oxidative stress. J. Neurochem. 81, 1061-1072.

Piacentini, M., D'Eletto, M., Falasca, L., Farrace, M.G., and Rodolfo, C. (2011). Transglutaminase 2 at the crossroads between cell death and survival. Adv. Enzymol. Relat. Areas Mol. Biol. 78, 197-246.

Pinkas DM, Strop P, Brunger AT, Khosla C. (2007) Transglutaminase 2 undergoes a large conformational change upon activation. (http://www.ncbi.nlm.nih.gov/pubmed/18092889) PLoS Biol. 5(12):e327.

Rodolfo, C., Mormone, E., Matarrese, P., Ciccosanti, F., Farrace, M.G., Garofano, E., Piredda, L., Fimia, G.M., Malorni, W., and Piacentini, M. (2004). Tissue transglutaminase is a multifunctional BH3-only protein. J Biol Chem. 279, 54783-54792.

Rossin, F., D'Eletto, M., Macdonald, D., Farrace, M.G., and Piacentini, M. (2011). TG2 transamidating activity acts as a reostat controlling the interplay between apoptosis and autophagy. Amino Acids 42, 1793-1802.

Rossin F, D'Eletto M, Falasca L, Sepe S, Cocco S, Fimia GM, Campanella M, Mastroberardino PG, Farrace MG, Piacentini M. (2014)Transglutaminase 2 ablation leads to mitophagy impairment associated with a metabolic shift towards aerobic glycolysis. (http://www.ncbi.nlm.nih.gov/pubmed/25060553) Cell Death Differ. Jul 25. doi: 10.1038/cdd.2014.106.

Song Y, Kirkpatrick LL, Schilling AB, Helseth DL, Chabot N, Keillor JW, Johnson GV, Brady ST (2013) Transglutaminase and polyamination of tubulin: posttranslational modification for stabilizing axonal microtubules. Neuron. 78(1), 109-23.

Vezza, R., Habib, A., and FitzGerald, G.A. (1999). Differential signaling by the thromboxane receptor isoforms via the novel GTP-binding protein, Gh. J. Biol. Chem. 274, 12774-12779.

Chapter 5: Caspases: helpers and killers (Zhivotovsky)

Boris Zhivotovsky

Institute of Environmental Medicine, Karolinska Institutet, Stockholm, Sweden; Faculty of Medicine, Lomonosov Moscow State University, Moscow, Russia

Abstract:

We now know over a dozen primary types of caspases, and potential caspase substrates now include at least 500 different proteins. These functionally different caspases are activated by different means including dimerization, cleavage, and binding to specific multiprotein platforms. Based on structural differences, mammalian caspases were classified into two groups: those with long prodomains (initiators of apoptosis) and those with short prodomains (effectors of apoptosis). Caspase-2 may function as both. In other situations caspases appear to be activated for the differentiation and do not destroy the cells. These roles for caspases beyond their function in apoptosis limit the potential therapeutic value of using pan-caspase inhibitors to protect cells. However, with selection of new compounds and consideration of how they might be used, they still have potential for protecting injured organs and organs subject to transplant. For example, one of the greatest challenges is to understand what drives the same protein (for example, caspase-8) to be a part of different complexes and how to influence these processes. There is cross-talk not only within apoptosis (receptor-mediated and mitochondria-mediated pathways), but also between various cell death modalities, and caspases are involved in this cross-talk but one of the most impressive links between caspases and apoptosis hallmarks is the caspase-dependent activation of endonucleases.

Introduction

When caspases were first discovered in the mid-1990s, they were recognized as the primary effectors of apoptosis, and cell death was assumed to be their primary function. However, we now know over a dozen primary types of caspases, and the story has become far more complex. One group regulates inflammation via processing of cytokines; one caspase functions in keratinocyte differentiation, and another group functions in apoptosis, subdivided into effectors of apoptosis, and membrane-bound or apoptosome-activated initiators of apoptosis. Caspases can digest inhibitors such as the inhibitor of caspase-activated DNase (ICAD) and they can themselves be inhibited by proteins such as inhibitors of apoptosis proteins (IAPs). Potential caspase substrates now include at least 500 different proteins. These functionally different caspases are activated by different means including dimerization, cleavage, and binding to specific multiprotein platforms. In other situations such as differentiation of erythroblasts, activation of T-cells, and regulation of cell cycle of B-cells, caspases appear to be activated for the differentiation and do not destroy the cells. These roles for caspases beyond their function in apoptosis limit the potential therapeutic value of using pan-caspase inhibitors to protect cells. However, with selection of new compounds and consideration of how they might be used, they still have potential for protecting injured organs and organs subject to transplant. For example, one of the greatest challenges is to understand what drives the same protein (for example, caspase-8) to be a part of different complexes and how to influence these processes. The recently introduced protein-protein database for the DD superfamily will be helpful to study the DD superfamily-mediated formation of cell-death-activating complexes. Beyond being solely of academic interest, data obtained might allow us to therapeutically manipulate cell death in various pathological processes. The fact that caspases have roles other than as effectors of cell death leads to unexpected complications in situations where caspase-dependent cell death is considered desirable, such as in response to cancer chemotherapy. Although the results of therapeutic use of caspase inhibitors are intriguing, additional carefully designed studies with adequate methodology, patient sample size, and follow-up need to be performed before any of these medications may be recommended for the treatment of different diseases. Thus, caspases can help to maintain various physiological processes in cells fulfilling non-apoptotic functions; they can act as killers during apoptosis. The near future will provide us with many more surprises in understanding their functions.

My interest in the field of cell death started in the mid-1970s. At that time radiobiology was dominated by the dogma that proliferating and undifferentiated cells are the most sensitive to radiation treatment. Consequently, our laboratory focused on understanding why lymphoid cells, especially thymocytes, being terminally differentiated and non-proliferating, were so sensitive to ionizing radiation. We hypothesized that radiation-induced lymphoid cell death represents an example of a more general phenomenon, programmed cell death (PCD) (Khanson 1979). This hypothesis was later accepted by the scientific community (Yamada and Ohyama, 1988).

The first apoptosis-related biochemical event was reported by Skalka *et al.* (1976) when the DNA of lymphocytes irradiated *in vivo* was shown to be degraded into oligonucleosomal-length fragments. This process was subsequently linked to endonuclease activity (Wyllie, 1980; Zhivotovsky et al., 1980) and has since become a prominent and routinely used biochemical hallmark of apoptosis. However, a question about the precise role of nuclear events in apoptosis, *i.e.* whether they were causally related to this process, was raised when Nuc-1, a protein that is essential for DNA degradation in the nematode *Caenorhabditis elegans,* was shown to act downstream of ced-3 and ced-4 (Ellis and Horvitz, 1986). Unfortunately, at that time the mechanism of action of ced-3 in the development of PCD was not clear. Many researchers in the field were sure that one of the initial events of cellular execution was activation of endonuclease(s).

Early evidence for the role of protease activation in apoptosis came from observations made by Lockshin (1969, 1981) and Bowen (1984) when they analyzed PCD in the intersegmental muscles of insects during metamorphosis. In these studies, the authors observed elevation of autophagic lysosome activity in areas of active cell death. It is now generally believed that the release of acid hydrolases from lysosomes plays a secondary role in PCD, since in several different experimental systems lysosomes appear intact until the final stages of cell disruption.

Later investigations into proteolytic mechanisms identified the granule exocytosis pathway of lymphocyte-mediated cytotoxicity and provided evidence that granule proteases, otherwise known as the granzymes (cytotoxic cell proteases (CCP)/fragmentins) contribute to the lethal damage inflicted upon target cells (Lobe et al., 1986; Masson et al., 1986; Klein et al., 1989; Shi et al., 1992). Kaufmann (1989) subsequently demonstrated that proteolysis of several nuclear substrates was an early characteristic of apoptosis in human myeloid cells exposed to various chemotherapeutic agents. The necessity of a stage of proteolysis was also implied in radiation-induced chromatin fragmentation (Soldatenko et al., 1991). It was reported that several distinct proteases could induce the morphological and biochemical features of apoptosis on their introduction into different cell types (Williams and Henkart, 1994). However, interest in the role of intracellular proteases in apoptosis was markedly stimulated by the observation that both the *C. elegans* cell death gene, ced-3, and its mammalian homolog, interleukin-Iβ-converting enzyme (ICE), contain a conserved pentapeptide domain at the active site and share a strong structural homology (Miura et al., 1993; Yuan et al., 1993). Following this observation, additional members of the ICE gene family, with cysteine protease properties, have been cloned. The frenetic pace of identification of new homologs has led to inconsistent and multiple names for many of these enzymes. As a consequence, the general scientific community is finding it increasingly difficult to follow this provocative and rapidly moving field. Therefore, a group of scientists who were deeply involved in this study has proposed to use the name 'caspase' as a root for serial names for all family members (Alnemri et al., 1996). As stated by the authors, "the selection of caspase was based on two catalytic properties of these enzymes. The 'c' is intended to reflect a cysteine protease mechanism, and 'aspase' refers to their ability to cleave after aspartic acid, the most distinctive catalytic feature of this protease family".

On the first AACR Special Conference "Cell Death and Cancer" (1993) several presentations were focused on the role of proteases in PCD (Eastman et al., 1994). In an excellent presentation Dr. H. Robert Horvitz clearly showed that one of the key genes that act positively to control cell death in *C. elegans* is ced-3, which is a cysteine protease. During the coffee break, the keynote speaker at the conference, Dr. Andrew Wyllie, told me that development of cell death research was leading to more and more complications: "even proteases are implied in the sequence of events leading to cell death. It looks as though the field of PCD will soon be as complicated as the field of the cell cycle". In fact, Dr. Wyllie was correct. However, successful development of both fields led to new discoveries, which were recognized by two Nobel Prizes in physiology or medicine in 2001 and 2002.

Several seminal discoveries on the role of proteases in cell death revolutionized the field and provided understanding of molecular mechanisms of apoptosis as well as links among various previous observations. First, it was shown that although overexpression of any caspases led to apoptotic cell death, not all of them were directly involved in apoptosis. Thus, caspase-1, -4, -5, -11 and -12 were implicated in regulation of inflammation, via processing of cytokines, whereas caspase-2, -3, -6, -7, -8, -9, and -10 were deeply involved in apoptosis and caspase-14 was found to be critical for keratinocyte differentiation.

Second, based on structural differences, mammalian caspases were classified into two groups: those with long prodomains (caspases-1,-2,-4,-5,-8,-9, -10, -11, and 12) and those with short prodomains (caspases-3,-6, -7, and -14). Short prodomain caspases (caspase-3,-6, and -7) were termed effectors of apoptosis. Long prodomain apoptotic caspases (caspase-2,-8, -9, and -10) were termed initiators of apoptosis. Interestingly, caspase-2 may function as both an initiator and an effector, similar to ced-3 in *C. elegans*; ontogenically, caspase-2 is most closely related to ced-3 (Figure 1) (Galluzzi et al., 2008).

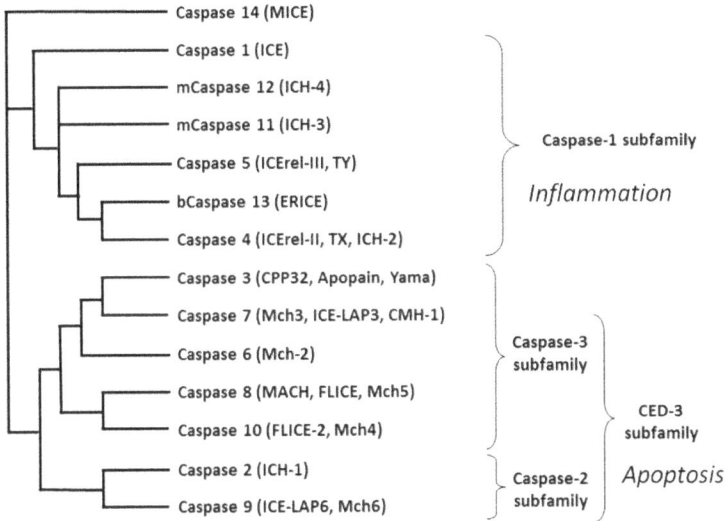

Figure 1. Family of Caspases - Cysteinyl Aspartate-Specific Proteases.

Caspases are synthesized as inactive precursors (zymogens, procaspases). The procaspase molecule (32–57 kDa) contains several domains: N-terminal prodomain and large (17–21 kDa) and small (10–13 kDa) subunits (for review, see Vaculova and Zhivotovsky, 2008). Short prodomain effector caspases are constitutively produced in cells as dimers and proteolytic processing by an initiator enzyme is required to trigger their activity. Processing occurs at an aspartate residue in the intersubunit linker, which is specific for targeting by initiator caspases as well as by granzyme B. When active, effector caspases target a broad spectrum of cellular proteins, ultimately leading to cell death. In contrast to effector caspases, initiator caspases are translated as monomeric zymogens, except caspase-2 which is dimeric, and their activation is more complex. The initial activation step for long prodomain caspases is proximity-induced dimerization. The dimerization occurs through the binding of adaptor proteins to the prodomain caspase activation and recruitment domain (CARD; caspase-1, -2, -4, -5, -9, -11 and -12) or death effector domain (DED;

caspase-8 and -10) motifs. There are specific adaptor proteins for each long prodomain caspase.

It has been shown that formation of high molecular weight (HMW) complexes is an essential step for activation of all long prodomain caspases. In 1995 Dr. Peter Krammer and colleagues showed that cytotoxicity-dependent APO-1 (Fas/CD95)-associated proteins form a death-inducing signaling complex (DISC) with the receptor (Kischkel et al., 1995). Four proteins, CAP1–4, were suggested to be APO-1 apoptosis-transducing molecules *in vivo*. However, it was unclear how this complex was activated and which enzymatic activity it performed. A year later, using different experimental approaches, two groups were able to identify a protein that provides the missing link both functionally and physically. Thus, the joint work of the groups of Drs. Krammer and Vishva Dixit using nanoelectrospray tandem mass spectrometry has shown that CAP4 is, in fact, a procaspase-8 (FLICE according to the original name) (Muzio et al., 1996). The group of Dr. David Wallach identified the same protein (originally called MACH) by a yeast two-hybrid screen using FADD/MORT1 as bait (Boldin et al., 1996). These teams clearly indicated that dominant-negative isoforms of caspase-8 block both CD95- and TNF-induced apoptosis, implying a role for caspase-8 in both signaling systems. Accordingly, the so-called receptor-mediated apoptotic pathway was deciphered, in which extracellular ligands stimulate receptor oligomerization and DISC assembly with caspase-8 (or as was found later, caspase-10) as the most upstream activated proteolytic enzyme, leading to the sequence of events that completely destroys the dead cell. A number of models were suggested to uncover dynamics of death-receptor signaling and DISC, and recently the stoichiometry of CD95 and TRAIL-R DISC complexes was revealed by quantitative and systems biological approaches (Dickens et al., 2012; Schleich et al., 2012).

Using a complex biochemical approach, Dr. Xiaodong Wang and colleagues were able to decipher the second molecular complex that is activated during mitochondria-mediated apoptosis (Li et al., 1997). In the presence of dATP, cytochrome c, released from the mitochondria, interacts with Apaf-1 and procaspase-9 leading to activation of this HMW complex. Dr. Michael Hengartner introduced the term 'apoptosome', to name this complex (Hengartner, 1997). Release of cytochrome c is regulated by various mechanisms, which involve the function of several pro- and anti-apoptotic members from the Bcl-2 family of proteins that preserve or disrupt mitochondrial integrity. Inhibitors of apoptosis proteins (IAPs) inhibit the proteolytic activity of mature caspase-9 in the apoptosome, as well as execution of caspase-3 activity. In turn, IAPs are inactivated and caspase activity is restored by proteins such as SMAC/Diablo or HtrA2/Omi, which are released from the mitochondria. Interplay among these proteins is complex but mathematical modeling of this complexity has been able to dissect spatiotemporal activation of the apoptosome-dependent pathway (Rehm et al., 2009). Importantly, a direct connection exists between functions of DISC and apoptosome complexes, which requires BH3-only proteins of Bcl-2 family (Li et al., 1998).

The third protein complex, the PIDDosome, was described by Dr. Jurg Tschopp and colleagues as a multiprotein complex that can act as a procaspase-2 activation platform upon genotoxic stress, by enabling autocatalytic procaspase-2 activation through induced proximity (Tinel and Tschopp, 2004). Assembly of the complex is provided by interactions between the C-terminal fragment of PIDD (p53-induced protein with a death domain, DD), RAIDD (receptor-interacting protein-associated ICH-1/CED-3 homologous protein with a DD), and procaspase-2 through their CARD and DD interactions. RAIDD plays the role of a bridge that binds PIDD via DD:DD interactions and procaspase-2 via CARD:CARD interactions. Interestingly, PIDDosome formation depends on the proteolysis of PIDD. The shortest PIDD fragment, PIDD-CC, can interact with RAIDD leading to the formation of a 'death-related' multiprotein complex. The PIDD-C fragment is known to be a positive regulator of NFκB signaling (Tinel et al., 2007). Therefore, PIDD can play the role of a molecular switch between cell survival and death. Intriguingly, activation of caspase-2 may occur in the absence of PIDD and RAIDD. We found that downregulation of PIDD and RAIDD by siRNA failed to influence processing and activation of caspase-2 after treatment with 5-fluoruracil (Vakifahmetoglu et al., 2006). Moreover, there is evidence that mouse embryonic fibroblasts (MEFs) from PIDD-deficient mice undergo PIDD-independent apoptosis that occurs normally in response to various stress signals. The formation of HMW complex, which contains caspase-2, in wild-type, *pidd-/-*, or *raidd-/-* SV40-immortalized MEFs after a temperature shift has been reported (Manzl et al., 2009). The HMW complex containing caspase-2, but not PIDD or RAIDD, can also be formed upon cisplatin treatment (our unpublished data). The initiator role of caspase-2 in pore-forming toxin-mediated apoptosis was demonstrated and HMW complex containing caspase-2 was isolated from cells in these experimental conditions (Imre et al., 2012). It seems that within this complex caspase-2 is activated in a PIDD/RAIDD-independent manner. The data indicate the existence of alternative cellular platforms for caspase-2 activation. However, the protein components of these complexes and the molecular mechanisms are still unknown and demand further investigation.

Accumulating evidence has revealed cross-talk not only within apoptosis (receptor-mediated and mitochondria-mediated pathways), but also between various cell death modalities, and caspases are involved in this cross-talk. Thus, natural (mutations) or artificial (chemical inhibitors) inactivation of caspases might cause the shift from apoptosis to necrosis or necroptosis, or to the mixture of these cell death modes. Cleavage of Atg5 by calpain causes a shift from autophagy to apoptosis via activation of mitochondria-mediated caspase-dependent mechanism (Yousefi et al., 2006).

As described above, a molecular switch between cell survival and death is present and involves NFκB. Under conditions of induction of apoptosis, active caspase-8 cleaves RIP1 and inactivates it, but when caspase activity is blocked necroptosis is triggered. On induction of necroptosis, RIP1 interacts with kinase RIP3 via RHIM domains, forming amyloid fibers according to the X-ray structure of this complex (Cho et al., 2009; Ofengeim and Yuan, 2013). Since caspase-8 is inhibited, RIP1 and RIP3 participate in the formation of the necroptosis-inducing complex, the so-called 'necrosome'. This complex includes FADD, caspase-8, RIP1, and RIP3; however, recent studies have shown that RIP3 might also induce necroptosis by a RIP1-independent mechanism (Feoktistova et al., 2011). Thus, the mechanism of molecular execution of necroptosis remains unclear and requires further detailed investigation.

Another important necroptosis-inducing multiprotein platform is the 'RIPoptosome' (Feoktistova et al., 2011; Tenev et al., 2011). Its formation depends on the activity of cIAPs1/2. The downregulation of cIAPs by SMAC mimetics or genotoxic stress triggers formation of the RIP1/FADD/caspase-8 complex independent of the involvement of TNF, TRAIL, CD95L, or mitochondria, and induces caspase-8-mediated apoptosis or caspase-independent necroptosis. Interestingly, it has been shown that the RIPoptosome can be recruited to other signaling platforms, such as the Toll-like receptor-3, and therefore, could initiate necroptosis upon immune or pro-inflammatory signals (Feoktistova et al., 2011).

The cross-talk between necroptosis and apoptosis is controlled by the balance of caspase-8 and c-FLIP isoforms (Feoktistova et al., 2011). It has been proposed that formation of caspase-8 homodimers within the RIPoptosome results in its catalytic activity and induces apoptosis. In contrast, formation of caspase-8–c-FLIP$_L$ heterodimers results in limited catalytic activity, which is not sufficient to trigger apoptosis but could cleave RIP1. This, in turn, leads to RIPoptosome disassembly and cell survival. Although the core components of the RIPoptosome and the necrosome have been identified, the stoichiometry of these complexes is still unknown. Further investigation of complex components will help to fully understand the mechanism of PCD regulation.

As previously mentioned, formation of HMW complexes is essential for activation of apoptotic caspases. However, once again it is necessary to mention the existence of caspases involved in regulation of inflammation, including caspase-1 and -5 (Lamkanfi and Dixit, 2012). Similar to the activation process of apoptotic caspases, activation of inflammatory caspases occurs through the formation of HMW complexes, called inflammasomes. Although the principle of organization and activation of inflammasomes is similar to the DISC, apoptosome, and PIDDosome, it is unclear whether the structures of these complexes are also similar.

To date, the growing list of caspase substrates involves at least 500 different protein molecules (www.casbah.ie). While some of them are considered to be common targets for caspase-mediated cleavage in all cells, cleavage of the others may be cell type-specific.

One of the most impressive links between caspases and apoptosis hallmarks is the caspase-dependent activation of endonucleases. As was mentioned above, one of the first described apoptotic events was fragmentation of DNA. However, which endonuclease is activated and the mechanism behind this event were unknown. In 1997–1998 two groups described a mechanism for disassembly of chromatin during apoptosis (Liu et al., 1997, 1998; Enari et al., 1998; Sakahira et al., 1998). They found that chromatin is cleaved by a nuclease that is activated by caspases exclusively in apoptosis. It also became clear that at least some elements of this mechanism are highly conserved among species. The endonuclease activity was shown to depend on two interacting molecules. The first was an unknown protein named ICAD (inhibitor of caspase-activated DNase), or DFF45 (DNA fragmentation factor), respectively, in mice and humans. This factor is expressed in many types of cells in an inactive, latent form. The second molecule of the partnership is CAD (caspase-activated endonuclease), or DFF40. The first protein (ICAD/DFF45) binds to, stabilizes and inactivates CAD/DFF40. During apoptosis ICAD/DFF45 releases

CAD/DFF40 after caspase-3 digestion, and it is ICAD/DFF45, not CAD/DFF40, that possesses caspase cleavage sites.

Although the significance of cleavage of some caspase substrates has been identified, the functional consequences of cleavage of many of these substrates still remain to be elucidated. Moreover, the association of cleavage of these substrates with other apoptotic events requires further investigation.

It has also emerged that apoptosis-associated caspases play non-apoptotic roles and that they are not simply destructive. Examples of these are cell differentiation, embryonic development, motility, compensatory proliferation, or tumor suppression (for review, see Connolly et al., 2014). The first example of a non-apoptotic function of caspases was presented in 1999 (De Maria et al., 1999). DeMaria and colleagues showed that exposure of erythroid progenitors to mature erythroblasts or death-receptor ligands resulted in caspase-mediated degradation of the transcription factor GATA-1, which is associated with impaired erythroblast development. The enzyme responsible for this process was identified as caspase-3. Thus, caspase-mediated cleavage of GATA-1 represented an important negative control mechanism in erythropoiesis. Transient activation of caspase-3 was also observed during T-cell stimulation and allogenic mixed lymphocyte reaction. Moreover, caspase-3 plays a role in cell cycle regulation of B-cells. Within these and related processes the 'killer' caspases clearly do not cause cell elimination, suggesting an entirely new role for caspases in determining the fate or behavior of cells.

At the end of the 1990s several biotechnology and pharmaceutical companies made attempts to develop caspase inhibitors for therapeutic uses. Unfortunately, almost all of these attempts failed, since in addition to apoptotic functions, the activity of caspases during normal physiological processes was inhibited. Several preclinical and early clinical studies suggest a potential benefit of caspase inhibitors in liver injury. Caspase inhibitors for human use were developed by IDUN

Pharmaceuticals, Gilead Sciences, Inc, and Vertex Pharmaceuticals. However, only one proof-of-concept trial has been performed in patients with Hepatitis C virus (HCV). This study employed IDN-6556, which was developed by IDUN Pharmaceuticals (currently part of Pfizer) and is an irreversible pan-caspase inhibitor. In the HCV patient population, all doses of IDN-6556 significantly lowered alanine aminotransferase and aspartate aminotransferase (Pockros et al., 2007). A human trial with IDN-6556 was also conducted in organ preservation injury. IDN-6556, when administered in cold storage and flush solutions during liver transplantation, seems to offer local therapeutic protection against cold ischemia/warm reperfusion liver injury (Baskin-Bey et al., 2007). Unfortunately, significant methodological issues including small sample size, short follow-up, and use of surrogate markers, mar these studies. In particular, none of these agents has been shown to consistently decrease the critical outcomes of mortality or time to transplant.

Concluding remarks

Since the discovery of caspases in the mid-1990s and the description of their role in cell death, significant progress has been made that is essential in the context of normal cell physiology and disease. We now know how and in which complex(es) caspases are activated and the consequences of their activation, and have validated many of their substrates. However, many unanswered questions remain to be addressed. For example, one of the greatest challenges is to understand what drives the same protein (for example, caspase-8) to be a part of different complexes and how to influence these processes. The recently introduced protein-protein database for the DD superfamily will be helpful to study the DD superfamily-mediated formation of cell-death-activating complexes. Beyond being solely of academic interest, data obtained might allow us to therapeutically manipulate cell death in various pathological processes.

Regardless of how caspases are regulated, it is clear that they do have roles other than as effectors of cell death. This also leads to unexpected complications in situations where caspase-dependent cell death is considered desirable, such as in response to cancer chemotherapy. The newfound alternative roles of caspases present the possibility that chemotherapy drugs may induce a wide range of cell behaviors such as increased migration and compensatory proliferation of cancer cells that are both unexpected and unwanted.

Although the results of therapeutic use of caspase inhibitors are intriguing, additional carefully designed studies with adequate methodology, patient sample size, and follow-up need to be performed before any of these medications may be recommended for the treatment of different diseases. Thus, caspases can help to maintain various physiological processes in cells fulfilling non-apoptotic functions; they can act as killers during apoptosis. The near future will provide us with many more surprises in understanding their functions.

References:

Alnemri ES, Livingston DJ, Nicholson DW, Salvesen G, Thornberry NA, Wong WW, Yuan J. Human ICE/CED-3 protease nomenclature. Cell. 1996; 87: 171.

Baskin-Bey ES, Washburn K, Feng S, Oltersdorf T, Shapiro D, Huyghe M, Burgart L, Garrity-Park M, van Vilsteren FG, Oliver LK, Rosen CB, Gores GJ. Clinical trial of the pan-caspase inhibitor, IDN-6556, in human liver preservation injury. Am J Transplant. 2007; 7:218–25.

Boldin, M. P., Goncharov, T.M., Goltsev, E.V., Wallach, D. Involvement of MACH, a Novel MORT1/FADD-Interacting Protease, in Fas/APO-1- and TNF Receptor–Induced Cell Death. Cell. 1996; 85: 803-815.

Bowen I. D. Laboratory techniques for demonstrating cell death. In: Cell Ageing and Cell Death, 1984; vol. 25, pp. 1 38, Davis I. and Sijee D (eds), Cambridge University Press, Cambridge.

Cho YS1, Challa S, Moquin D, Genga R, Ray TD, Guildford M, Chan FK. Phosphorylation-driven assembly of the RIP1-RIP3 complex regulates programmed necrosis and virus-induced inflammation. Cell. 2009; 137:1112-23.

Connolly PF, Jäger R, Fearnhead HO. New roles for old enzymes: killer caspases as the engine of cell behavior changes. Front Physiol. 2014; 5:149.

De Maria R, Zeuner A, Eramo A, Domenichelli C, Bonci D, Grignani F, Srinivasula SM, Alnemri ES, Testa U, Peschle C. Negative regulation of erythropoiesis by caspase-mediated cleavage of GATA-1. Nature. 1999; 401:489-93.

Dickens LS, Boyd RS, Jukes-Jones R, Hughes MA, Robinson GL, Fairall L, Schwabe JW, Cain K, Macfarlane M. A death effector domain chain DISC model reveals a crucial role for caspase-8 chain assembly in mediating apoptotic cell death. Mol Cell. 2012; 47: 291-305.

Eastman A., Grant S., Lock R., Tritton T., Van Houten N., Yuan J. Cell Death in Cancer and Development: AACR Special Conference in Cancer Research. Cancer Res. 1994; 54: 2812-8.

Ellis HM, Horvitz HR. Genetic control of programmed cell death in the nematode C. elegans. Cell. 1986; 44: 817-29.

Enari M, Sakahira H, Yokoyama H, Okawa K, Iwamatsu A, Nagata S. A caspase-activated DNase that degrades DNA during apoptosis, and its inhibitor ICAD. Nature. 1998; 391:43-50.

Feoktistova M1, Geserick P, Kellert B, Dimitrova DP, Langlais C, Hupe M, Cain K, MacFarlane M, Häcker G, Leverkus M. cIAPs block Ripoptosome formation, a RIP1/caspase-8 containing intracellular cell death complex differentially regulated by cFLIP isoforms. Mol Cell. 2011; 43:449-63.

Galluzzi L, Joza N, Tasdemir E, Maiuri MC, Hengartner M, Abrams JM, Tavernarakis N, Penninger J, Madeo F, Kroemer G. No death without life: vital functions of apoptotic effectors. Cell Death Differ. 2008; 15: 1113-23.

Hengartner MO. Apoptosis. CED-4 is a stranger no more. Nature. 1997; 388: 714-5.

Imre G1, Heering J, Takeda AN, Husmann M, Thiede B, zu Heringdorf DM, Green DR, van der Goot FG, Sinha B, Dötsch V, Rajalingam K. Caspase-2 is an initiator caspase responsible for pore-forming toxin-mediated apoptosis. EMBO J. 2012; 31: 2615-28.

Kaufmann S. H. Induction of endonucleolytic DNA cleavage in human acute myelogenous leukemia cells by etoposide, camptothecin and other cytotoxic anticancer drugs: a cautionary note. Cancer Res. 1989; 49: 5870-8.

Khanson KP. Molecular mechanisms of the interphase death of lymphoid cells. Radiobiologiia. 1979; 19: 814-20 (Russian).

Kischkel FC1, Hellbardt S, Behrmann I, Germer M, Pawlita M, Krammer PH, Peter ME. Cytotoxicity-dependent APO-1 (Fas/CD95)-associated proteins form a death-inducing signaling complex (DISC) with the receptor. EMBO J. 1995; 14: 5579-88.

Klein JL, Shows TB, Dupont B, Trapani JA. Genomic organization and chromosomal assignment for a serine protease gene (CSPB) expressed by human cytotoxic lymphocytes. Genomics. 1989; 5: 110-7.

Lamkanfi M, Dixit VM. Inflammasomes and Their Roles in Health and Disease. Annu Rev Cell Dev Biol 2012;28:137–161.

Li H1, Zhu H, Xu CJ, Yuan J. Cleavage of BID by caspase 8 mediates the mitochondrial damage in the Fas pathway of apoptosis. Cell. 1998; 94: 491-501.

Li P, Nijhawan D, Budihardjo I, Srinivasula SM, Ahmad M, Alnemri ES, Wang X. Cytochrome c and dATP-dependent formation of Apaf-1/caspase-9 complex initiates an apoptotic protease cascade. Cell. 1997; 91: 479-89.

Liu X, Li P, Widlak P, Zou H, Luo X, Garrard WT, Wang X. The 40-kDa subunit of DNA fragmentation factor induces DNA fragmentation and chromatin condensation during apoptosis. Proc Natl Acad Sci U S A. 1998; 95:8461-6.

Liu X, Zou H, Slaughter C, Wang X. DFF, a heterodimeric protein that functions downstream of caspase-3 to trigger DNA fragmentation during apoptosis. Cell. 1997; 89:175-84.

Lobe CG, Finlay BB, Paranchych W, Paetkau VH, Bleackley RC. Novel serine proteases encoded by two cytotoxic T lymphocyte-specific genes. Science. 1986; 232: 858-61.

Lockshin R. A. Cell death in metamorphosis. In: Cell Death in Biology and Pathology, 1981, pp. 79-121, Bowen I. D. and Lockshin R. A (eds), Chapman and Hall.

Lockshin R. A. Lysosomes in insects. In: Lysosomes in Biology and Pathology, 1969, pp. 363 391. Dingle J. T. and Fell H. B. (eds), North-Holland, Amsterdam.

Manzl C, Krumschnabel G, Bock F, Sohm B, Labi V, Baumgartner F, Logette E, Tschopp J, Villunger A. Caspase-2 activation in the absence of PIDDosome formation. J Cell Biol 2009;185: 291–303.

Masson D, Zamai M, Tschopp J. Identification of granzyme A isolated from cytotoxic T-lymphocyte-granules as one of the proteases encoded by CTL-specific genes. FEBS Lett. 1986; 208: 84-8.

Miura M., Zhu H., Rotello R., Hartwieg E. A. and Yuan J. Induction of apoptosis in fibroblasts by IL-lb-converting enzyme, a mammalian homolog of the *C. elegans* cell death gene ced-3. Cell 1993; 75: 653-60.

Muzio M, Chinnaiyan AM, Kischkel FC, O'Rourke K, Shevchenko A, Ni J, Scaffidi C, Bretz JD, Zhang M, Gentz R, Mann M, Krammer PH, Peter ME, Dixit VM. FLICE, a novel FADD-homologous ICE/CED-3-like protease, is recruited to the CD95 (Fas/APO-1) death--inducing signaling complex. Cell. 1996; 85: 817-27.

Ofengeim D, Yuan J. Regulation of RIP1 kinase signalling at the crossroads of inflammation and cell death. Nat Rev Mol Cell Biol 2013; 14: 727–36.

Pockros PJ, Schiff ER, Shiffman ML, McHutchison JG, Gish RG, Afdhal NH, Makhviladze M, Huyghe M, Hecht D, Oltersdorf T, Shapiro DA. Oral IDN-6556, an antiapoptotic caspase inhibitor, may lower aminotransferase activity in patients with chronic hepatitis C. Hepatology. 2007; 46:324–9.

Rehm M1, Huber HJ, Hellwig CT, Anguissola S, Dussmann H, Prehn JH. Dynamics of outer mitochondrial membrane permeabilization during apoptosis. Cell Death Differ. 2009; 16: 613-23.

Sakahira H, Enari M, Nagata S. Cleavage of CAD inhibitor in CAD activation and DNA degradation during apoptosis. Nature. 1998; 391:96-9.

Schleich K, Warnken U, Fricker N, Oztürk S, Richter P, Kammerer K, Schnölzer M, Krammer PH, Lavrik IN. Stoichiometry of the CD95 death-inducing signaling complex: experimental and modeling evidence for a death effector domain chain model. Mol Cell. 2012; 47: 306-19.

Shi L1, Kraut RP, Aebersold R, Greenberg AH. A natural killer cell granule protein that induces DNA fragmentation and apoptosis. J Exp Med. 1992; 175: 553-66.

Skalka M, Matyásová J, Cejková M. DNA in chromatin of irradiated lymphoid tissues degrades in vivo into regular fragments. FEBS Lett. 1976; 72: 271-4.

Soldatenko VA, Denisenko MF, Alferova TM, Filippovich IV. Chromatin degradation in the death of thymic lymphocytes induced by radiation or dexamethasone: the necessity for a stage of preliminary proteolysis. Radiobiologiia. 1991; 31: 180-7 (Russian).

Tenev T1, Bianchi K, Darding M, Broemer M, Langlais C, Wallberg F, Zachariou A, Lopez J, MacFarlane M, Cain K, Meier P. The Ripoptosome, a signaling platform that assembles in response to genotoxic stress and loss of IAPs. Cell. 2011; 43:432-48.

Tinel A, Janssens S, Lippens S, et al. Autoproteolysis of PIDD marks the bifurcation between pro-death caspase-2 and pro-survival NF-kappaB pathway. EMBO J 2007; 26:197–208.

Tinel A, Tschopp J. The PIDDosome, a protein complex implicated in activation of caspase-2 in response to genotoxic stress. Science 2004; 304:843–6.

Vaculova, A., and Zhivotovsky, B. Caspases: determination of their activities in apoptotic cells. Meth. Enzymol. 2008; 442: 157-181.

Vakifahmetoglu H, Olsson M, Orrenius S, Zhivotovsky B. Functional connection between p53 and caspase-2 is essential for apoptosis induced by DNA damage. Oncogene 2006; 25: 5683–92.

Williams M. S. and Henkart P. A. Apoptotic cell death induced by intracellular proteolysis. J. Immunol. 1994; 153:4247-55.

Wyllie AH. Glucocorticoid-induced thymocyte apoptosis is associated with endogenous endonuclease activation. Nature. 1980; 284: 555-6.

Yamada T, Ohyama H. Radiation-induced interphase death of rat thymocytes is internally programmed (apoptosis). Int J Radiat Biol Relat Stud Phys Chem Med. 1988; 53: 65-75.

Yousefi S1, Perozzo R, Schmid I, Ziemiecki A, Schaffner T, Scapozza L, Brunner T, Simon HU. Calpain-mediated cleavage of Atg5 switches autophagy to apoptosis. Nat Cell Biol. 2006; 8: 1124-32.

Yuan J., Shaham S., Ledoux S., Ellis H. M. and Horvitz H. R. The C. Elegans cell death gene ced-3 encodes a protein similar to mammalian interleukin-lb-converting enzyme. Cell 1993; 75:641-52.

Zhivotovsky BD, Zvonareva NB, Voskoboïnikov GV, Khanson KP. Molecular mechanisms of the interphase death of lymhoid cells. 2. The comparative characteristics of the postradiation degradation and nuclease digestion products of rat thymus chromatin. Radiobiologiia. 1980; 20: 502-7 (Russian).

Chapter 6: Seeking day jobs for all apoptosis-related factors – inside one perspective (Hardwick)

J. Marie Hardwick

W. Harry Feinstone Department of Molecular Microbiology and Immunology, Johns Hopkins University Bloomberg School of Public Health, Baltimore, MD 21205 USA

Running title: The outside perspective on cell death

Correspondence: email: hardwick@jhu.edu

Keywords: apoptosis, programmed cell death, mitochondria, Bcl-2, model organisms

Abstract

Cell death factors were once thought to be sequestered in a dormant state in healthy cells, but are now being discovered to have alternative functions in healthy cells. These healthy-cell functions are likely related biochemically to their death-jobs. Here I summarize my laboratory's unique contributions and perspectives that led us to pursue the day-jobs of death factors, with the overall hypothesis that evolution linked their key day-jobs to cell death to ensure that the billions of deaths per day per individual remain balanced. This search also led to new insights into the power of cell death in driving genome evolution, which has important implications for cancer.

Gaining a perspective on cell death

Powerful insights that shaped the cell death field

To understand genetically controlled cell death, each investigator develops a unique perspective shaped by their own experiments and the insights of others. Particularly influential to my laboratory were the early acumens provided by Hamburger, Levi-Montalcini, Lockshin, Horvitz and others who had observed these events and conceptualized deliberate cell death during animal development (Ellis and Horvitz, 1986; Hamburger and Levi-Montalcini, 1949; Lockshin and Williams, 1964). The cell death meme began after John Isaacs, Sten Orrenius, Doug Green, John Hickman, Peter Krammer, Barbara Osborne, Stan Korsmeyer and many others popularized the previously coined term "apoptosis" through elegant studies in tumor cells and lymphocytes. The core apoptosis machineries and their pathways were mapped by the pioneering genetic screens from Horvitz and from Steller working on invertebrate model organisms (Ellis and Horvitz, 1986; Steller, 1995). A foundation for biochemical dissection of these pathways was laid by independent studies from Lazebnik (Lazebnik et al, 1994) and from Newmeyer (Newmeyer et al, 1994), who recapitulated the dying process using cell-free extracts. The assignment of a biochemical function to the first cell death effector molecule, a protease that would eventually be called a caspase, by Pickup (Ray et al, 1992) and by Yuan (Yuan et al, 1993), and the caspase specificities, activation and regulatory mechanisms from Thornberry, Dixit, Salvesen, Xiaodong Wang and others (Komiyama et al, 1994; Li et al, 1997; Salvesen and Dixit, 1997; Tewari et al, 1995; Thornberry et al, 1992), launched an enormous momentum. On another front, Krammer and Debatin, Osborne and Schwartz, Driscoll, Nagata, Henson and others revealed early evidence for cell death signaling in neurons, lymphocytes and the engulfment of apoptotic cells (Driscoll, 1992; Fadok et al, 1992; Itoh et al, 1991; Schwartz and Osborne, 1993; Trauth et al, 1989). Definitive evidence that apoptosis governs viral pathogenesis was revealed by Clem and Miller in insects (Clem et al, 1991), and by Levine and Huang in our group for mammals (Levine et al, 1993). A revolutionary discovery was the B-cell lymphoma gene BCL-2, and that BCL-2 and its viral mimics might promote tumorigenesis by suppressing apoptosis, instead of stimulating cell division, as demonstrated by Vaux, Korsmeyer, Croce and Tsujimoto, Eileen White, Rickinson and others (Henderson et al, 1993; Korsmeyer, 1992; Tsujimoto et al, 1984; Vaux and Korsmeyer, 1999; White et al, 1991). As a consequence, many scientists began turning their attention to the matter of dying purposefully. This core concept that cell death inhibition (not just cell division) is a component of tumorigenesis, as well as developmental processes, now permeates biology. Cell death is an essential ongoing process that when unbalanced likely underlies many human diseases beyond cancer and neurodegenerative disorders. While extensive knowledge has been gained from the insightful works of thousands of investigators, major fundamental questions remain. For example, how do BCL-2 family proteins really work, what are the day-jobs of caspases that may directly link survival and death, and how do cells know when to reveal their weaknesses that trigger suicide and their disposal by other cells? Knowing these and other fundamentals of cell death will help devise more effective tools for the clinic, and possibly also inform complex societal questions.

Troubling nomenclature – evolved versus accidental cell death

While nomenclature is essential for communication, there is the inevitable struggle to keep pace with new discoveries, and such struggles continue in the cell death field. Richard Lockshin originally applied the term programmed cell death to describe developmental processes (Lockshin and Williams, 1964). Subsequently, the term apoptosis was coined with the intention of describing deliberate programmed cell death in both homeostatic and pathological events (Kerr et al, 1972). However, because this original definition of apoptosis was supported only by morphological evidence, the term apoptosis has come to be commonly reserved for death mediated predominantly by mammalian caspases-9, -8 and-3 and their orthologous counterparts that cause classical apoptotic morphologies. By contrast, the literature commonly applies the term programmed cell death with an even broader definition that extends to all gene-dependent death, while others avoid potential miscommunication by dropping the word programmed, referring simply to cell death (Dondelinger et al, 2014; Moriwaki and Chan, 2014).

Recently, the Nomenclature Committee on Cell Death (NCCD) recommended that programmed cell death (PCD) be applied to developmental plus homeostatic processes, which constitute only a subset of a new larger category referred to as regulated cell death (RCD) (Galluzzi et al, 2015). Regulated cell death is roughly defined as evolutionarily selected (gene-dependent) cell death, and therefore could be manipulated genetically or therapeutically by altering the functions of those gene products. One arguable point is the Nomenclature Committee's definition of accidental cell death, suggested to be equivalent to cell death by direct assault (gene-independent cell death in which the dying cell makes no contribution to its own death (Galluzzi et al, 2015). The Nomenclature Committee's apparent dismissal of the possibility that genes might occasionally cause cell death unintentionally (genes that did not evolve for this purpose), together with their many cell death metaphors, has set themselves up for some sparring with evolutionary biologists (Aouacheria et al, 2013; Reynolds, 2014).

Turning death factors on their heads – truth is stranger than fiction

Pro-survival and pro-death activities housed in the same BCL-2 proteins

In the earlier years of the molecular cell death field, through the work of Emily Cheng, Rollie Clem, Chris Karp, David Kirsch, Christopher Karp, Soyoung Seo, David Bellows and Victor Nava in my group, we discovered that BCL-2 family proteins, including the anti-apoptotic BCL-2 and BCL-xL proteins and pro-apoptotic BAD, but not viral BCL-2 homologs, are cleaved by caspases, both in reconstituted in vitro systems and in apoptotic cells (Bellows et al, 2000; Bellows et al, 2002; Cheng et al, 1997a; Cheng et al, 1997b; Clem et al, 1998; Kirsch et al, 1999; Nava et al, 1997; Nava et al, 1998; Seo et al, 2004). This cleavage event releases a deadly C-terminal fragment, converting anti-death factors into pro-death factors. In contrast, the equivalent engineered truncations of most all viral BCL-2 homologs fail to kill cells (Cheng et al, 1997a; Clem et al, 1998; Seo et al, 2004).

Soon after, pro-apoptotic BID was shown by others to be cleaved by caspase-8, releasing the well-studied truncated BID (tBID) that directly activates BAX and suppresses BCL-2/BCL-xL to permeabilize the outer mitochondrial membrane and cause apoptosis (Bellows et al, 2000; Li et al, 1998; Luo et al, 1998). This remains the famed crosstalk mechanism connecting the death receptor-signaling (extrinsic) caspase-8 pathway with the mitochondrial (intrinsic) caspase-9 death pathway.

At first we assumed that the function of BCL-2/BCL-xL prior to cleavage was limited to their anti-apoptotic activity (inhibiting BAX/BAK-induced apoptosis). However, work by Sarah Berman, Yingbei Chen, Yihru Fannjiang and Brian Polster in our lab, and additional studies with long-term collaborators Liz Jonas and Len Kaczmarek (Yale University) and Suzanne Zukin (Einstein), demonstrated that BCL-2 family proteins have apoptosis-unrelated roles prior to activation of the cell death pathway, including the regulation of mitochondrial dynamics (fission/fusion/biogenesis) and energetics, at least in part by limiting excess flux of ions across the inner mitochondrial membrane, where at least some endogenous BCL-2 family proteins reside, contrary to conventional wisdom (Hardwick et al, 2012). Many assumed that pro-apoptotic proteins such as BAD and BID were simply inert inactive killers prior to apoptosis. The alternative possibility is that, like BCL-2/BCL-xL, caspases also abolish the largely unknown day-job functions of these BH3-only proteins BAD and BID in normal healthy cells. In fact, BAD has critical functions in regulating glucose metabolism (Stanley et al, 2014). Perhaps even BID has another yet uncharacterized function before conversion to a killer, such as normal DNA damage responses. Its affinity for cardiolipin, found primarily on the mitochondrial inner membranes, may also be an important clue.

The eight canonical BH3-only proteins, and potentially the dozens of other factors reported to contain BH3 motifs, have suggested roles as sentinels of cell stress. This is consistent with the selective preferences for specific caspase cleavage sites in BAD. Cleavage at any particular site is highly dependent on the specific type of death stimulus (Seo et al, 2004). Cleavage at one caspase site in BAD is required for cell death following growth factor withdrawal but is irrelevant following gamma irradiation or virus infection, while a different BAD cleavage site is required for gamma irradiation and virus-induced apoptosis, but not growth factor withdrawal (Seo et al, 2004).

We uncovered even more unexpected results studying the brains of Bax and Bak knockout mice, suggesting that even these direct effectors of apoptosis could also have day-jobs in healthy cells. In contrast to expectation and going against current dogma, Yihru Fannjiang, Jennifer Lewis and George Oyler in our group found arguably the most pronounced phenotype observed to date for Bax or Bak knockout mice. Compared to control littermates, the knockout mice are highly susceptible to fatal infections with the acute neuronotropic Sindbis virus, accompanied by increased neuronal death (Fannjiang et al, 2003; Lewis et al, 1999). Even more remarkable, using Sindbis viruses as Bax/Bak expression vectors to reconstitute Bax and Bak knockout mice, also rescued these knockout mice from lethal infections (Fannjiang et al, 2003; Lewis et al, 1999). These protective effects of Bax and Bak are not limited to virus infections. Our extensive study of Bak knockouts revealed that Bak could be strongly pro-death or anti-death in a range of other neurological disease models (e.g. Parkinson's disease, epilepsy and stroke), dependent on the neuronal subtype, developmental stage and specific stimulus. Protection from cell death by Bax and Bak was also observed in cultured brain slices but never in dissociated neuron cultures, implying network involvement (Fannjiang et al, 2003; Lewis et al, 1999).

Particularly intriguing was that Bak knockouts had more severe seizures within a few minutes after kainic acid injection, long before neuronal death was detected (Fannjiang et al, 2003).

This was the first clue leading us to investigate the role of BCL-2 family proteins in synaptic activity of healthy neurons through a long-standing collaboration with Jonas, Kaczmarek and Hickman carried out in the summer labs at the Marine Biological Laboratory in Woods Hole.

To investigate the in vivo consequences of caspase cleavage of Bcl-xL, Yingbei Chen in my lab constructed a knockin mouse with point mutations in the two caspase cleavage sites of Bcl-xL, D61A and D76A, rendering Bcl-xL uncleavable by caspases (Chen et al; Fujita et al, 1998; Ofengeim et al, 2012). Ofengeim and Zukin showed that these knockin mice were resistant to ischemic brain injury (Ofengeim et al, 2012). They found that ABT-737, a small molecule specific inhibitor of BCL-xL, protected rats from ischemic injury, the opposite to expected for an pro-death cancer therapeutic (Ofengeim et al, 2012). Although ABT-737 promotes cell death in tumor cells, its ability to suppress neuronal death following transient ischemia suggests that ABT-737 may also inhibit the pro-death fragment of BCL-xL. Thus cleavage of BCL-xL by caspases may not only inactivate its anti-death function but also activate a death function in whole animals. These mice also have interesting thymic phenotypes currently under study.

Day-jobs for caspases

We were puzzled in the early 1990's to find that we were unable to create stable cell lines expressing the direct caspase inhibitors CrmA of cowpox virus or P35 of baculovirus (unpublished). We consistently obtained hundreds of cell colonies under drug selections with control empty vectors, but not with these caspase inhibitors, apparently because caspases are important for cell growth and/or survival, or possibly by inhibiting non-apoptotic necroptosis (Degterev et al, 2008). In either case, these findings spurred our interest in the day-jobs of caspases.

Caspases are phylogenetically conserved cysteine proteases, some of which can promote apoptosis (e.g. caspases-9, -8, and -

3), while others promote non-apoptotic necrotic cell death by pyroptosis (caspases-1 and -11) (Lamkanfi and Dixit, 2014). Caspases also have additional functions, for example caspase-8 inhibits RIPK-dependent necroptotic cell death, and caspase-1 is a key activator of innate immune responses to infection (de Rivero Vaccari et al, 2014; Hyman and Yuan, 2012; Li and Yuan, 2008). In addition, caspase-1 has also been suggested to have alternative substrates, including Bid (Pham et al, 2012) and Bcl-xL (Cheng et al, 1997a; Clem et al, 1998), and ongoing efforts seek to identify additional targets (Pham et al, 2014). Yet other mammalian and invertebrate caspases have poorly defined or unknown functions (Nutt et al, 2005; Yang et al, 2010).

Particularly in mammalian and *Drosophila* neurons, caspases that induce apoptosis in other cell types were found to be critical for normal neuronal activity by pruning neuronal endings where synapses form to communicate between neurons. The pruning of some synapses is required to strengthen other synapses in normal brain (Erturk et al, 2014; Koto et al, 2009; Lee et al, 2013). The day-job mechanisms of caspases are currently under investigation, but may not be limited to apoptosis-like dismantling of synaptic endings. Bcl-xL is an anti-apoptotic Bcl-2 family protein critical for neuron survival, but our collective works have found that Bcl-xL and other Bcl-2 family proteins appear to have additional functions in regulating cellular energetics (Cheng et al, 1997a; Cheng et al, 1996; Cheng et al, 1997b) and controlling synaptic activity (Alavian et al, 2011; Chen et al, 2011; Fannjiang et al, 2003; Hickman et al, 2008; Jonas et al, 2003; Li et al, 2008).

We recently embarked on a new approach to study the day-jobs of caspases using *Drosophila*, inspired by the earlier work of Hogan (Ho Lam) Tang and Ming Chiu Fung (Hong Kong University) (Tang et al, 2012; Tang et al, 2009). Hogan Tang and Holly Tang in my lab constructed a new caspase biosensor, designated CaspaseTracker (Tang et al, 2015a). CaspaseTracker combines a natural caspase cleavage site from DIAP1 (*Drosophila* IAP/inhibitor of apoptosis 1) (Bardet et al, 2008), which was mutated to prevent it from inhibiting caspases (Li et al, 2011). This caspase-cleavable sensor was combined with the GAL4-inducible *Drosophila* G-Trace system (Evans et al, 2009). CaspaseTracker permanently marks any cell that has experienced caspase activity at any time in its past (Tang et al, 2015a). Biosensor activity is inducible in *Drosophila* egg chambers by starvation and cold-shock, indicating caspase activation as expected (Tang et al, 2015a). Specificity of the biosensor for caspases is demonstrated by a control biosensor lacking any basal or inducible biosensor activity because of a single point mutation in the caspase cleavage site. More remarkable, and similar to earlier experiments with cultured cells (Tang et al, 2015b; Tang et al, 2012; Tang et al, 2009), if flies are allowed to recover from starvation and cold shock for 3 days, healthy CaspaseTracker egg chambers reappear indicating recovery from apoptosis, termed "anastasis" (Tang et al, 2012).

However, the most obvious phenotype of the CaspaseTracker fly is the basal biosensor activity in untreated flies raised under optimized laboratory conditions. That is, many morphologically healthy cells throughout the internal organs of the fly exhibit caspase biosensor activity. Biosensor-positive cells are particularly abundant in the brain and optic lobes, but are also prominent throughout the digestive track, the kidney tubules and other organ systems. Somewhat geometric patterns of biosensor-positive cells in some tissues suggest a functional organization of cells that have ongoing or past caspase activity (Tang et al, 2015a).

Do yeast and mammals share molecular cell death programs?

Yeast as a model for cell death

My desire to take advantage of the "awesome power" of yeast genetics to study gene-dependent cell death was resisted by the members of my lab until the day Madeo and colleagues reported the pro-death function of metacaspase *MCA1/YCA1* encoded by *Saccharomyces cerevisiae* (Madeo et al, 2002). The skepticism expressed by thoughtful lab members turned out to be mild, even invigorating by comparison to the vehement rejection by the cell death and genetics fields regarding the possible existence of genetically encoded death programs in unicellular organisms, despite earlier support for this idea (Ameisen, 2002). However, a sequential series of students in my lab, Yihru Fannjiang, Wen-Chih Cheng, Xinchen Teng and Margaret Dayhoff-Brannigan, drove the point home. Our initial efforts using *Saccharomyces cereivsiae* provided compelling evidence that yeast may have a purpose in dying, for example in response to pathogenic yeast viruses that induce gene-dependent cell death (Cheng et al, 2008a; Ivanovska and Hardwick, 2005), similar to our earlier work when the roles of cleaved BCL-2 and BCL-xL in mammalian apoptosis were doubted (Cheng et al, 1997a; Cheng et al, 1996; Levine et al, 1993; Lewis et al, 1999; Ubol et al, 1994). Our early studies also identified a conserved protein that can promote cell death (after stress) in both yeast and mammalian cells, apparently by related mechanisms. The human mitochondrial fission factor Drp1, identified through studies in yeast (Bleazard et al, 1999; Ingerman et al, 2005; Smirnova et al, 2001), had recently been reported to collaborate with BAX at the mitochondrial outer membrane to promote mammalian cell death (Cassidy-Stone et al, 2008; Frank et al, 2001; Karbowski et al, 2002; Montessuit et al, 2010). We showed that deletion of the yeast homolog Dnm1 renders yeast significantly resistant to several death-inducing stresses, and cell death required its GTPase activity, similar to mammals (Berman et al, 2009; Fannjiang et al, 2004; Hardwick and Cheng, 2004; Li et al, 2008). Others have confirmed our results under different conditions in yeast (Bink et al, 2010; Palermo et al, 2007; Scheckhuber et al, 2007; Weinberger et al, 2010), worms (Breckenridge et al, 2003; Jagasia et al, 2005), flies (Abdelwahid et al, 2011; Goyal et al, 2007) and mice (Frank et al, 2001; Wakabayashi et al, 2009).

Yeast cell death studies reveal widespread mutation-driven selection for new mutations

Being accustomed to finding opposing anti-/pro-death functions of mammalian BCL-2 family proteins, we did not flinch when knockouts of the conserved Dnm1-binding mitochondrial receptor, Fis1 (Cerveny et al, 2007; Shaw and Nunnari, 2002), had the opposite phenotype – cell death sensitive. We collaborated with Blake Hill to solve the structure of Fis1, but like several other groups who also solved the structure of Fis1 from a range of species, this alpha-helical tail-anchor protein lacked a 'BCL-2' fold (Dohm et al, 2004; Koirala et al, 2010; Suzuki et al, 2003; Zhang and Chan, 2007). As we became experienced yeast geneticists, we made another surprising discovery. Several *FIS1* knockout yeast strains that were made independently and acquired from different sources, each also had acquired a unique second mutation in the gene (*WHI2*), which is responsible for the cell death phenotype (Cheng et al, 2008a; Cheng et al, 2008b; Teng et al, 2011; Teng et al, 2013; Teng and Hardwick, 2013). In contrast to some reviewers' suggestions that Fis1 was unrelated to the cell death phenotype, our ultimate conclusion was that Fis1 is indeed a critical pro-survival factor, such that compensatory mutations (e.g. in *WHI2* or *SIN3*) are nearly required to sustain long-term viability of *fis1*-deficient strains. Others have confirmed these secondary mutations (Bink et al, 2010; Lang et al, 2013; Mendl et al, 2011; Szamecz et al, 2014). Importantly, this was the first clue that would eventually lead to another controversial finding - the high prevalence of a single meaningful secondary mutation that arose at least partly in response to losing the function of the first deleted gene. Xinchen Teng in our lab demonstrated that an estimated 75% of knockout strains have acquired a single second site mutation affecting cell death and/or nutrient-sensing. That is, one gene mutation begets two gene mutations (Cheng et al, 2008b; Teng et al, 2013). Cell death studies using yeast by a growing community of investigators have not yet revealed a multi-step death pathway analogous to the early work in *C. elegans* by Horvitz and colleagues, but these are very early days.

Extended perspective

Our fascination with the unexpected has led us to develop uncommon model systems (including many not discussed here, such as mosquitoes, squid and others). By exploring the unknown mechanisms of viral pathogenesis and the role of apoptosis regulators in virus-infected mammalian brains, we uncovered day-jobs for pro-death BCL-2 family members and caspases. Additional factors that have been investigated by the other talented members of my lab whose works are not discussed here, are reserved for a future article. My background as a virologist heavily influenced my perspective on cell death, being accustomed to the knife's edge that separates protective versus detrimental host responses to infection. To understand the molecular events that link health and death will require dissecting the fulcrum point between the anti-death and pro-death functions belonging to the same protein. Estimates from Xinchen Teng's yeast genomic screens in our lab indicate that 800-2000 genes (~15-30% of the genome) strongly contribute to cell death following stress based on knockout studies following stress (Teng et al, 2013). Thus, there may be a multitude of initiation steps towards cell death, and potentially many conserved alternative yet undiscovered death pathways.

The established evidence that viruses and other pathogens exist in nature as complex quasispecies with considerable genomic adaptability provided us with an important perspective while uncovering the role of cell death (and avoidance thereof) in driving genome evolution. The evidence we have obtained indicates that almost any functional mutation may be capable of causing a genomic disturbance sufficient to drive the selection for a new specific compensatory mutation that affects cell growth and/or cell death. This seems both obvious and logical (suppressor mutations), but this possibility is difficult to address in human tumorigenesis. However, it is quite feasible to identify relevant pairs of genes using yeast, each pair of mutant genes consisting of the first mutation (knockout gene) and the compensatory second mutant gene, which subsequently can be identified by genetics and genome sequencing. Interestingly, the human homologs of these pairs of mutant yeast genes also significantly co-occur in specific human tumors, even though the frequency at which either one of these genes is mutated in cancer may be low. This has important implications for tumorigenesis that currently are far from mainstream hypotheses. While DNA mutations are initially acquired randomly, the next step in evolution is the selection for the small fraction of these mutations that are (partially) compensatory, and this selection process is likely not to be very random. However, the network of compensating mutations, based on our yeast studies, is likely to draw a new map, and will likely be permeated by cell death regulators with dual roles in healthy cells. This is the beginning of an exciting new era.

Acknowledgements

I sincerely thank the tireless efforts of all the past and current members of the Hardwick lab, who are responsible first hand for the studies described here. I especially thank Emily Cheng, Rollie Clem, Xinchen Teng and Wen-Chih Cheng for their thoughtful comments on this manuscript. The work described here was supported by multiple awards, and most recently by 5R01NS037402, 1R01NS083373 and 2RO1GM077875.

References

E. Abdelwahid, S. Rolland, X. Teng, B. Conradt, J.M. Hardwick, and K. White. Mitochondrial involvement in cell death of non-mammalian eukaryotes. Biochim Biophys Acta 1813 (2011), pp. 597-607

K.N. Alavian, H. Li, L. Collis, L. Bonanni, L. Zeng, S. Sacchetti, E. Lazrove, P. Nabili, B. Flaherty, M. Graham, et al. Bcl-xL regulates metabolic efficiency of neurons through interaction with the mitochondrial F1FO ATP synthase. Nat Cell Biol 13 (2011), pp. 1224-1233

J.C. Ameisen. On the origin, evolution, and nature of programmed cell death: a timeline of four billion years. Cell Death Differ 9 (2002), pp. 367-393

A. Aouacheria, V. Rech de Laval, C. Combet, and J.M. Hardwick. Evolution of Bcl-2 homology motifs: homology versus homoplasy. Trends Cell Biol 23 (2013), pp. 103-111

P.L. Bardet, G. Kolahgar, A. Mynett, I. Miguel-Aliaga, J. Briscoe, P. Meier, and J.P. Vincent. A fluorescent reporter of caspase activity for live imaging. Proc Natl Acad Sci U S A 105 (2008), pp. 13901-13905

D.S. Bellows, B.N. Chau, P. Lee, Y. Lazebnik, W.H. Burns, and J.M. Hardwick. Antiapoptotic herpesvirus Bcl-2 homologs escape caspase-mediated conversion to proapoptotic proteins. J Virol 74 (2000), pp. 5024-5031

D.S. Bellows, M. Howell, C. Pearson, S.A. Hazlewood, and J.M. Hardwick. Epstein-Barr virus BALF1 is a BCL-2-like antagonist of the herpesvirus antiapoptotic BCL-2 proteins. J Virol 76 (2002), pp. 2469-2479

S.B. Berman, Y.B. Chen, B. Qi, J.M. McCaffery, E.B. Rucker, 3rd, S. Goebbels, K.A. Nave, B.A. Arnold, E.A. Jonas, F.J. Pineda, et al. Bcl-x L increases mitochondrial fission, fusion, and biomass in neurons. J Cell Biol 184 (2009), pp. 707-719

A. Bink, G. Govaert, I.E. Francois, K. Pellens, L. Meerpoel, M. Borgers, G. Van Minnebruggen, V. Vroome, B.P. Cammue, and K. Thevissen. A fungicidal piperazine-1-carboxamidine induces mitochondrial fission-dependent apoptosis in yeast. FEMS Yeast Res 10 (2010), pp. 812-818

W. Bleazard, J.M. McCaffery, E.J. King, S. Bale, A. Mozdy, Q. Tieu, J. Nunnari, and J.M. Shaw. The dynamin-related GTPase Dnm1 regulates mitochondrial fission in yeast. Nat Cell Biol 1 (1999), pp. 298-304

D.G. Breckenridge, M. Stojanovic, R.C. Marcellus, and G.C. Shore. Caspase cleavage product of BAP31 induces mitochondrial fission through endoplasmic reticulum calcium signals, enhancing cytochrome c release to the cytosol. J Cell Biol 160 (2003), pp. 1115-1127

A. Cassidy-Stone, J.E. Chipuk, E. Ingerman, C. Song, C. Yoo, T. Kuwana, M.J. Kurth, J.T. Shaw, J.E. Hinshaw, D.R. Green, et al. Chemical inhibition of the mitochondrial division dynamin reveals its role in Bax/Bak-dependent mitochondrial outer membrane permeabilization. Dev Cell 14 (2008), pp. 193-204

K.L. Cerveny, Y. Tamura, Z. Zhang, R.E. Jensen, and H. Sesaki. Regulation of mitochondrial fusion and division. Trends Cell Biol 17 (2007), pp. 563-569

Y.B. Chen, M.A. Aon, Y.T. Hsu, L. Soane, X. Teng, J.M. McCaffery, W.C. Cheng, B. Qi, H. Li, K.N. Alavian, et al. Bcl-xL regulates mitochondrial energetics by stabilizing the inner membrane potential. J Cell Biol 195 (2011), pp. 263-276

Y.B. Chen, J. Huska, B.A. Roelofs, and J.M. Hardwick. Knockin caspase-uncleavable Bcl-xL mouse retains a normal thymus during aging. (unpublished)

E.H. Cheng, D.G. Kirsch, R.J. Clem, R. Ravi, M.B. Kastan, A. Bedi, K. Ueno, and J.M. Hardwick. Conversion of Bcl-2 to a Bax-like death effector by caspases. Science 278 (1997a), pp. 1966-1968

E.H. Cheng, B. Levine, L.H. Boise, C.B. Thompson, and J.M. Hardwick. Bax-independent inhibition of apoptosis by Bcl-XL. Nature 379 (1996), pp. 554-556

E.H. Cheng, J. Nicholas, D.S. Bellows, G.S. Hayward, H.G. Guo, M.S. Reitz, and J.M. Hardwick. A Bcl-2 homolog encoded by Kaposi sarcoma-associated virus, human herpesvirus 8, inhibits apoptosis but does not heterodimerize with Bax or Bak. Proc Natl Acad Sci U S A 94 (1997b), pp. 690-694

W.C. Cheng, K.M. Leach, and J.M. Hardwick. Mitochondrial death pathways in yeast and mammalian cells. Biochim Biophys Acta 1783 (2008a), pp. 1272-1279

W.C. Cheng, X. Teng, H.K. Park, C.M. Tucker, M.J. Dunham, and J.M. Hardwick. Fis1 deficiency selects for compensatory mutations responsible for cell death and growth control defects. Cell Death Differ 15 (2008b), pp. 1838-1846

R.J. Clem, E.H. Cheng, C.L. Karp, D.G. Kirsch, K. Ueno, A. Takahashi, M.B. Kastan, D.E. Griffin, W.C. Earnshaw, M.A. Veliuona, *et al.* Modulation of cell death by Bcl-XL through caspase interaction. Proc Natl Acad Sci U S A 95 (1998), pp. 554-559

R.J. Clem, M. Fechheimer, and L.K. Miller. Prevention of apoptosis by a baculovirus gene during infection of insect cells. Science 254 (1991), pp. 1388-1390

J.P. de Rivero Vaccari, W.D. Dietrich, and R.W. Keane. Activation and regulation of cellular inflammasomes: gaps in our knowledge for central nervous system injury. J Cereb Blood Flow Metab 34 (2014), pp. 369-375

A. Degterev, J. Hitomi, M. Germscheid, I.L. Ch'en, O. Korkina, X. Teng, D. Abbott, G.D. Cuny, C. Yuan, G. Wagner, *et al.* Identification of RIP1 kinase as a specific cellular target of necrostatins. Nat Chem Biol 4 (2008), pp. 313-321

J.A. Dohm, S.J. Lee, J.M. Hardwick, R.B. Hill, and A.G. Gittis. Cytosolic domain of the human mitochondrial fission protein fis1 adopts a TPR fold. Proteins 54 (2004), pp. 153-156

Y. Dondelinger, W. Declercq, S. Montessuit, R. Roelandt, A. Goncalves, I. Bruggeman, P. Hulpiau, K. Weber, C.A. Sehon, R.W. Marquis, *et al.* MLKL Compromises Plasma Membrane Integrity by Binding to Phosphatidylinositol Phosphates. Cell Rep 7 (2014), pp. 971-981

M. Driscoll. Molecular genetics of cell death in the nematode Caenorhabditis elegans. J Neurobiol 23 (1992), pp. 1327-1351

H.M. Ellis, and H.R. Horvitz. Genetic control of programmed cell death in the nematode C. elegans. Cell 44 (1986), pp. 817-829

A. Erturk, Y. Wang, and M. Sheng. Local pruning of dendrites and spines by caspase-3-dependent and proteasome-limited mechanisms. J Neurosci 34 (2014), pp. 1672-1688

C.J. Evans, J.M. Olson, K.T. Ngo, E. Kim, N.E. Lee, E. Kuoy, A.N. Patananan, D. Sitz, P. Tran, M.T. Do, et al. G-TRACE: rapid Gal4-based cell lineage analysis in Drosophila. Nat Methods 6 (2009), pp. 603-605

V.A. Fadok, D.R. Voelker, P.A. Campbell, J.J. Cohen, D.L. Bratton, and P.M. Henson. Exposure of phosphatidylserine on the surface of apoptotic lymphocytes triggers specific recognition and removal by macrophages. J Immunol 148 (1992), pp. 2207-2216

Y. Fannjiang, W.C. Cheng, S.J. Lee, B. Qi, J. Pevsner, J.M. McCaffery, R.B. Hill, G. Basañez, and J.M. Hardwick. Mitochondrial fission proteins regulate programmed cell death in yeast. Genes Dev 18 (2004), pp. 2785-2797

Y. Fannjiang, C.H. Kim, R.L. Huganir, S. Zou, T. Lindsten, C.B. Thompson, T. Mito, R.J. Traystman, T. Larsen, D.E. Griffin, et al. BAK alters neuronal excitability and can switch from anti- to pro-death function during postnatal development. Dev Cell 4 (2003), pp. 575-585

S. Frank, B. Gaume, E.S. Bergmann-Leitner, W.W. Leitner, E.G. Robert, F. Catez, C.L. Smith, and R.J. Youle. The role of dynamin-related protein 1, a mediator of mitochondrial fission, in apoptosis. Dev Cell 1 (2001), pp. 515-525

N. Fujita, A. Nagahashi, K. Nagashima, S. Rokudai, and T. Tsuruo. Acceleration of apoptotic cell death after the cleavage of Bcl-XL protein by caspase-3-like proteases. Oncogene 17 (1998), pp. 1295-1304

L. Galluzzi, J.M. Bravo-San Pedro, I. Vitale, S.A. Aaronson, J.M. Abrams, D. Adam, E.S. Alnemri, L. Altucci, D. Andrews, M. Annicchiarico-Petruzzelli, et al. Essential versus accessory aspects of cell death: recommendations of the NCCD 2015. Cell Death Differ 22 (2015), pp. 58-73

G. Goyal, B. Fell, A. Sarin, R.J. Youle, and V. Sriram. Role of mitochondrial remodeling in programmed cell death in Drosophila melanogaster. Dev Cell 12 (2007), pp. 807-816

V. Hamburger, and R. Levi-Montalcini. Proliferation, differentiation and degeneration in the spinal ganglia of the chick embryo under normal and experimental conditions. J Exp Zool 111 (1949), pp. 457-501

J.M. Hardwick, Y.B. Chen, and E.A. Jonas. Multipolar functions of BCL-2 proteins link energetics to apoptosis. Trends Cell Biol 22 (2012), pp. 318-328

J.M. Hardwick, and W.C. Cheng. Mitochondrial programmed cell death pathways in yeast. Dev Cell 7 (2004), pp. 630-632

S. Henderson, D. Huen, M. Rowe, C. Dawson, G. Johnson, and A. Rickinson. Epstein-Barr virus-coded BHRF1 protein, a viral homologue of Bcl-2, protects human B cells from programmed cell death. Proc Natl Acad Sci U S A 90 (1993), pp. 8479-8483

J.A. Hickman, J.M. Hardwick, L.K. Kaczmarek, and E.A. Jonas. Bcl-xL inhibitor ABT-737 reveals a dual role for Bcl-xL in synaptic transmission. J Neurophysiol 99 (2008), pp. 1515-1522

B.T. Hyman, and J. Yuan. Apoptotic and non-apoptotic roles of caspases in neuronal physiology and pathophysiology. Nat Rev Neurosci 13 (2012), pp. 395-406

E. Ingerman, E.M. Perkins, M. Marino, J.A. Mears, J.M. McCaffery, J.E. Hinshaw, and J. Nunnari. Dnm1 forms spirals that are structurally tailored to fit mitochondria. J Cell Biol 170 (2005), pp. 1021-1027

N. Itoh, S. Yonehara, A. Ishii, M. Yonehara, S. Mizushima, M. Sameshima, A. Hase, Y. Seto, and S. Nagata. The polypeptide encoded by the cDNA for human cell surface antigen Fas can mediate apoptosis. Cell 66 (1991), pp. 233-243

I. Ivanovska, and J.M. Hardwick. Viruses activate a genetically conserved cell death pathway in a unicellular organism. J Cell Biol 170 (2005), pp. 391-399

R. Jagasia, P. Grote, B. Westermann, and B. Conradt. DRP-1-mediated mitochondrial fragmentation during EGL-1-induced cell death in C. elegans. Nature 433 (2005), pp. 754-760

E.A. Jonas, D. Hoit, J.A. Hickman, T.A. Brandt, B.M. Polster, Y. Fannjiang, E. McCarthy, M.K. Montanez, J.M. Hardwick, and L.K. Kaczmarek. Modulation of synaptic transmission by the BCL-2 family protein BCL-xL. J Neurosci 23 (2003), pp. 8423-8431

M. Karbowski, Y.J. Lee, B. Gaume, S.Y. Jeong, S. Frank, A. Nechushtan, A. Santel, M. Fuller, C.L. Smith, and R.J. Youle. Spatial and temporal association of Bax with mitochondrial fission sites, Drp1, and Mfn2 during apoptosis. J Cell Biol 159 (2002), pp. 931-938

J.F. Kerr, A.H. Wyllie, and A.R. Currie. Apoptosis: a basic biological phenomenon with wide-ranging implications in tissue kinetics. Br J Cancer 26 (1972), pp. 239-257

D.G. Kirsch, A. Doseff, B.N. Chau, D.S. Lim, N.C. de Souza-Pinto, R. Hansford, M.B. Kastan, Y.A. Lazebnik, and J.M. Hardwick. Caspase-3-dependent cleavage of Bcl-2 promotes release of cytochrome c. J Biol Chem 274 (1999), pp. 21155-21161

S. Koirala, H.T. Bui, H.L. Schubert, D.M. Eckert, C.P. Hill, M.S. Kay, and J.M. Shaw. Molecular architecture of a dynamin adaptor: implications for assembly of mitochondrial fission complexes. J Cell Biol 191 (2010), pp. 1127-1139

T. Komiyama, C.A. Ray, D.J. Pickup, A.D. Howard, N.A. Thornberry, E.P. Peterson, and G. Salvesen. Inhibition of interleukin-1 beta converting enzyme by the cowpox virus serpin CrmA. An example of cross-class inhibition. J Biol Chem 269 (1994), pp. 19331-19337

S.J. Korsmeyer. Bcl-2: an antidote to programmed cell death. Cancer Surv 15 (1992), pp. 105-118

A. Koto, E. Kuranaga, and M. Miura. Temporal regulation of Drosophila IAP1 determines caspase functions in sensory organ development. J Cell Biol 187 (2009), pp. 219-231

M. Lamkanfi, and V.M. Dixit. Mechanisms and Functions of Inflammasomes. Cell 157 (2014), pp. 1013-1022

G.I. Lang, D.P. Rice, M.J. Hickman, E. Sodergren, G.M. Weinstock, D. Botstein, and M.M. Desai. Pervasive genetic hitchhiking and clonal interference in forty evolving yeast populations. Nature (2013), pp.
Y.A. Lazebnik, S.H. Kaufmann, S. Desnoyers, G.G. Poirier, and W.C. Earnshaw. Cleavage of poly(ADP-ribose) polymerase by a proteinase with properties like ICE. Nature 371 (1994), pp. 346-347

G.G. Lee, K. Kikuno, S. Nair, and J.H. Park. Mechanisms of postecdysis-associated programmed cell death of peptidergic neurons in Drosophila melanogaster. J Comp Neurol 521 (2013), pp. 3972-3991

B. Levine, Q. Huang, J.T. Isaacs, J.C. Reed, D.E. Griffin, and J.M. Hardwick. Conversion of lytic to persistent alphavirus infection by the bcl-2 cellular oncogene. Nature 361 (1993), pp. 739-742

J. Lewis, G.A. Oyler, K. Ueno, Y.R. Fannjiang, B.N. Chau, J. Vornov, S.J. Korsmeyer, S. Zou, and J.M. Hardwick. Inhibition of virus-induced neuronal apoptosis by Bax. Nat Med 5 (1999), pp. 832-835

H. Li, Y. Chen, A.F. Jones, R.H. Sanger, L.P. Collis, R. Flannery, E.C. McNay, T. Yu, R. Schwarzenbacher, B. Bossy, et al. Bcl-xL induces Drp1-dependent synapse formation in cultured hippocampal neurons. Proc Natl Acad Sci U S A 105 (2008), pp. 2169-2174

H. Li, H. Zhu, C.J. Xu, and J. Yuan. Cleavage of BID by caspase 8 mediates the mitochondrial damage in the Fas pathway of apoptosis. Cell 94 (1998), pp. 491-501

J. Li, and J. Yuan. Caspases in apoptosis and beyond. Oncogene 27 (2008), pp. 6194-6206

P. Li, D. Nijhawan, I. Budihardjo, S.M. Srinivasula, M. Ahmad, E.S. Alnemri, and X. Wang. Cytochrome c and dATP-dependent formation of Apaf-1/caspase-9 complex initiates an apoptotic protease cascade. Cell 91 (1997), pp. 479-489

X. Li, J. Wang, and Y. Shi. Structural mechanisms of DIAP1 auto-inhibition and DIAP1-mediated inhibition of drICE. Nat Commun 2 (2011), pp. 408

R.A. Lockshin, and C.M. Williams. Programmed cell death—II. Endocrine potentiation of the breakdown of the intersegmental muscles of silkmoths. J Insect Physiol 10 (1964), pp. 643-659

X. Luo, I. Budihardjo, H. Zou, C. Slaughter, and X. Wang. Bid, a Bcl2 interacting protein, mediates cytochrome c release from mitochondria in response to activation of cell surface death receptors. Cell 94 (1998), pp. 481-490

F. Madeo, E. Herker, C. Maldener, S. Wissing, S. Lachelt, M. Herlan, M. Fehr, K. Lauber, S.J. Sigrist, S. Wesselborg, et al. A caspase-related protease regulates apoptosis in yeast. Mol Cell 9 (2002), pp. 911-917

N. Mendl, A. Occhipinti, M. Muller, P. Wild, I. Dikic, and A.S. Reichert. Mitophagy in yeast is independent of mitochondrial fission and requires the stress response gene WHI2. J Cell Sci 124 (2011), pp. 1339-1350

S. Montessuit, S.P. Somasekharan, O. Terrones, S. Lucken-Ardjomande, S. Herzig, R. Schwarzenbacher, D.J. Manstein, E. Bossy-Wetzel, G. Basanez, P. Meda, et al. Membrane remodeling induced by the dynamin-related protein Drp1 stimulates Bax oligomerization. Cell 142 (2010), pp. 889-901

K. Moriwaki, and F.K. Chan. Necrosis-dependent and independent signaling of the RIP kinases in inflammation. Cytokine Growth Factor Rev 25 (2014), pp. 167-174

V.E. Nava, E.H. Cheng, M. Veliuona, S. Zou, R.J. Clem, M.L. Mayer, and J.M. Hardwick. Herpesvirus saimiri encodes a functional homolog of the human bcl-2 oncogene. J Virol 71 (1997), pp. 4118-4122

V.E. Nava, A. Rosen, M.A. Veliuona, R.J. Clem, B. Levine, and J.M. Hardwick. Sindbis virus induces apoptosis through a caspase-dependent, CrmA-sensitive pathway. J Virol 72 (1998), pp. 452-459

D.D. Newmeyer, D.M. Farschon, and J.C. Reed. Cell-free apoptosis in Xenopus egg extracts: inhibition by Bcl-2 and requirement for an organelle fraction enriched in mitochondria. Cell 79 (1994), pp. 353-364

L.K. Nutt, S.S. Margolis, M. Jensen, C.E. Herman, W.G. Dunphy, J.C. Rathmell, and S. Kornbluth. Metabolic regulation of oocyte cell death through the CaMKII-mediated phosphorylation of caspase-2. Cell 123 (2005), pp. 89-103

D. Ofengeim, Y.B. Chen, T. Miyawaki, H. Li, S. Sacchetti, R.J. Flannery, K.N. Alavian, F. Pontarelli, B.A. Roelofs, J.A. Hickman, et al. N-terminally cleaved Bcl-xL mediates ischemia-induced neuronal death. Nat Neurosci 15 (2012), pp. 574-580

V. Palermo, C. Falcone, and C. Mazzoni. Apoptosis and aging in mitochondrial morphology mutants of S. cerevisiae. Folia Microbiol 52 (2007), pp. 479-483

V.C. Pham, V.G. Anania, Q.T. Phung, and J.R. Lill. Complementary methods for the identification of substrates of proteolysis. Methods Enzymol 544 (2014), pp. 359-380

V.C. Pham, R. Pitti, V.G. Anania, C.E. Bakalarski, D. Bustos, S. Jhunjhunwala, Q.T. Phung, K. Yu, W.F. Forrest, D.S. Kirkpatrick, et al. Complementary proteomic tools for the dissection of apoptotic proteolysis events. J Proteome Res 11 (2012), pp. 2947-2954

C.A. Ray, R.A. Black, S.R. Kronheim, T.A. Greenstreet, P.R. Sleath, G.S. Salvesen, and D.J. Pickup. Viral inhibition of inflammation: cowpox virus encodes an inhibitor of the interleukin-1 beta converting enzyme. Cell 69 (1992), pp. 597-604

A.S. Reynolds. The deaths of a cell: how language and metaphor influence the science of cell death. Stud Hist Philos Biol Biomed Sci 48 Pt B (2014), pp. 175-184

G.S. Salvesen, and V.M. Dixit. Caspases: intracellular signaling by proteolysis. Cell 91 (1997), pp. 443-446

C.Q. Scheckhuber, N. Erjavec, A. Tinazli, A. Hamann, T. Nystrom, and H.D. Osiewacz. Reducing mitochondrial fission results in increased life span and fitness of two fungal ageing models. Nat Cell Biol 9 (2007), pp. 99-105

L.M. Schwartz, and B.A. Osborne. Programmed cell death, apoptosis and killer genes. Immunol Today 14 (1993), pp. 582-590

S.Y. Seo, Y.B. Chen, I. Ivanovska, A.M. Ranger, S.J. Hong, V.L. Dawson, S.J. Korsmeyer, D.S. Bellows, Y. Fannjiang, and J.M. Hardwick. BAD is a pro-survival factor prior to activation of its pro-apoptotic function. J Biol Chem 279 (2004), pp. 42240-42249

J.M. Shaw, and J. Nunnari. Mitochondrial dynamics and division in budding yeast. Trends Cell Biol 12 (2002), pp. 178-184

E. Smirnova, L. Griparic, D.L. Shurland, and A.M. van der Bliek. Dynamin-related protein Drp1 is required for mitochondrial division in mammalian cells. Mol Biol Cell 12 (2001), pp. 2245-2256

I.A. Stanley, S.M. Ribeiro, A. Gimenez-Cassina, E. Norberg, and N.N. Danial. Changing appetites: the adaptive advantages of fuel choice. Trends Cell Biol 24 (2014), pp. 118-127

H. Steller. Mechanisms and genes of cellular suicide. Science 267 (1995), pp. 1445-1449

M. Suzuki, S.Y. Jeong, M. Karbowski, R.J. Youle, and N. Tjandra. The solution structure of human mitochondria fission protein Fis1 reveals a novel TPR-like helix bundle. J Mol Biol 334 (2003), pp. 445-458

B. Szamecz, G. Boross, D. Kalapis, K. Kovacs, G. Fekete, Z. Farkas, V. Lazar, M. Hrtyan, P. Kemmeren, M.J. Groot Koerkamp, et al. The genomic landscape of compensatory evolution. PLoS Biol 12 (2014), pp. e1001935

H.L. Tang, H.M. Tang, M.C. Fung, and J.M. Hardwick. In vivo CaspaseTracker biosensor system for detecting anastasis and non-apoptotic caspase activity. Sci Rep 5 (2015a), pp. 9015

H.L. Tang, H.M. Tang, J.M. Hardwick, and M.C. Fung. Strategies for Tracking Anastasis, A Cell Survival Phenomenon that Reverses Apoptosis. J Vis Exp (2015b), pp.
H.L. Tang, H.M. Tang, K.H. Mak, S. Hu, S.S. Wang, K.M. Wong, C.S. Wong, H.Y. Wu, H.T. Law, K. Liu, et al. Cell survival, DNA damage, and oncogenic transformation after a transient and reversible apoptotic response. Mol Biol Cell 23 (2012), pp. 2240-2252

H.L. Tang, K.L. Yuen, H.M. Tang, and M.C. Fung. Reversibility of apoptosis in cancer cells. Br J Cancer 100 (2009), pp. 118-122

X. Teng, W.C. Cheng, B. Qi, T.X. Yu, K. Ramachandran, M.D. Boersma, T. Hattier, P.V. Lehmann, F.J. Pineda, and J.M. Hardwick. Gene-dependent cell death in yeast. Cell Death Dis 2 (2011), pp. 1-9 [e188]

X. Teng, M. Dayhoff-Brannigan, W.C. Cheng, C.E. Gilbert, C.N. Sing, N.L. Diny, S.J. Wheelan, M.J. Dunham, J.D. Boeke, F.J. Pineda, et al. Genome-wide consequences of deleting any single gene. Mol Cell 52 (2013), pp. 485-494

X. Teng, and J.M. Hardwick. Quantification of genetically controlled cell death in budding yeast. Methods Mol Biol 1004 (2013), pp. 161-170

M. Tewari, L.T. Quan, K. O'Rourke, S. Desnoyers, Z. Zeng, D.R. Beidler, G.G. Poirier, G.S. Salvesen, and V.M. Dixit. Yama/CPP32 beta, a mammalian homolog of CED-3, is a CrmA-inhibitable protease that cleaves the death substrate poly(ADP-ribose) polymerase. Cell 81 (1995), pp. 801-809

N.A. Thornberry, H.G. Bull, J.R. Calaycay, K.T. Chapman, A.D. Howard, M.J. Kostura, D.K. Miller, S.M. Molineaux, J.R. Weidner, J. Aunins, et al. A novel heterodimeric cysteine protease is required for interleukin-1 beta processing in monocytes. Nature 356 (1992), pp. 768-774

B.C. Trauth, C. Klas, A.M. Peters, S. Matzku, P. Moller, W. Falk, K.M. Debatin, and P.H. Krammer. Monoclonal antibody-mediated tumor regression by induction of apoptosis. Science 245 (1989), pp. 301-305

Y. Tsujimoto, L.R. Finger, J. Yunis, P.C. Nowell, and C.M. Croce. Cloning of the chromosome breakpoint of neoplastic B cells with the t(14;18) chromosome translocation. Science 226 (1984), pp. 1097-1099

S. Ubol, P.C. Tucker, D.E. Griffin, and J.M. Hardwick. Neurovirulent strains of Alphavirus induce apoptosis in bcl-2-expressing cells: role of a single amino acid change in the E2 glycoprotein. Proc Natl Acad Sci U S A 91 (1994), pp. 5202-5206

D.L. Vaux, and S.J. Korsmeyer. Cell death in development. Cell 96 (1999), pp. 245-254

J. Wakabayashi, Z. Zhang, N. Wakabayashi, Y. Tamura, M. Fukaya, T.W. Kensler, M. Iijima, and H. Sesaki. The dynamin-related GTPase Drp1 is required for embryonic and brain development in mice. J Cell Biol 186 (2009), pp. 805-816

M. Weinberger, A. Mesquita, T. Caroll, L. Marks, H. Yang, Z. Zhang, P. Ludovico, and W.C. Burhans. Growth signaling promotes chronological aging in budding yeast by inducing superoxide anions that inhibit quiescence. Aging (Albany NY) 2 (2010), pp. 709-726

E. White, R. Cipriani, P. Sabbatini, and A. Denton. Adenovirus E1B 19-kilodalton protein overcomes the cytotoxicity of E1A proteins. J Virol 65 (1991), pp. 2968-2978

C.S. Yang, M.J. Thomenius, E.C. Gan, W. Tang, C.D. Freel, T.J. Merritt, L.K. Nutt, and S. Kornbluth. Metabolic regulation of Drosophila apoptosis through inhibitory phosphorylation of Dronc. Embo J 29 (2010), pp. 3196-3207

J. Yuan, S. Shaham, S. Ledoux, H.M. Ellis, and H.R. Horvitz. The C. elegans cell death gene ced-3 encodes a protein similar to mammalian interleukin-1 beta-converting enzyme. Cell 75 (1993), pp. 641-652

Y. Zhang, and D.C. Chan. Structural basis for recruitment of mitochondrial fission complexes by Fis1. Proc Natl Acad Sci U S A 104 (2007), pp. 18526-18530

PART 2: PROCESSES OF APOPTOSIS

Chapter 7: Modern history of the study of cell death: 1964...1994...2014 (Zakeri and Lockshin)

Zahra Zakeri and Richard A Lockshin[1,2]

[1]Department of Biology, Queens College and Graduate Center of the City University of New York, Flushing, NY 11367 USA

[2]Professor Emeritus, Department of Biological Sciences, St. John's University, Jamaica, NY 11439 USA

Abstract

In the last twenty years we have learned an enormous amount about the mechanisms of cell death, including the effectors of apoptosis and the many membrane-bound, mitochondria-derived, and diffusible adjusters of the threshold of apoptosis. We have also come to appreciate that cells have many options and do not necessarily die by apoptosis. Depending to a large extent on the energetics, dynamics, and kinetics of stresses vs resources, cells can undergo substantial amounts of autophagy, necroptosis, or even necrosis. Most importantly, cells in even the most homogeneous culture respond differently in terms of timing and threshold to noxious stimuli, and some may even resist the stimulus. Thus, whether we wish to protect cells such as myocardial myocytes or neurons; or to kill cancer cells, we need to know much more specifically what determines the threshold at which cell death becomes inevitable. As is indicated by several authors in this volume, this next phase of research is now rapidly developing.

1964-1994

When we first described developmental cell death as "programmed," we were referring to the observation that the death of cells at metamorphosis was a developmental event like any other, and that in the situation we were studying it was controlled by endocrine, neural, and neurosecretory processes and effected by....lysosomes. [2]Apoptosis was not yet on the

---

[2]The doctoral thesis entitled "Programmed cell death in an insect" was accepted in 1963, and the publications from that thesis appeared in 1964 and 1965. The idea was relatively obvious, but had not been previously directly addressed.

horizon, and the exciting topic of the time was the function of lysosomes, discovered a few years earlier. Christian de Duve and his collaborators had realized that lysosomes "burst" or leak enzymes in the highly toxic situation that they modeled, carbon tetrachloride hepatotoxicity, and they had concluded that leakage of lysosomes was a primary cause of cell death. Although many laboratories within short order determined that lysosomes did not routinely rupture, the literature in the 1960's strongly reflected everyone's interest in lysosomes.

Meanwhile, John Kerr was leading another school of thought, based on the observation that many cells condensed, and their chromatin coalesced, as they died. Such a behavior, which Kerr described as "shrinkage necrosis," was not explicable by the proposition that respiration and hence energy for ion pumps would fail, and the cell would accumulate the osmolyte lactic acid, which would draw water into the cell and cause it to lyse in the classic pattern of necrosis. In 1972, Kerr, along with Andrew Wyllie and Alastair Currie, described the pattern as a generalized one, which they named "apoptosis"(Kerr, Wyllie et al. 1972, Wyllie, Kerr et al. 1980). Other researchers began to look for, and publish, other instances in which apoptotic cells could be seen. The history of this period has been covered by several authors (Clarke and Clarke 1995, Clarke and Clarke 1996, Lockshin and Zakeri 2001, Kerr 2002, Vaux 2002, Diamantis, Magiorkinis et al. 2008, Zakeri and Lockshin 2008, Clarke and Clarke 2012) and will not be re-reviewed here.

By the early 1990's interest in cell death had expanded owing to at least three major breakthroughs. First, Wyllie's group, by establishing that the coalescence of chromatin resulted from the internucleosomal hydrolysis of DNA, provided an easy and cheap means of documenting at least instances of substantial apoptosis--electrophoresis of DNA to reveal a typical "ladder" of DNA fragments differing in size by 180 base pairs. Many laboratories quickly documented active apoptosis where cell death had previously been missed. Second, Brenner, Sulston, and Horvitz identified specific genes that controlled a block of

191 predictable deaths during the embryogenesis of *Caenorhabditis elegans*, thus concretely documenting the rather intuitive assumptions of the earlier developmental biologists. When Junying Yuan and H. Robert Horvitz established a few years later that an essential killer gene was a protease conserved from roundworms to humans, the race to understand the new form of cell death was on. Third, the rise of molecular genetics had led to the identification of several genes that were commonly mutated in many types of cancer. Several laboratories quickly recognized that at least three groups of these genes were associated with patterns of cell death rather than control of mitosis. These groups of genes included p53, as analyzed by Bert Vogelstein, Tyler Jacks, and Scott Lowe; Fas/Apo-1, as separately studied by Shigekazu Nagata in Japan and Peter Krammer in Germany; and an inhibitor of cell death, Bcl-2, as described by Stan Korsmeyer, Gabriel Nuñez, Suzanne Cory, and David Vaux and as influenced by cytochrome c (Xiaodong Wang (Liu, Kim et al. 1996, Yang, Liu et al. 1997), Donald Newmeyer (Newmeyer, Farschon et al. 1994, Kluck, Bossy-Wetzel et al. 1997, Kluck, Martin et al. 1997)).

Because of this flurry of attention, in short order different societies held special sessions on "apoptosis" or "programmed cell death," special meetings were convened: by pathologists in Sardinia, and a Banbury Conference in Cold Spring Harbor; conference series were launched as Cold Spring Harbor Symposia, Keystone Conferences, and Gordon Conferences; and the European Cell Death Society and the International Cell Death Society (also called originally the Death Poet's Society) were founded.

1994-2014
A general pattern in science is that a new idea is greeted with disinterest or disbelief; it catches hold, becomes a certainty, a paradigm, and the only explanation of a phenomenon; and then doubt creeps in, with some difficulty, until finally the idea is softened and rendered more inclusive, less rigid but more

adaptable. This is the pattern of programmed cell death/apoptosis.

In 1994 the brilliance of the genetic, technical, and medical advances had built a very heady atmosphere, in which apoptosis explained everything and would soon be controlled to treat many diseases. Mammalian and roundworm cell death sequences were nearly identical. Cells contained within themselves the capacity to commit suicide, in the form of a caspase (ced 3 in roundworms, caspases 3 and 7 in mammalian cells); the caspases were held in abeyance by inhibitors of cell death (anti-apoptotic bcl-2 family in mammals, ced 9 in worms), from which they were released by other factors such as ced 4 (worms) and Apaf-1; and overall there were genes that regulated in which cells the caspases would be released (ces genes in worms, many in mammals--Horvitz 2003). The mammalian system was more complicated, in that there were initiator and effector caspases and at least two routes of activation (extrinsic, involving plasma membrane-bound receptors; and intrinsic, involving depolarization and permeabilization of mitochondria) but basically the systems were straightforward, self-contained, and comprehensible. Cancers resisted apoptosis and could be treated by activating apoptosis. To prevent neurodegeneration or cure AIDS, one should be able to block apoptosis. Many publications announced, even triumphantly, that they demonstrated that "cells died by apoptosis"--a curious statement at best, since it essentially is equivalent to saying, "he died because his heart stopped beating". Very few researchers addressed apoptosis for what it was, essentially the post-mortem of a cell, the manifestation of its death, as opposed to the source and process of its death.

This certainty (and confidence) dominated the research atmosphere, to some extent to the exclusion of other ideas. We published a paper in 1993 entitled "Delayed internucleosomal DNA fragmentation in programmed cell death," in which we argued that much of the destruction of some cells took place long before we saw signs of apoptosis (Zakeri, Quaglino et al. 1993), echoing and expanding our earlier reports of what we had interpreted to be "lysosomal cell death"(Lockshin and Beaulaton 1974, Beaulaton and Lockshin 1977, Lockshin and Beaulaton 1979, Beaulaton and Lockshin 1982). The publication of this paper was delayed by the vigorous protests of one reviewer, who insisted that the death had to be apoptotic, and that our techniques were insufficiently sensitive. (They were 10 fold more sensitive than the techniques he or she proposed.) When we went to a meeting shortly after the paper was published, several people sought us out and thanked us for publishing it, saying, "we have seen similar results, but cannot get a paper accepted that says so".

The initial apoptosis-based therapies failed, and the companies that offered to "measure apoptosis" as a diagnostic tool did not survive. The failure of the first therapies, combined with a growing ambiguity typified by that 1993 paper, led to a new phase in which there was room for alternatives to apoptosis.

The apoptosis-based therapies failed for at least three major reasons. First, although several cancers are apparently driven by mutations in the primary controllers of apoptosis, such as bcl-2 or p53, only very rarely is the proximate mechanism of apoptosis (the effector caspases) altered. Caspase-3 is only one of the genes lost in the MCF-7 line of breast cancer cells ((Janicke, Sprengart et al. 1998); see also http://cancer.sanger.ac.uk/cosmic/search?q=mcf7&domain=cosmic), and in many other cancers, the threshold of activation is altered, with the sequence of controllers being otherwise intact. Second, even under the most stringent of laboratory conditions, not all cells die, nor is their death synchronous, even when the cell line is essentially homogeneous. We have no explanation for this variability. In a living organism or human patient we have much less ability to target or discriminate among cells. Third, as has become obvious now that we acknowledge other pathways to death, one does not need an instruction manual to die. If a cell is under sufficient pressure, it will fail. If apoptosis is blocked, a cell can fail by other means. To return to the analogy above, if a swimmer has a heart attack while swimming, getting him out of the water does not guarantee his survival. In fact, in many publications, a reported prevention of death was more appropriately described as delay of death. It should have been obvious that, in laboratory experiments where apoptosis was induced by administration of toxins such as cycloheximide or staurosporine, often at concentrations 10X the amount needed to produce their specific biochemical effects; even where apoptosis was inhibited the cells would ultimately die.

The 20 years between 1994 and 2014 were marked by a consolidation of our understanding of the mechanics of apoptosis, including most notably the expansion of our knowledge of how mitochondria affect the potential for and activation of apoptosis, and our growing tolerance of alternatives. The former aspect has engendered a new and more promising therapeutic approach, at least for cancers, in the targeting of specific cells to activate their death-receptor pathways to apoptosis or, relying on the difference in metabolism between malignant and normal cells, targeting the mitochondria with BH3 mimetics. The latter has led to our appreciation, most notably, of the importance of autophagy and of necroptosis, with other potentially relevant forms of death at least considered, if not fully accepted. Finally, today we at least acknowledge the importance of understanding the history and metabolism of a cell, and the stresses on it, but we have not yet learned how to protect cells, as we must do to address neurodegenerative diseases and cell-killing infections such as HIV.

How mitochondria affect the potential for apoptosis

Analysis of the role of bcl-2 and its related molecules led to an understanding that the interaction and competition of the pro- and anti-apoptotic variants took place on the mitochondria, with evidence from the laboratories of Donald Newmeyer, Douglas Green, and Xiaodong Wang calling attention to the leakage of cytochrome c into the cytoplasm, a story that quickly turned to the identification of the apoptosome and the mechanism for activation of caspase-9, triggering the final phase of metabolically-activated apoptosis. Kroemer identified the depolarization of the mitochondria (Kroemer, Petit et al. 1995, Zamzami, Marchetti et al. 1995), leading to the development of an understanding of the elaborate nature of the mitochondria as a sensor of many metabolic perturbations and a common final arbiter of the decision to undergo apoptosis (Green, Galluzzi et al. 2014). However, the picture became more complicated with the recognition of the importance of autophagy.(Galluzzi, Bravo-San Pedro et al. 2014, Green, Galluzzi et al. 2014, Marino, Pietrocola et al. 2014).

Autophagy and alternative forms of death

Since the early days of cell death literature, autophagy has never truly faded from our images of cell death, but our understanding of its role has evolved considerably. De Duve and Wattiaux' interpretation of the role of lysosomes failed because they did not understand that the properties of carbon tetrachloride that made it a hepatotoxin--its solubility in and ability to dissolve lipid membranes--were the same as the properties that allowed it to rupture lysosomes (Appelmans, Wattiaux et al. 1955, De Duve and Wattiaux 1956). Nevertheless, autophagy as a mode of cell death made sense in some circumstances. The death of a large, cytoplasm-rich, poorly mitotic or postmitotic cell is much more than an incident. It entails the clearance of a substantial amount of material. Thus, if a cell such as one from an involuting mammary gland or a large secretory gland in a metamorphosing insect is to die in a controlled and biologically efficient manner, it must first divest itself of substantial cytoplasm(Zakeri, Quaglino et al. 1993, Lockshin and Zakeri 2001, 2002). Most commonly, this is done by massive autophagy. In other instances, the tail muscles of metamorphosing tadpoles break into sarcolytes, or chunks of muscle consisting of a few sarcomeres (Weber 1964); atrophying muscles lose most of their sarcoplasm, through ubiquitin-driven proteolysis, perhaps influenced by changing solubility of myofibrillar proteins as they depolarize; and whole intestinal epithelial cells are released into the intestinal lumen. In contrast, many cells of the reticuloendothelial system, such as lymphocytes and thymocytes, are relatively small, cytoplasm-restricted cells that are either highly mitotic or derive from highly mitotic lines. For them, rapid destruction of the chromatin is a biologically intelligent approach. Thus researchers considering "autophagic cell death" and "apoptosis" tended to look at different cells. Similarly, neurons depend very much on their substratum and on their interactions with other cells, whereas blood-borne cells do not. Thus neurologists were far more concerned with growth factors, cell-to-cell communication, and interactions with substratum. In the 1970's and 1980's there were many excellent publications, primarily relying on electron microscopy, documenting substantial expansion of the lysosomal and autophagosomal systems in dying cells (see, for instance, (Helminen and Ericsson 1968, Helminen and Ericsson 1968, Helminen and Ericsson 1968, Helminen, Ericsson et al. 1968).

Elucidation of the extrinsic and intrinsic pathways of apoptosis made it clear that the lysosomal system was not an intrinsic or linear component of most forms of cell death, and several emerging techniques made study of the lysosomal system more accessible. First, yeast had lysosomes and were genetically malleable. Primarily through the laboratory of Dan Klionsky, knockout of several genes established what is now known as the ATG (autophagosome) pathway (Wang, Zhao et al. 1996, Mizushima, Noda et al. 1998), which Beth Levine and others determined to be highly homologous to human genes controlling autophagocytosis (Liang, Jackson et al. 1999, Liang, Yu et al. 2001, Levine and Klionsky 2004). These genes and their products could be measured by the standard techniques of molecular biology. Second, the maturation of techniques allowing coupling of fluors to proteins led to the production of fluor-labeled components of autophagosomes, such as LC-3, while high resolution fluorescence and confocal microscopy made it possible to visualize autophagosomes, autolysosomes, and autophagic vesicles, and to track their formation and disappearance in cultured cells. Third, many laboratories found that analysis of the basis of toxicity of fungal and bacterial toxins was fertile ground, creating an armamentarium of drugs that specifically activated or inhibited autophagic destruction of cytoplasm. The findings of many laboratories using these techniques led to a consensus, most eloquently argued by Kroemer and his colleagues, but generally appreciated by many researchers, that autophagy is most commonly not an alternative path to death, and that "autophagic cell death," while possible, is at best uncommon and in any case raises many questions.

It now appears that autophagy most commonly protects cells. Autophagy and apoptosis pathways can intersect at various points--for instance, autophagy may remove mitochondria, lowering energy resources and triggering apoptosis; or autophagy may sequester damaged mitochondria, preventing leaks of cytochrome c and AIF, protecting against apoptosis; and beclin-3, a component of the autophagy cycle, can be degraded by caspases. However, in most situations autophagy appears to be activated when a cell is stressed and it serves to remove damaged organelles or consume cellular reserves to supplement the energy available to the cell. Autophagy therefore protects the cell against transient hard times. In these situations, the activation of autophagy forestalls apoptosis and may prevent it altogether. In unusual situations, such as late in the cycle of viral infection, the virus may instigate an overwhelming autophagy, which drains all the resources from the cell, killing it and allowing the virus to escape(Lockshin and Zakeri 2007).

In developmental situations, such as in metamorphosing insect tissues or involuting post-lactational mammary epithelium, autophagy consumes the bulk of the cytoplasm before, at the last moment, clear signs of apoptosis destroy the remaining skeleton of the cell (Facey and Lockshin 2010). This situation causes us to reflect on the meaning of "autophagic cell death".

In the situations that we know, the death is triggered by appearance and withdrawal of hormones (high ecdysterone and low juvenile hormone in insects, withdrawal of prolactin in mammals) as well as potentially by physical signals related to accumulation of milk) but simple atrophy is not so dramatic. However we have no evidence that the autophagy that leads to the eventual death of the cell differs in any manner from that used to maintain homeostasis, with the exception that it appears to relentless and unlimited. The question is what has provoked the autophagy, and whether or not even in this instance it simply represents the attempt of a cell to survive a difficulty, albeit here without a brake. Thus apoptosis would ensue only when the last efforts to salvage the cell have failed.

In addition to autophagy, several researchers have noted forms of cell death that cannot be readily characterized as autophagy or apoptosis (such as necroptosis (described in Chapter 15) and paraptosis (Sperandio et al., 2004; Schneider et al., 2004; Wang et al., 2004), as well as cell activities that include aspects of apoptosis but do not end with the disappearance of the cell, such as the loss of cytoplasm from differentiating spermatozoa and the loss of organelles from differentiating lens fibers, erythrocytes, and keratinocytes. Regression of dendrites from neurons deprived of growth factors or contact with other neurons may involve caspases but not necessarily the death of the cell. In other instances, such as when osteoblasts or intestinal epithelial cells die, they die in an environment in which phagocytosis of apoptotic remnants is unlikely. Thus these latter cells frequently initiate a controlled cell death but terminate in something that more resembles necrosis, as do some cells that are injured seriously enough that they cannot complete apoptosis before exhausting their energy resources. The death of a cell very much depends on the dynamics and history of its metabolism, the intensity and dynamics of the stress on the cell, and the cell's relationship to other cells and growth factors(Loos, Engelbrecht et al. 2013). In fact, one might with some justification argue that there are as many ways to die as there are types of cells. It is perhaps more reasonable to focus on the main pathways of death, remembering that, if the stress is not removed or if a vital function, such as mitochondrial respiration, is seriously compromised, the cell will ultimately die whether or not one specific pathway is blocked.

Protecting or destroying cells

Virtually all current efforts to exploit our knowledge of cell death for therapeutic purposes focus on the increased resistance of malignant cells to apoptosis. Major clinical trials attempt to target cells to initiate apoptosis or resensitize cells to triggers of apoptosis by inveigling specific cells to take up BH3 mimetics or competitors of Apoptosis Inhibiting Factor, or to direct to specific cells agonists related to the TNF-α family of activators of externally driven apoptosis. These techniques show some promise since, as long as the target is precise, the goal is to elicit a usually still-present but silent means of self-destruction. However, another huge concern is the protection of cells at risk, most critically neurons faced with whatever, certainly metabolic, pressures that lead to metabolic disease, but also in more acute situations such as infarct and inflammatory destruction. While ultimately the goal would be to anticipate and prevent or relieve the pressures, there is certainly merit in protecting cells, at least transiently until the biological situation has been stabilized, and thus staving off death. Here we most need insight into the metabolic problems and the most critical vulnerabilities of the cell, and most importantly, what determines the threshold or thresholds at which cells commit to death. Such information would not only help us to protect vulnerable cells, it would help us to more successfully kill them. It may be emotionally satisfying, and it can produce joy and hope, to induce apoptosis in 90% of malignant cells, but the last 10% that are resistant to apoptosis are very dangerous. If we can rely on other biological activities, such as the capability of the immune system (Zitvogel, Casares et al. 2004, Zitvogel, Tesniere et al. 2006, Zitvogel, Apeton et al. 2008, Zitvogel and Kroemer 2008, Green, Ferguson et al. 2009),

90% may be enough, but otherwise we will have to learn how cells partition above and below the threshold.

References

Appelmans, F., R. Wattiaux and C. De Duve (1955). "Tissue fractionation studies. 5. The association of acid phosphatase with a special class of cytoplasmic granules in rat liver." Biochem J 59(3): 438-445.

Beaulaton, J. and R. A. Lockshin (1977). "Ultrastructural study of the normal degeneration of the intersegmental muscles of Anthereae polyphemus and Manduca sexta (Insecta, Lepidoptera) with particular reference of cellular autophagy." J Morphol 154(1): 39-57.

Beaulaton, J. and R. A. Lockshin (1982). "The relation of programmed cell death to development and reproduction: comparative studies and an attempt at classification." Int Rev Cytol 79: 215-235.

Clarke, P. G. and S. Clarke (1995). "Historic apoptosis." Nature 378(6554): 230.

Clarke, P. G. and S. Clarke (1996). "Nineteenth century research on naturally occurring cell death and related phenomena." Anat Embryol (Berl) 193(2): 81-99.

Clarke, P. G. and S. Clarke (2012). "Nineteenth century research on cell death." Exp Oncol 34(3): 139-145.

De Duve, C. and R. Wattiaux (1956). "Tissue fractionation studies. VII. Release of bound hydrolases by means of triton X-100." Biochem J 63(4): 606-608.

Diamantis, A., E. Magiorkinis, G. H. Sakorafas and G. Androutsos (2008). "A brief history of apoptosis: from ancient to modern times." Onkologie 31(12): 702-706.

Facey, C. O. and R. A. Lockshin (2010). "The execution phase of autophagy associated PCD during insect metamorphosis." Apoptosis 15(6): 639-652.

Galluzzi, L., J. M. Bravo-San Pedro and G. Kroemer (2014). "Organelle-specific initiation of cell death." Nat Cell Biol 16(8): 728-736.

Green, D. R., T. Ferguson, L. Zitvogel and G. Kroemer (2009). "Immunogenic and tolerogenic cell death." Nat Rev Immunol 9(5): 353-363.

Green, D. R., L. Galluzzi and G. Kroemer (2014). "Cell biology. Metabolic control of cell death." Science 345(6203): 1250256.

Helminen, H. J. and J. L. Ericsson (1968). "Studies on mammary gland involution. 3. Alterations outside auto- and heterophagocytic pathways for cytoplasmic degradation." J Ultrastruct Res 25(3): 228-239.

Helminen, H. J. and J. L. Ericsson (1968). "Studies on mammary gland involution. I. On the ultrastructure of the lactating mammary gland." J Ultrastruct Res **25**(3): 193-213.

Helminen, H. J. and J. L. Ericsson (1968). "Studies on mammary gland involution. II. Ultrastructural evidence for auto- and heterophagocytosis." J Ultrastruct Res **25**(3): 214-227.

Helminen, H. J., J. L. Ericsson and S. Orrenius (1968). "Studies on mammary gland involution. IV. Histochemical and biochemical observations on alterations in lysosomes and lysosomal enzymes." J Ultrastruct Res **25**(3): 240-252.

Horvitz, H. R. (2003). "Nobel lecture. Worms, life and death." Biosci Rep **23**(5-6): 239-303.

Janicke, R. U., M. L. Sprengart, M. R. Wati and A. G. Porter (1998). "Caspase-3 is required for DNA fragmentation and morphological changes associated with apoptosis." J Biol Chem **273**(16): 9357-9360.

Kerr, J. F. (2002). "History of the events leading to the formulation of the apoptosis concept." Toxicology **181-182**: 471-474.

Kerr, J. F., A. H. Wyllie and A. R. Currie (1972). "Apoptosis: a basic biological phenomenon with wide-ranging implications in tissue kinetics." Br J Cancer **26**(4): 239-257.

Kluck, R. M., E. Bossy-Wetzel, D. R. Green and D. D. Newmeyer (1997). "The release of cytochrome c from mitochondria: a primary site for Bcl-2 regulation of apoptosis." Science **275**(5303): 1132-1136.

Kluck, R. M., S. J. Martin, B. M. Hoffman, J. S. Zhou, D. R. Green and D. D. Newmeyer (1997). "Cytochrome c activation of CPP32-like proteolysis plays a critical role in a Xenopus cell-free apoptosis system." Embo j **16**(15): 4639-4649.

Kroemer, G., P. Petit, N. Zamzami, J. L. Vayssiere and B. Mignotte (1995). "The biochemistry of programmed cell death." Faseb J **9**(13): 1277-1287.

Levine, B. and D. J. Klionsky (2004). "Development by self-digestion: molecular mechanisms and biological functions of autophagy." Dev Cell **6**(4): 463-477.

Liang, X. H., S. Jackson, M. Seaman, K. Brown, B. Kempkes, H. Hibshoosh and B. Levine (1999). "Induction of autophagy and inhibition of tumorigenesis by beclin 1." Nature **402**(6762): 672-676.

Liang, X. H., J. Yu, K. Brown and B. Levine (2001). "Beclin 1 contains a leucine-rich nuclear export signal that is required for its autophagy and tumor suppressor function." Cancer Res **61**(8): 3443-3449.

Liu, X., C. N. Kim, J. Yang, R. Jemmerson and X. Wang (1996). "Induction of apoptotic program in cell-free extracts: requirement for dATP and cytochrome c." Cell 86(1): 147-157.

Lockshin, R. A. and J. Beaulaton (1974). "Programmed cell death. Cytochemical appearance of lysosomes when death of the intersegmental muscles is prevented." J Ultrastruct Res 46(1): 63-78.

Lockshin, R. A. and J. Beaulaton (1979). "Programmed cell death. electrophysiological and ultrastructural correlations in metamorphosing muscles of lepidopteran insects." Tissue Cell 11(4): 803-819.

Lockshin, R. A. and Z. Zakeri (2001). "Programmed cell death and apoptosis: origins of the theory." Nat Rev Mol Cell Biol 2(7): 545-550.

Lockshin, R. A. and Z. Zakeri (2002). "Caspase-independent cell deaths." Curr Opin Cell Biol 14(6): 727-733.

Lockshin, R. A. and Z. Zakeri (2007). "Cell death in health and disease." J Cell Mol Med 11(6): 1214-1224.

Loos, B., A. M. Engelbrecht, R. A. Lockshin, D. J. Klionsky and Z. Zakeri (2013). "The variability of autophagy and cell death susceptibility: Unanswered questions." Autophagy 9(9): 1270-1285.

Marino, G., F. Pietrocola, T. Eisenberg, Y. Kong, S. A. Malik, A. Andryushkova, S. Schroeder, T. Pendl, A. Harger, M. Niso-Santano, N. Zamzami, M. Scoazec, S. Durand, D. P. Enot, A. F. Fernandez, I. Martins, O. Kepp, L. Senovilla, C. Bauvy, E. Morselli, E. Vacchelli, M. Bennetzen, C. Magnes, F. Sinner, T. Pieber, C. Lopez-Otin, M. C. Maiuri, P. Codogno, J. S. Andersen, J. A. Hill, F. Madeo and G. Kroemer (2014). "Regulation of autophagy by cytosolic acetyl-coenzyme A." Mol Cell 53(5): 710-725.

Mizushima, N., T. Noda, T. Yoshimori, Y. Tanaka, T. Ishii, M. D. George, D. J. Klionsky, M. Ohsumi and Y. Ohsumi (1998). "A protein conjugation system essential for autophagy." Nature 395(6700): 395-398.

Newmeyer, D. D., D. M. Farschon and J. C. Reed (1994). "Cell-free apoptosis in Xenopus egg extracts: inhibition by Bcl-2 and requirement for an organelle fraction enriched in mitochondria." Cell 79(2): 353-364.

Schneider D1, Gerhardt E, Bock J, Müller MM, Wolburg H, Lang F, Schulz JB. (2004) "Intracellular acidification by inhibition of the Na+/H+-exchanger leads to caspase-independent death of cerebellar granule neurons resembling paraptosis." Cell Death Differ. 11(7):760-70.

Sperandio S, Poksay K, de Belle I, Lafuente MJ, Liu B, Nasir J, Bredesen DE (2004) "Paraptosis: mediation by MAP kinases and inhibition by AIP-1/Alix."Cell Death Differ. 2004 11(10):1066-75.

Vaux, D. L. (2002). "Apoptosis timeline." Cell Death Differ 9(4): 349-354.

Wang, Y. X., H. Zhao, T. M. Harding, D. S. Gomes de Mesquita, C. L. Woldringh, D. J. Klionsky, A. L. Munn and L. S. Weisman (1996). "Multiple classes of yeast mutants are defective in vacuole partitioning yet target vacuole proteins correctly." Mol Biol Cell **7**(9): 1375-1389.

Wang Y1, Li X, Wang L, Ding P, Zhang Y, Han W, Ma D.(2004). "An alternative form of paraptosis-like cell death, triggered by TAJ/TROY and enhanced by PDCD5 overexpression."J Cell Sci. **117** (8):1525-32.

Weber, R. (1964). "Ultrastructural changes in regressing tail muscles of Xenopus larvae at metamorphosis." J Cell Biol **22**: 481-487.

Wyllie, A. H., J. F. Kerr and A. R. Currie (1980). "Cell death: the significance of apoptosis." Int Rev Cytol **68**: 251-306.

Yang, J., X. Liu, K. Bhalla, C. N. Kim, A. M. Ibrado, J. Cai, T. I. Peng, D. P. Jones and X. Wang (1997). "Prevention of apoptosis by Bcl-2: release of cytochrome c from mitochondria blocked." Science **275**(5303): 1129-1132.

Zakeri, Z. and R. A. Lockshin (2008). "Cell death: history and future." Adv Exp Med Biol **615**: 1-11.

Zakeri, Z. F., D. Quaglino, T. Latham and R. A. Lockshin (1993). "Delayed internucleosomal DNA fragmentation in programmed cell death." Faseb j **7**(5): 470-478.

Zamzami, N., P. Marchetti, M. Castedo, C. Zanin, J. L. Vayssiere, P. X. Petit and G. Kroemer (1995). "Reduction in mitochondrial potential constitutes an early irreversible step of programmed lymphocyte death in vivo." J Exp Med **181**(5): 1661-1672.

Zitvogel, L., L. Apetoh, F. Ghiringhelli and G. Kroemer (2008). "Immunological aspects of cancer chemotherapy." Nat Rev Immunol **8**(1): 59-73.

Zitvogel, L., N. Casares, M. O. Pequignot, N. Chaput, M. L. Albert and G. Kroemer (2004). "Immune response against dying tumor cells." Adv Immunol **84**: 131-179.

Zitvogel, L. and G. Kroemer (2008). "The immune response against dying tumor cells: avoid disaster, achieve cure." Cell Death Differ **15**(1): 1-2.

Zitvogel, L., A. Tesniere and G. Kroemer (2006). "Cancer despite immunosurveillance: immunoselection and immunosubversion." Nat Rev Immunol **6**(10): 715-727.

Chapter 8: Efferocytosis: Molecular Mechanisms and Immune Signaling from Dying Cells (Kumar, Smith, and Birge)

Sushil Kumar[1], Brendan Smith[1], and Raymond B. Birge [1,2]

[1]Department of Microbiology, Biochemistry, and Molecular Genetics, Rutgers School of Biomedical and Health Sciences – Cancer Center, 205 South Orange Avenue, Newark, NJ 07103

[2]Corresponding author: Department of Biochemistry and Molecular Biology, Rutgers School of Biomedical and Health Sciences – Cancer Center, 205 South Orange Avenue, Newark, NJ 07103

Abstract:

Apoptosis is an evolutionarily conserved and tightly regulated cell death modality. It serves important roles in the biology of life, by removing unwanted cells that have reached advanced age or whose genomes have been irreparably damaged, as well as by sculpting complex structures during embryogenesis. Apoptosis culminates in the rapid and decisive removal of cell corpses by neighboring or recruited viable cells, a term called efferocytosis to distinguish the engulfment of apoptotic cells from other phagocytic processes. Over the past 20 years, the molecular and cell biological events associated with efferocytosis have been rigorously studied, and many of the genes and pathways have now been delineated. Moreover, the detection and elimination of corpses under physiological conditions promotes an anti-inflammatory response at the cellular and tissue level, as well as immunological tolerance at the organism level. Consequently, defects in efferocytosis have been linked to various pathological conditions that include systemic lupus erythematosis (SLE) and atherosclerosis. Conversely, under certain conditions, the killing of tumor cells with chemotherapy and radiation can induce immunogenic cell death, that culminates in the ability of the host to mount an effective anti-tumor response. In this chapter, we review some of the historical milestones inthe field of efferocytosis, and speculate where the field may be heading in future years.

Introduction:

By the early 1990's, the field of apoptosis was in a rapid expansive phase following the identification of a set of cell death defective genes (commonly abbreviated CED genes) in

Caenorhabditis elegans (*C. elegans*) that defined at the molecular and genetic level how apoptosis was arranged in metazoans (Yuan and Horvitz 2004). These discoveries helped establish the central dogma of apoptotic cell death, where apoptosis was positively regulated by CED3 (caspase) and CED4 (APAF1) (Yuan and Horvitz 1990, Yuan and Horvitz 1992, Hengartner and Horvitz 1994), and negatively regulated by CED9 (Bcl-2) (Hengartner and Horvitz 1994), a body of work that, in part, led to the Nobel Prize in Physiology and Medicine to H. Robert Horvitz and his fellow researchers in *C. elegans* in 2002 (**Fig. 1**) (Marx 2002, Horvitz 2003). Perhaps initially less universally recognized, but conceptually of equal importance, these genetic studies in worms also identified a second set of genes (indeed a larger array of genes) comprising CED1, CED2, CED5, CED6, CED7, CED8, CED10, and CED12 that regulated the engulfment of apoptotic corpses (Ellis, Jacobson et al. 1991). The identification of these gene products unequivocally demonstrated that clearance (like apoptosis) was genetically programmed in multicellular organisms to ensure rapid and decisive removal of cell corpses by neighboring viable cells.

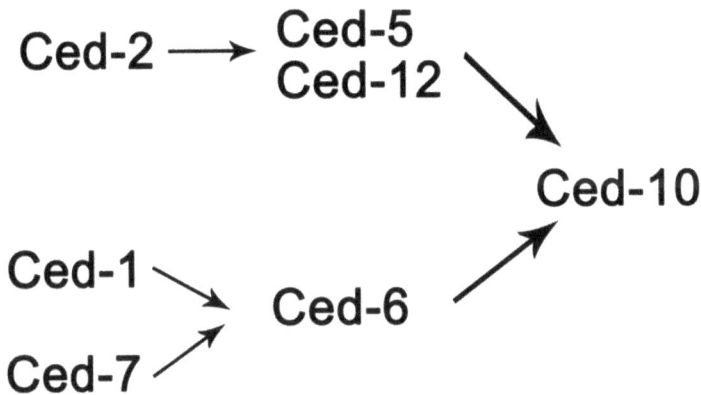

Fig. 1. **The genetic pathway for efferocytosis in C. elegans:** Genetic analysis has led to the identification of several different mutations that affect engulfment of apoptotic cells (efferocytosis). These genes include CED1 (ATP binding cassette ABC1 transporter), CED2 (Crk II), CED5 (DOCK180/DOCK1),

CED6 (GULP), CED7 (CD91/Scavenger receptor from endothelial cells), CED8 (Xkr8), CED10 (Rac1), and Ced12 (ELMO). Arrows indicate positive regulatory elements.

Complementation studies revealed that the aforementioned genes that regulated engulfment encoded two evolutionarily conserved modules, the first defined by a complementation group involving CED2 (Crk) (Reddien and Horvitz 2000), CED5 (DOCK180) (Wu and Horvitz 1998), CED12 (ELMO) (Zhou, Caron et al. 2001) (Gumienny, Brugnera et al. 2001) (biochemically, CED2/CED5/and CED12 actually form a ternary protein complex and can be co-immunoprecipitated) that was upstream of CED10 (Rac1) (Reddien and Horvitz 2000), and a second complementation group comprised of CED1 (CD91/Scavenger receptor from endothelial cells (SREC)-like protein) (Zhou, Hartwieg et al. 2001), CED6 (GULP) (Liu and Hengartner 1998) (Su, Nakada-Tsukui et al. 2002), and CED7 (ATP binding cassette ABC1 transporter) (Wu and Horvitz 1998). Like the apoptosis pathways, the engulfment pathways could be mapped by genetic rescue experiments; For example a CED2 or CED5 loss of function mutant could be rescued by a CED10 gain of function mutant. Interestingly, while the initial studies identified two complementation groups, later studies showed that CED10 gain of function mutants could rescue both pathways, implicating the Rho family GTPase, Rac1, as the master switch signal for the clearance of apoptotic cells in multicellular organisms (Kinchen, Cabello et al. 2005). Classical cell biological experiments showed that Rac1 was also essential for lamellipodia and actin-dependent polarized cell migration, thereby linking polarized motility and efferocytosis as a common cellular and biochemical event (Gumienny, Brugnera et al. 2001) (Kinchen and Ravichandran 2007).

Alterations in the surface of dying cells; Apoptotic Cell Associated Molecular Patterns:

While the aforementioned genetic studies in C. *elegans* provided a logical conceptual framework for how apoptotic cells were engulfed and delivered to phagolysosomes, perhaps unexpectedly, these studies provided less insight into how apoptotic cells are recognized as effete, as only two CED mutants (CED1 and CED7) were identified as cell surface receptors expected to interact with exposed surfaces on the apoptotic cells. Indeed, unlike the regulation of efferocytosis in the relatively simple organism *C. elegans*, efferocytosis in higher metazoans appears to represent an exceedingly complex biology, and over the past two decades, over 40 cell surface receptors or apoptotic cell bridging molecules (called opsonins) have been identified (Savill, Dransfield et al. 2002) (Poon, Lucas et al. 2014). Clearly, while prioritizing this complexity and appreciating the nature of the apparent redundancy in apoptotic cell receptors remains an ongoing challenge, a few general rules may provide some conceptual relief to this field.

First, it appears that the surface of the apoptotic cell possesses molecular patterns that are partially shared with molecular components on the surface of pathogens (the so called "pathogen-associated molecular patterns" or PAMPs) that initiate innate immune responses towards pathogens (Poon, Hulett et al. 2010). These components on the apoptotic cell surface, aptly referred to as "apoptotic cell-associated molecular patterns" (ACAMPs), recruit early components of the classical complement pathway (C1q) (Mevorach, Mascarenhas et al. 1998, Vandivier, Ogden et al. 2002, Galvan, Greenlee-Wacker et al. 2012), the alternative complement pathway (C3) (Nauta, Daha et al. 2003, Nauta, Castellano et al. 2004), and components of the mannose-binding lectin (MBL) pathway (Ogden, deCathelineau et al. 2001, Savill, Dransfield et al. 2002), and by doing so, provide "eat-me" cues for phagocytes. Notably however, these complexes, which form on the surface of the apoptotic cell membrane, never activate late components of complement or lead to the formation of a membrane attack complex (MAC). Moreover, in stark contrast to the pro-inflammatory responses of PAMPs on pathogen surfaces, ACAMPs provide a non-inflammatory signal and promote immune tolerance.

A second and related consideration predicts that the surface of the apoptotic cell is biochemically and molecularly modified to serve as constituents of "altered-self". Indeed, Savill and colleagues originally described this phenomenon, which implicated irreversible modified endogenous resident macromolecules, and now it is clear that both lipids and glycoproteins become molecularly modified at the cell surface during apoptosis (Savill, Dransfield et al. 2002). Although the exact nature of altered self-epitopes is far from completely understood, the removal of sialic acid from glycoproteins (Meesmann, Fehr et al. 2010) and the oxidation of certain phospholipids provide such signals (Chang, Bergmark et al. 1999, Kagan, Gleiss et al. 2002, Kagan, Borisenko et al. 2003). Oxidized lipids and low density lipoproteins (LDLs), in turn, interact with a variety of scavenger receptors on phagocytes such as CD68 (Platt, Suzuki et al. 1996, Platt, Suzuki et al. 2000), CD36 (Ren, Silverstein et al. 1995, Stern, Savill et al. 1996), LOX-1 (Murphy, Tacon et al. 2006) and SCARF1 (Ramirez-Ortiz, Pendergraft et al. 2013), which can contribute substantially to apoptotic cell clearance. Indeed, in *Drosophila melanogaster*, the CD36-related protein Croquemort expressed on hemocytes (macrophages) appears to be one of the principal engulfment receptors, as a knockout of Croquemort result in a failure to clear apoptotic cells during embryogenesis (Franc, Heitzler et al. 1999). Likewise, in mammals, scavenger receptors including CD36 and SCARF1 are also critical for efferocytosis, and recent studies have linked SCARF1 (possibly as a co-receptor for C1q) with failed clearance and autoimmunity (Kimani, Geng et al. 2014).

In addition to their role as eat-me signals for efferocytosis, oxidized lipids and LDLs also have important intracellular roles in phagocytic cells by interacting with lipid sensing receptors of the LXR and Peroxisome Proliferator Activated Receptor gamma (PPARγ families (Mukundan, Odegaard et al. 2009, Roszer, Menendez-Gutierrez et al. 2011). These receptors sense increases in lipid mass in phagocytes and contribute by up-regulating additional engulfment receptors that control eating in a feed-forward loop (N, Bensinger et al. 2009), but also limit the amount of inflammation by directing the expression of anti-inflammatory phenotypes (Bensinger, Bradley et al. 2008).

Re-localization of PS and intracellular proteins to the surface of the apoptotic cell:

Concomitant with biochemical modifications of endogenous molecules that serve as cues for efferocytosis, many cellular macromolecules also become re-localized from an intracellular locale to a final determined destiny at the extracellular surface of the plasma membrane. In many cases, these re-localization events require both caspase activation and ATP, implicating externalization as part of the apoptotic program, and ensuring that apoptosis and engulfment are temporally coupled by common checkpoints. While the complete repertoire of plasma membrane changes is not fully understood, the externalization of phosphatidylserine (PS) is arguably the most emblematic event associated with apoptosis, and PS clearly serves as recognition module for a variety of receptors and serum proteins that contribute to efferocytosis (Krahling, Callahan et al. 1999, Schlegel, Callahan et al. 2000) (Fadok, Voelker et al. 1992) (Fig. 2). Indeed, many of the PS receptors described in Fig. 2, such as MFG-E8, TIM-1, TIM-4, MERTK, C1q, and SCARF1, when targeted by knockout in mice, result in an impaired capacity to clear apoptotic cells in vivo, validating the centrality of PS externalization as a central recognition cue for efferocytosis in metazoans.

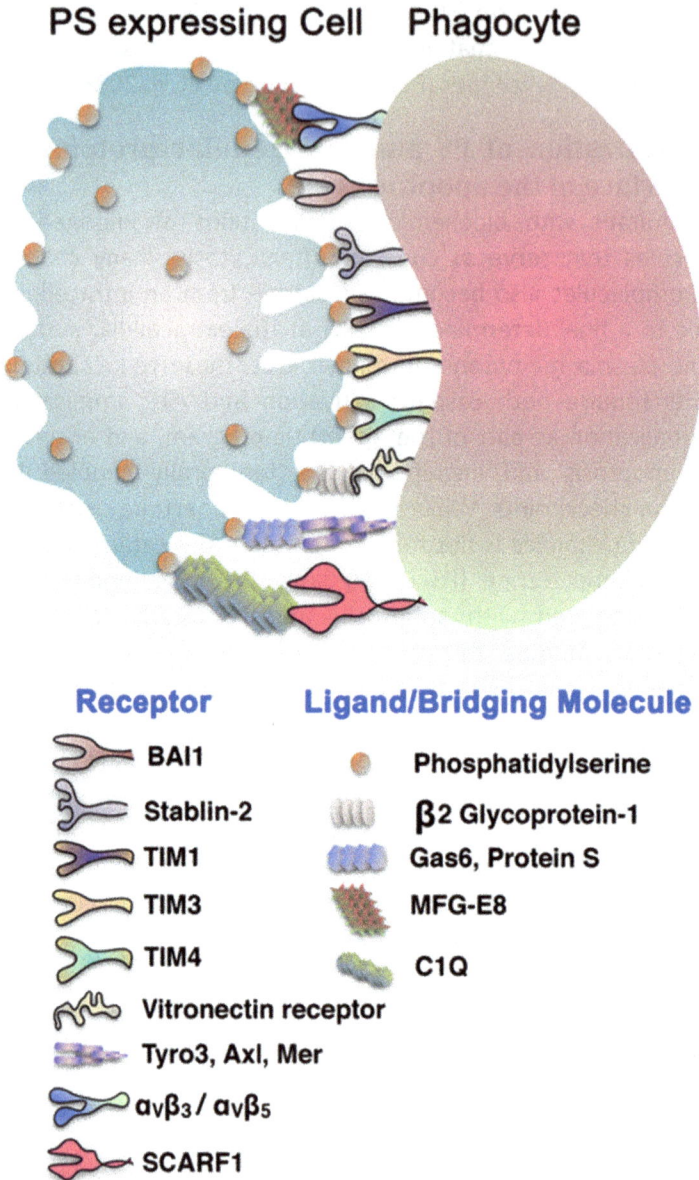

Fig.2. PS-dependent recognition of apoptotic cells: Apoptotic cells expressing PS are recognized by various receptors present on phagocytes that include BAI1, Stablin-2, TIM1, 3, 4, Tyro, Axl, Mer, SCARF1 and Vitronectin receptors

among others. These PS receptors either bind PS directly or indirectly through bridging molecules such as β2 glycoprotein1, Gas6, protein S, MFG-E8 and C1q.

While historically the field of PS biology has been predominantly occupied with the identification and characterization of a vast array of PS receptors and opsonins (Wu, Tibrewal et al. 2006), over the past few years there has been a renewed focus on the mechanisms by which PS is externalized, and a realization that (i) different scramblases externalize PS by different mechanisms, and (ii) not all the externalized PS has the same biological function (Suzuki, Umeda et al. 2010, Suzuki, Denning et al. 2013). For example, externalization of PS by the calcium-activated scramblase TMEM16 is readily reversible, and remarkably, does not serve as a signal for engulfment (Segawa, Suzuki et al. 2011). Rather, the PS that is externalized by TMEM16 creates a nucleation surface for the binding and activation of clotting factors (Yang, Kim et al. 2012). In contrast, externalization of PS by the caspase-activated scramblase Xkr8 (CED8) is largely irreversible and induces efferocytosis (Suzuki, Imanishi et al. 2014). Under these conditions, the PS that is externalized by Xkr8 is recognized by conventional PS receptors and opsonins such as Gas6/MERTK and MFG-E8αvβ3 integrin (Kimani, Geng et al. 2014) (Fig. 3). Clearly, an obvious and important question concerns the distinction between the PS that is externalized by TMEM16 and by Xkr8, and what molecular signatures explain these different itineraries. Presently, at least two distinct possibilities exist, and recent evidence supports both scenarios. In the first, it is possible that the PS externalized by TMEM16 versus Xkr8 results in different PS saturation density at the cell surface or that PS exists in different lipid microdomains. This could occur if TMEM16 and Xkr8 themselves are activated at different subcellular localizations that immediately redistribute PS to distinct microdomains. Alternatively, the different fates of PS could also result if re-localization mechanisms back to the inner membrane (i.e. floppase activity) are different as a result of cell activation (TMEM16) versus apoptosis (Xkr8). Indeed, during apoptosis, a second PS-dependent enzyme (ATP11C) is also cleaved by caspase 3, inactivating floppase activity, and therefore sustaining the exposure of PS, possibility explaining the irreversibility of this event (Segawa, Kurata et al. 2014). The fact that TMEM16 causes reversible PS exposure, while Xkr8 causes irreversible PS exposure, likely contributes, at least in part, to the different effects on efferocytosis (Fig. 3).

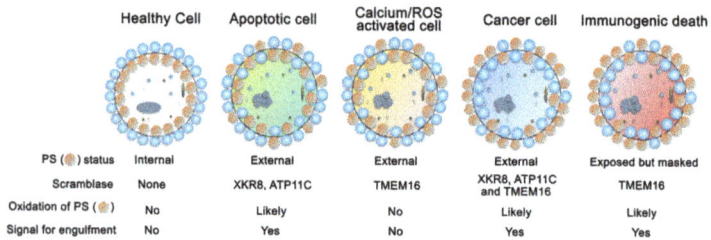

	Healthy Cell	Apoptotic cell	Calcium/ROS activated cell	Cancer cell	Immunogenic death
PS ( ) status	Internal	External	External	External	Exposed but masked
Scramblase	None	XKR8, ATP11C	TMEM16	XKR8, ATP11C and TMEM16	TMEM16
Oxidation of PS ( )	No	Likely	No	Likely	Likely
Signal for engulfment	No	Yes	No	Yes	Yes

Fig. 3. Differential roles of PS scramblases in distinct types of cell deaths: Externalization of PS to the outer leaflet of cell membrane is a characteristic feature of cell death. However, apoptotic cells, calcium/ROS activated cells, cancer cells and immunogenically dead cells employ distinct PS scramblases that further lead to differential distribution and oxidation of externalized PS.

In addition to the issue of PS surface density, driven by activation of different scramblases, another important conceptual idea that has gained traction in recent years is that the PS externalized that serves as an eat-me signal is covalently modified by oxidation on one or more the PS acyl chains, resulting in oxPS and a more "palatable" signal for efferocytic receptors (Kagan, Gleiss et al. 2002, Tyurin, Balasubramanian et al. 2014). Indeed, in some cases, such as in the binding of Gas6, MFG-E8, and TIM1, these proteins bind with higher affinity to oxPS than non-oxidized PS, thereby providing a biochemical rationale (Tyurin, Balasubramanian et al. 2014). Moreover, from a mechanistic level, the oxidation of PS during apoptosis has an additional important implication, as one of the proposed mechanisms for PS oxidation involves cytochrome c-dependent PS oxidation, with cytochrome c acquiring a gain-of-function peroxidase activity once released from the mitochondria (Jiang,

Kini et al. 2004). In this model, the cytochrome c released from the mitochondria during mitochondrial outer membrane permeabilization (MOMP) would serve two interrelated functions, first as a central component of the apoptosome, and second, to concomitantly catalyze the oxidation of PS to ensure that apoptotic cells are swiftly and decisively cleared by phagocytes (Jiang, Kini et al. 2004, Kagan, Tyurin et al. 2005) (Fig. 3).

Finally, in addition to the re-localization the PS, many resident cellular proteins are re-localized at the plasma membrane during apoptosis. These events can result in the deposition of neo-epitopes on the surface of apoptotic cells, such as cytosolic glycolytic enzymes TPI and Enolase (Ucker, Jain et al. 2012), or the endoplasmic reticulum (ER)-luminal protein Calreticulin (CRT) (Gardai, McPhillips et al. 2005). On the other hand, the internalization of proteins away from the membrane, such as CD47, CD300a, CD31 can also occur during apoptosis (Gardai, McPhillips et al. 2005, Gardai, Bratton et al. 2006) (Brown, Heinisch et al. 2002) (Simhadri, Andersen et al. 2012). CD47 is a member of a very interesting class of receptors that function as "don't eat-me signals" that must be incapacitated for the cells to be properly recognized for engulfment. Interestingly, CD47 is up-regulated on certain tumor cells in order to avoid efferocytosis, suggesting that both positive and negative regulation exists (Jaiswal, Jamieson et al. 2009). Together, while many epitopes have now been defined that provide recognition cues for binding or internalization of apoptotic cells, there is still a great deal to be learned in terms of the carbohydrate, proteome, and lipidome of the apoptotic cell membrane.

Post-efferocytosis immune responses, signals from beyond the grave:

While efferocytosis is clearly recognized as a waste-disposal management system that carries out the removal and degradation of apoptotic cells in phagolysosomes, over the past two decades it has also become quite apparent that apoptotic cell/phagocyte interactions have profound immunological consequences in higher metazoans (Savill, Dransfield et al. 2002) (Poon, Lucas et al. 2014) (Birge and Ucker 2008), both at the level of professional phagocytes (macrophages and dendritic cells) (Voll, Herrmann et al. 1997, Huynh, Fadok et al. 2002), as well as non-professional phagocytes such as epithelial cells (Juncadella, Kadl et al. 2013) and astrocytes (Loov, Hillered et al. 2012). Indeed, it is likely that efferocytosis serves as a global immune surveillance system to autopsy the quality and mechanisms of death within the tissue environment. Along these same lines, it is equally likely that the supreme goal of apoptosis, despite all the intracellular and organelle changes that occur, is to transmit appropriate information via the plasma membrane lipidome/proteome to communicate signals to immune effector cells. Depending on the nature and content of the lipidome and proteome of the apoptotic cell surface, information would be provided about the past history of the cell, whether it died a natural or unnatural death, and whether it was stressed, transformed, senescent, or infected by virus. Mediated by the subsequent production of specific cytokines and chemokines, and processing of antigens derived from the apoptotic cells (Albert, Sauter et al. 1998, Kroemer, Galluzzi et al. 2013), both innate and adaptive immune responses would ensue to harness immunological outcomes **(Fig. 4)**. As discussed below, such immunological responses maintain homeostasis under physiological conditions (tolerance), but can lead to inflammation and autoimmunity under pathophysiological conditions (Poon, Lucas et al. 2014).

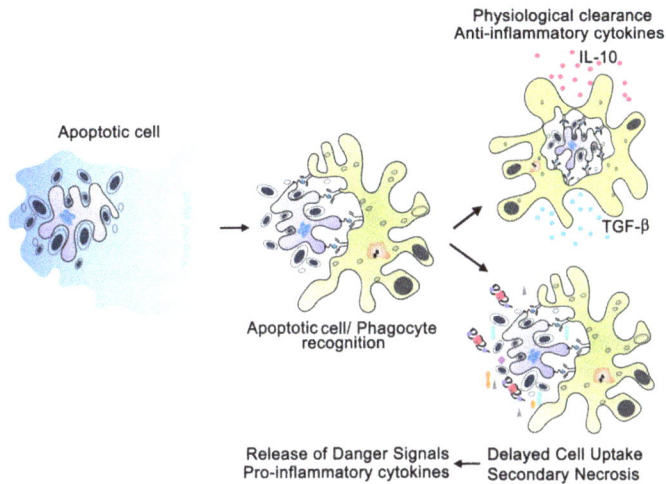

Fig. 4. Efferocytosis: A choice between physiological clearance and secondary necrosis: Apoptotic cells secrete "Find me" signals that induce its phagocytic recognition. During physiological clearance, the apoptotic cell is efficiently engulfed by the phagocyte in a process called as efferocytosis and leads to production of anti-inflammatory cytokines such as IL-10 and TGF- β. In an event of delayed cell uptake or efferocytosis, secondary necrosis followed by release of various danger signal and pro-inflammatory cytokines take place.

Under physiological conditions, the interaction between apoptotic cells and phagocytes is anti-inflammatory and tolerogenic, and spares the tissue microenvironment of inflammation and fibrosis (Voll, Herrmann et al. 1997, Steinman, Turley et al. 2000, Birge and Ucker 2008) (Gaipl, Beyer et al. 2003). Of the molecular determinants on the apoptotic cell surface that direct tolerogenic death, exposed PS is arguably the pre-eminent signal for immunosuppression (as it is for engulfment) although other neo-epitopes undoubtedly also contribute. Externalized PS has pleiotropic immune responses that (i) inhibit NF-kB and p38 MAPK activation on human DCs (Doffek, Chen et al. 2011), (ii) down-regulate TLR4 signaling and dampen pro-inflammatory cytokine production via TAM receptors (Lu and Lemke 2001, Cohen, Caricchio et al. 2002), (iii) prevent the activation of macrophages (Gaipl, Beyer et al. 2003) and DCs (He, Yin et al. 2009), and (iv) and polarize macrophages towards M2 and prevent the maturation dendritic cells to cross-presents antigens in a class-I restricted manner (Kimani, Geng et al. 2014). Moreover, phagocytes that have encountered PS-positive apoptotic cells produce immunosuppressive IL-10 and TGF-β, factors that create a tolerized microenvironment that prevents immune activation (Fadok, Bratton et al. 1998) (Huynh, Fadok et al. 2002) (Juncadella, Kadl et al. 2013). For these reasons PS externalized on apoptotic cells has been considered a global immunosuppressive signaling checkpoint by preventing the activation of auto-reactive cells that contribute to autoimmunity.

By contrast to the maintenance of tolerance when apoptotic cells are cleared under physiological conditions, when efferocytosis is compromised, either when apoptosis exceeds clearance capacity or when engulfment receptors are inadequate, apoptotic cells undergo secondary necrosis in peripheral tissues, secondary lymphoid tissues, or germinal centers and can release intracellular components as a sterile infection (Baumann, Kolowos et al. 2002, Munoz, Lauber et al. 2010, Shao and Cohen 2011). Under these conditions, auto-reactive intracellular constituents such as self-nucleic acids and modified histones are released from cells, which can lead to auto-antibody production and a pathology highly reminiscent of systemic lupus erythematosis (SLE) (Munoz, Lauber et al. 2010) **(Fig. 4)**. While the relationship between defective apoptotic clearance and SLE is strongly predicted by studies in knockout mice, whereby several of the PS receptor knockout mice and complement receptor knockout mice develop an SLE-like phenotype, the relationship between SLE and human SLE is still somewhat enigmatic (Kimani, Geng et al. 2014). In this respect, while there are some rare genetic associations between clearance factors (i.e. C1q and MFG-8) and SLE (Botto, Dell'Agnola et al. 1998, Yamaguchi, Fujimoto et al. 2010, Yamamoto, Yamaguchi et al. 2014) (Hanayama, Tanaka et al. 2004, Hu, Wu et al. 2009), these events are relatively uncommon compared to the incidence of human lupus in the general population. Presently, the genetics between failed efferocytosis and SLE is a topical and ongoing area of investigation in the field of cell death.

In between the extremes of classical apoptosis (efficient clearance) and secondary necrosis (inefficient clearance) lies a range of cell death pathways with unique molecular features (Kroemer, Galluzzi et al. 2009). For example, programmed necrosis, characterized by RIP1/RIP3/MLKL-dependent necrosome formation, and calcium/ROS-dependent mitochondrial permeability transition (MPT), are both considered pro-inflammatory based on the interaction of yet unknown constituents on the dying cells with phagocytes (Kroemer, Galluzzi et al. 2009, Linkermann and Green 2014, Tait, Ichim et al. 2014). Indeed, given the frequency of new and unique forms of death (Kroemer, Galluzzi et al. 2009), a challenge to the field will be to employ the "phagocytic response output" as a tool to assess similarities and differences in cell death programs. In this capacity, the gene expression patterns and cytokine arrays could be profiled from macrophages or DCs during their interaction with apoptotic or necrotic cells **(Fig. 5).**

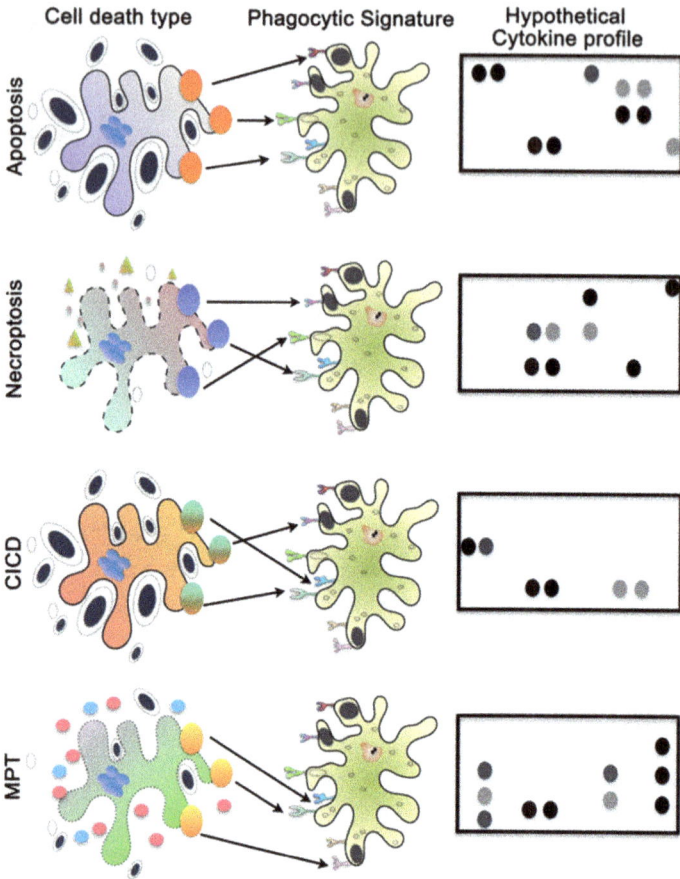

Fig. 5. Engagement of different dying cells with phagocytic receptors may trigger differential cytokine profile: Cells that have undergone cell death through various mechanisms including apoptosis, necroptosis, caspase-independent cell death (CICD) and mitochondrial permeability transition (MPT) may expose different ligands that in turn may engage distinct receptors on phagocytes. A hypothetical cytokine profile shown here indicates the possibility that these distinct dead cell-phagocyte recognitions may trigger differential secretion of cytokines.

Immunogenic death of tumor cells and anti-tumor responses:

The above-mentioned scenario in which the surface of the apoptotic cell can determine the post-efferocytic immunological outcomes is elegantly demonstrated by studies from Kroemer and colleagues showing that certain tumor cells treated with anthracyclines or irradiation can elicit immune stimulatory activities that lead to a durable anti-tumor response, a process called immunogenic cell death (Eggermont, Kroemer et al. 2013, Kroemer, Galluzzi et al. 2013, Ladoire, Hannani et al. 2014, Sukkurwala, Martins et al. 2014, Yamazaki, Hannani et al. 2014). Mechanistically, such studies have shown that if certain "lock-and-key" molecular determinants can exist, DCs can adapt a mature phenotype and cross-present antigens to anti-tumor T cells and mount an anti-tumor response. While the complete molecular determinants are not completely defined, several of the features of immunogenic death have been elucidated and validated with *in vivo* tumor models and immunocompetent mice. For example, chemotherapeutics that induce the release of high-mobility group box 1 (HMGB1), CRT, and ATP from dying tumor cells satisfy the minimal requirements for immunogenic death, although there are likely to be additional components. ATP appears to be required not only to recruit DCs to the vicinity of apoptotic tumor cells via purinoreceptor-1 (P2Y2) receptors (Elliott, Chekeni et al. 2009), but also to activate the inflammasome and induce DC maturation and prime DC for anti-tumor adaptive immunity (Ghiringhelli, Apetoh et al. 2009, Aymeric, Apetoh et al. 2010). CRT, on the other hand, acts as a danger signal and traffics apoptotic cargo into a cross-presentation competent itinerary (Chaput, De Botton et al. 2007, Obeid, Tesniere et al. 2007).

Secreted HMGB1 also appears to have a critical role in immunogenic death, although presently the mechanisms by which it impinges on tumor immunity are not completely understood. On the one hand, HMGB1 may act as a Toll-like receptor 4 (TLR4) agonist, which can lead to activation and maturation of DCs as a pre-requisite of cross-presentation (Apetoh, Ghiringhelli et al. 2007, Yamazaki, Hannani et al. 2014). On the other hand, recent studies also suggest that HMGB1 can act as a mask for PS (Liu, Wang et al. 2008), thereby indirectly blocking the anti-inflammatory and tolerogenic activity of PS. Indeed, it is well known that blocking PS in the tumor microenvironment, either with blocking antibodies (Bavituximab) (He, Yin et al. 2009, Yin, Huang et al. 2013) or recombinant proteins (Annexin 1) (Rovere, Sabbadini et al. 1999, Bondanza, Zimmermann et al. 2004), can markedly enhance the ability to mount an anti-tumor response.

Summary:

Concomitant to the major advances in our understanding of apoptosis and programed cell death and necrosis over the past 20 years, similar advances have occurred in the field of efferocytosis. Not only do we have a much clearer understanding of the genes and pathways that govern apoptotic cell clearance, but recent evidence supports the idea that the biology of efferocytosis can be therapeutically manipulated in diseases such as autoimmunity and cancer. Clearly, the next challenge will be to harness the knowledge of the apoptotic cell clearance pathways for a clinical benefit.

Final remarks:

The formation of the Cell Death Society in 1994 at The Rockefeller University (at that time called the Death Poets Society) represented a collective embracement of the new field of cell death that brought together researchers from incredibly diverse interests, ranging from developmental biology, genetics, immunology, biochemistry, cancer biology, and neuroscience. Some people brushed up against this most peculiar field and quickly turned away. Others jumped on and off at different times depending on the circumstances of their research projects. But a sizable number of scientists, we'll never know how many, got it big time and embraced the field of cell death to call it their own. Over the past 20 years, the International Cell Death Society has swept up 1000's of investigators and students and developed into a club with no requirements. But chances are, if your reading this with interest, you already know that.

References:

Albert, M. L., B. Sauter and N. Bhardwaj (1998). "Dendritic cells acquire antigen from apoptotic cells and induce class I-restricted CTLs." Nature 392(6671): 86-89.

Apetoh, L., F. Ghiringhelli, A. Tesniere, A. Criollo, C. Ortiz, R. Lidereau, C. Mariette, N. Chaput, J. P. Mira, S. Delaloge, F. Andre, T. Tursz, G. Kroemer and L. Zitvogel (2007). "The interaction between HMGB1 and TLR4 dictates the outcome of anticancer chemotherapy and radiotherapy." Immunol Rev 220: 47-59.

Aymeric, L., L. Apetoh, F. Ghiringhelli, A. Tesniere, I. Martins, G. Kroemer, M. J. Smyth and L. Zitvogel (2010). "Tumor cell death and ATP release prime dendritic cells and efficient anticancer immunity." Cancer Res 70(3): 855-858.

Baumann, I., W. Kolowos, R. E. Voll, B. Manger, U. Gaipl, W. L. Neuhuber, T. Kirchner, J. R. Kalden and M. Herrmann (2002). "Impaired uptake of apoptotic cells into tingible body macrophages in germinal centers of patients with systemic lupus erythematosus." Arthritis Rheum 46(1): 191-201.

Bensinger, S. J., M. N. Bradley, S. B. Joseph, N. Zelcer, E. M. Janssen, M. A. Hausner, R. Shih, J. S. Parks, P. A. Edwards, B. D. Jamieson and P. Tontonoz (2008). "LXR signaling couples sterol metabolism to proliferation in the acquired immune response." Cell 134(1): 97-111.

Birge, R. B. and D. S. Ucker (2008). "Innate apoptotic immunity: the calming touch of death." Cell Death Differ 15(7): 1096-1102.

Bondanza, A., V. S. Zimmermann, P. Rovere-Querini, J. Turnay, I. E. Dumitriu, C. M. Stach, R. E. Voll, U. S. Gaipl, W. Bertling, E. Poschl, J. R. Kalden, A. A. Manfredi and M. Herrmann (2004). "Inhibition of phosphatidylserine recognition heightens the immunogenicity of irradiated lymphoma cells in vivo." J Exp Med 200(9): 1157-1165.

Botto, M., C. Dell'Agnola, A. E. Bygrave, E. M. Thompson, H. T. Cook, F. Petry, M. Loos, P. P. Pandolfi and M. J. Walport (1998). "Homozygous C1q deficiency causes glomerulonephritis associated with multiple apoptotic bodies." Nat Genet 19(1): 56-59.

Brown, S., I. Heinisch, E. Ross, K. Shaw, C. D. Buckley and J. Savill (2002). "Apoptosis disables CD31-mediated cell detachment from phagocytes promoting binding and engulfment." Nature 418(6894): 200-203.

Chang, M. K., C. Bergmark, A. Laurila, S. Horkko, K. H. Han, P. Friedman, E. A. Dennis and J. L. Witztum (1999). "Monoclonal antibodies against oxidized low-density lipoprotein bind to apoptotic cells and inhibit their phagocytosis by elicited macrophages: evidence that oxidation-specific epitopes mediate macrophage recognition." Proc Natl Acad Sci U S A 96(11): 6353-6358.

Chaput, N., S. De Botton, M. Obeid, L. Apetoh, F. Ghiringhelli, T. Panaretakis, C. Flament, L. Zitvogel and G. Kroemer (2007). "Molecular determinants of immunogenic cell death: surface exposure of calreticulin makes the difference." J Mol Med (Berl) 85(10): 1069-1076.

Cohen, P. L., R. Caricchio, V. Abraham, T. D. Camenisch, J. C. Jennette, R. A. Roubey, H. S. Earp, G. Matsushima and E. A. Reap (2002). "Delayed apoptotic cell clearance and lupus-like autoimmunity in mice lacking the c-mer membrane tyrosine kinase." J Exp Med 196(1): 135-140.

Doffek, K., X. Chen, S. L. Sugg and J. Shilyansky (2011). "Phosphatidylserine inhibits NFkappaB and p38 MAPK activation in human monocyte derived dendritic cells." Mol Immunol 48(15-16): 1771-1777.

Eggermont, A. M., G. Kroemer and L. Zitvogel (2013). "Immunotherapy and the concept of a clinical cure." Eur J Cancer 49(14): 2965-2967.

Elliott, M. R., F. B. Chekeni, P. C. Trampont, E. R. Lazarowski, A. Kadl, S. F. Walk, D. Park, R. I. Woodson, M. Ostankovich, P. Sharma, J. J. Lysiak, T. K. Harden, N. Leitinger and K. S. Ravichandran (2009). "Nucleotides released by apoptotic cells act as a find-me signal to promote phagocytic clearance." Nature 461(7261): 282-286.

Ellis, R. E., D. M. Jacobson and H. R. Horvitz (1991). "Genes required for the engulfment of cell corpses during programmed cell death in Caenorhabditis elegans." Genetics 129(1): 79-94.

Fadok, V. A., D. L. Bratton, A. Konowal, P. W. Freed, J. Y. Westcott and P. M. Henson (1998). "Macrophages that have ingested apoptotic cells in vitro inhibit proinflammatory cytokine production through autocrine/paracrine mechanisms involving TGF-beta, PGE2, and PAF." J Clin Invest 101(4): 890-898.

Fadok, V. A., D. R. Voelker, P. A. Campbell, J. J. Cohen, D. L. Bratton and P. M. Henson (1992). "Exposure of phosphatidylserine on the surface of apoptotic lymphocytes triggers specific recognition and removal by macrophages." J Immunol 148(7): 2207-2216.

Franc, N. C., P. Heitzler, R. A. Ezekowitz and K. White (1999). "Requirement for croquemort in phagocytosis of apoptotic cells in Drosophila." Science 284(5422): 1991-1994.

Gaipl, U. S., T. D. Beyer, I. Baumann, R. E. Voll, C. M. Stach, P. Heyder, J. R. Kalden, A. Manfredi and M. Herrmann (2003). "Exposure of anionic phospholipids serves as anti-inflammatory and immunosuppressive signal-- implications for antiphospholipid syndrome and systemic lupus erythematosus." Immunobiology 207(1): 73-81.

Galvan, M. D., M. C. Greenlee-Wacker and S. S. Bohlson (2012). "C1q and phagocytosis: the perfect complement to a good meal." J Leukoc Biol 92(3): 489-497.

Gardai, S. J., D. L. Bratton, C. A. Ogden and P. M. Henson (2006). "Recognition ligands on apoptotic cells: a perspective." J Leukoc Biol 79(5): 896-903.

Gardai, S. J., K. A. McPhillips, S. C. Frasch, W. J. Janssen, A. Starefeldt, J. E. Murphy-Ullrich, D. L. Bratton, P. A. Oldenborg, M. Michalak and P. M. Henson (2005). "Cell-surface calreticulin initiates clearance of viable or apoptotic cells through trans-activation of LRP on the phagocyte." Cell 123(2): 321-334.

Ghiringhelli, F., L. Apetoh, A. Tesniere, L. Aymeric, Y. Ma, C. Ortiz, K. Vermaelen, T. Panaretakis, G. Mignot, E. Ullrich, J. L. Perfettini, F. Schlemmer, E. Tasdemir, M. Uhl, P. Genin, A. Civas, B. Ryffel, J. Kanellopoulos, J. Tschopp, F. Andre, R. Lidereau, N. M. McLaughlin, N. M. Haynes, M. J. Smyth, G. Kroemer and L. Zitvogel (2009). "Activation of the NLRP3 inflammasome in dendritic cells induces IL-1beta-dependent adaptive immunity against tumors." Nat Med 15(10): 1170-1178.

Gumienny, T. L., E. Brugnera, A. C. Tosello-Trampont, J. M. Kinchen, L. B. Haney, K. Nishiwaki, S. F. Walk, M. E. Nemergut, I. G. Macara, R. Francis, T. Schedl, Y. Qin, L. Van Aelst, M. O. Hengartner and K. S. Ravichandran (2001). "CED-12/ELMO, a novel member of the CrkII/Dock180/Rac pathway, is required for phagocytosis and cell migration." Cell 107(1): 27-41.

Hanayama, R., M. Tanaka, K. Miyasaka, K. Aozasa, M. Koike, Y. Uchiyama and S. Nagata (2004). "Autoimmune disease and impaired uptake of apoptotic cells in MFG-E8-deficient mice." Science 304(5674): 1147-1150.

He, J., Y. Yin, T. A. Luster, L. Watkins and P. E. Thorpe (2009). "Antiphosphatidylserine antibody combined with irradiation damages tumor blood vessels and induces tumor immunity in a rat model of glioblastoma." Clin Cancer Res 15(22): 6871-6880.

Hengartner, M. O. and H. R. Horvitz (1994). "C. elegans cell survival gene ced-9 encodes a functional homolog of the mammalian proto-oncogene bcl-2." Cell 76(4): 665-676.

Hengartner, M. O. and H. R. Horvitz (1994). "The ins and outs of programmed cell death during C. elegans development." Philos Trans R Soc Lond B Biol Sci 345(1313): 243-246.

Horvitz, H. R. (2003). "Worms, life, and death (Nobel lecture)." Chembiochem 4(8): 697-711.

Hu, C. Y., C. S. Wu, H. F. Tsai, S. K. Chang, W. I. Tsai and P. N. Hsu (2009). "Genetic polymorphism in milk fat globule-EGF factor 8 (MFG-E8) is associated with systemic lupus erythematosus in human." Lupus 18(8): 676-681.

Huynh, M. L., V. A. Fadok and P. M. Henson (2002). "Phosphatidylserine-dependent ingestion of apoptotic cells promotes TGF-beta1 secretion and the resolution of inflammation." J Clin Invest 109(1): 41-50.

Jaiswal, S., C. H. Jamieson, W. W. Pang, C. Y. Park, M. P. Chao, R. Majeti, D. Traver, N. van Rooijen and I. L. Weissman (2009). "CD47 is upregulated on circulating hematopoietic stem cells and leukemia cells to avoid phagocytosis." Cell 138(2): 271-285.

Jiang, J., V. Kini, N. Belikova, B. F. Serinkan, G. G. Borisenko, Y. Y. Tyurina, V. A. Tyurin and V. E. Kagan (2004). "Cytochrome c release is required for phosphatidylserine peroxidation during Fas-triggered apoptosis in lung epithelial A549 cells." Lipids 39(11): 1133-1142.

Juncadella, I. J., A. Kadl, A. K. Sharma, Y. M. Shim, A. Hochreiter-Hufford, L. Borish and K. S. Ravichandran (2013). "Apoptotic cell clearance by bronchial epithelial cells critically influences airway inflammation." Nature 493(7433): 547-551.

Kagan, V. E., G. G. Borisenko, B. F. Serinkan, Y. Y. Tyurina, V. A. Tyurin, J. Jiang, S. X. Liu, A. A. Shvedova, J. P. Fabisiak, W. Uthaisang and B. Fadeel (2003). "Appetizing rancidity of apoptotic cells for macrophages: oxidation, externalization, and recognition of phosphatidylserine." Am J Physiol Lung Cell Mol Physiol 285(1): L1-17.

Kagan, V. E., B. Gleiss, Y. Y. Tyurina, V. A. Tyurin, C. Elenstrom-Magnusson, S. X. Liu, F. B. Serinkan, A. Arroyo, J. Chandra, S. Orrenius and B. Fadeel (2002). "A role for oxidative stress in apoptosis: oxidation and externalization of phosphatidylserine is required for macrophage clearance of cells undergoing Fas-mediated apoptosis." J Immunol 169(1): 487-499.

Kagan, V. E., V. A. Tyurin, J. Jiang, Y. Y. Tyurina, V. B. Ritov, A. A. Amoscato, A. N. Osipov, N. A. Belikova, A. A. Kapralov, V. Kini, Vlasova, II, Q. Zhao, M. Zou, P. Di, D. A. Svistunenko, I. V. Kurnikov and G. G. Borisenko (2005). "Cytochrome c acts as a cardiolipin oxygenase required for release of proapoptotic factors." Nat Chem Biol 1(4): 223-232.

Kimani, S. G., K. Geng, C. Kasikara, S. Kumar, G. Sriram, Y. Wu and R. B. Birge (2014). "Contribution of Defective PS Recognition and Efferocytosis to Chronic Inflammation and Autoimmunity." Front Immunol 5: 566.

Kinchen, J. M., J. Cabello, D. Klingele, K. Wong, R. Feichtinger, H. Schnabel, R. Schnabel and M. O. Hengartner (2005). "Two pathways converge at CED-10 to mediate actin rearrangement and corpse removal in C. elegans." Nature 434(7029): 93-99.

Kinchen, J. M. and K. S. Ravichandran (2007). "Journey to the grave: signaling events regulating removal of apoptotic cells." J Cell Sci 120(Pt 13): 2143-2149.

Krahling, S., M. K. Callahan, P. Williamson and R. A. Schlegel (1999). "Exposure of phosphatidylserine is a general feature in the phagocytosis of apoptotic lymphocytes by macrophages." Cell Death Differ 6(2): 183-189.

Kroemer, G., L. Galluzzi, O. Kepp and L. Zitvogel (2013). "Immunogenic cell death in cancer therapy." Annu Rev Immunol 31: 51-72.

Kroemer, G., L. Galluzzi, P. Vandenabeele, J. Abrams, E. S. Alnemri, E. H. Baehrecke, M. V. Blagosklonny, W. S. El-Deiry, P. Golstein, D. R. Green, M. Hengartner, R. A. Knight, S. Kumar, S. A. Lipton, W. Malorni, G. Nunez, M. E. Peter, J. Tschopp, J. Yuan, M. Piacentini, B. Zhivotovsky, G. Melino and D. Nomenclature Committee on Cell (2009). "Classification of cell death: recommendations of the Nomenclature Committee on Cell Death 2009." Cell Death Differ 16(1): 3-11.

Ladoire, S., D. Hannani, M. Vetizou, C. Locher, L. Aymeric, L. Apetoh, O. Kepp, G. Kroemer, F. Ghiringhelli and L. Zitvogel (2014). "Cell-death-associated molecular patterns as determinants of cancer immunogenicity." Antioxid Redox Signal 20(7): 1098-1116.

Linkermann, A. and D. R. Green (2014). "Necroptosis." N Engl J Med 370(5): 455-465.

Liu, G., J. Wang, Y. J. Park, Y. Tsuruta, E. F. Lorne, X. Zhao and E. Abraham (2008). "High mobility group protein-1 inhibits phagocytosis of apoptotic neutrophils through binding to phosphatidylserine." J Immunol 181(6): 4240-4246.

Liu, Q. A. and M. O. Hengartner (1998). "Candidate adaptor protein CED-6 promotes the engulfment of apoptotic cells in C. elegans." Cell 93(6): 961-972.

Loov, C., L. Hillered, T. Ebendal and A. Erlandsson (2012). "Engulfing astrocytes protect neurons from contact-induced apoptosis following injury." PLoS One 7(3): e33090.

Lu, Q. and G. Lemke (2001). "Homeostatic regulation of the immune system by receptor tyrosine kinases of the Tyro 3 family." Science 293(5528): 306-311.

Marx, J. (2002). "Nobel Prize in Physiology or Medicine. Tiny worm takes a star turn." Science 298(5593): 526.

Meesmann, H. M., E. M. Fehr, S. Kierschke, M. Herrmann, R. Bilyy, P. Heyder, N. Blank, S. Krienke, H. M. Lorenz and M. Schiller (2010). "Decrease of sialic acid residues as an eat-me signal on the surface of apoptotic lymphocytes." J Cell Sci 123(Pt 19): 3347-3356.

Mevorach, D., J. O. Mascarenhas, D. Gershov and K. B. Elkon (1998). "Complement-dependent clearance of apoptotic cells by human macrophages." J Exp Med 188(12): 2313-2320.

Mukundan, L., J. I. Odegaard, C. R. Morel, J. E. Heredia, J. W. Mwangi, R. R. Ricardo-Gonzalez, Y. P. Goh, A. R. Eagle, S. E. Dunn, J. U. Awakuni, K. D. Nguyen, L. Steinman, S. A. Michie and A. Chawla (2009). "PPAR-delta senses and orchestrates clearance of apoptotic cells to promote tolerance." Nat Med 15(11): 1266-1272.

Munoz, L. E., K. Lauber, M. Schiller, A. A. Manfredi and M. Herrmann (2010). "The role of defective clearance of apoptotic cells in systemic autoimmunity." Nat Rev Rheumatol 6(5): 280-289.

Murphy, J. E., D. Tacon, P. R. Tedbury, J. M. Hadden, S. Knowling, T. Sawamura, M. Peckham, S. E. Phillips, J. H. Walker and S. Ponnambalam (2006). "LOX-1 scavenger receptor mediates calcium-dependent recognition of phosphatidylserine and apoptotic cells." Biochem J 393(Pt 1): 107-115.

N, A. G., S. J. Bensinger, C. Hong, S. Beceiro, M. N. Bradley, N. Zelcer, J. Deniz, C. Ramirez, M. Diaz, G. Gallardo, C. R. de Galarreta, J. Salazar, F. Lopez, P. Edwards, J. Parks, M. Andujar, P. Tontonoz and A. Castrillo (2009). "Apoptotic cells promote their own clearance and immune tolerance through activation of the nuclear receptor LXR." Immunity 31(2): 245-258.

Nauta, A. J., G. Castellano, W. Xu, A. M. Woltman, M. C. Borrias, M. R. Daha, C. van Kooten and A. Roos (2004). "Opsonization with C1q and mannose-binding lectin targets apoptotic cells to dendritic cells." J Immunol 173(5): 3044-3050.

Nauta, A. J., M. R. Daha, C. van Kooten and A. Roos (2003). "Recognition and clearance of apoptotic cells: a role for complement and pentraxins." Trends Immunol 24(3): 148-154.

Obeid, M., A. Tesniere, F. Ghiringhelli, G. M. Fimia, L. Apetoh, J. L. Perfettini, M. Castedo, G. Mignot, T. Panaretakis, N. Casares, D. Metivier, N. Larochette, P. van Endert, F. Ciccosanti, M. Piacentini, L. Zitvogel and G. Kroemer (2007). "Calreticulin exposure dictates the immunogenicity of cancer cell death." Nat Med 13(1): 54-61.

Ogden, C. A., A. deCathelineau, P. R. Hoffmann, D. Bratton, B. Ghebrehiwet, V. A. Fadok and P. M. Henson (2001). "C1q and mannose binding lectin engagement of cell surface calreticulin and CD91 initiates macropinocytosis and uptake of apoptotic cells." J Exp Med 194(6): 781-795.

Platt, N., H. Suzuki, T. Kodama and S. Gordon (2000). "Apoptotic thymocyte clearance in scavenger receptor class A-deficient mice is apparently normal." J Immunol 164(9): 4861-4867.

Platt, N., H. Suzuki, Y. Kurihara, T. Kodama and S. Gordon (1996). "Role for the class A macrophage scavenger receptor in the phagocytosis of apoptotic thymocytes in vitro." Proc Natl Acad Sci U S A 93(22): 12456-12460.

Poon, I. K., M. D. Hulett and C. R. Parish (2010). "Molecular mechanisms of late apoptotic/necrotic cell clearance." Cell Death Differ 17(3): 381-397.

Poon, I. K., C. D. Lucas, A. G. Rossi and K. S. Ravichandran (2014). "Apoptotic cell clearance: basic biology and therapeutic potential." Nat Rev Immunol 14(3): 166-180.

Ramirez-Ortiz, Z. G., W. F. Pendergraft, 3rd, A. Prasad, M. H. Byrne, T. Iram, C. J. Blanchette, A. D. Luster, N. Hacohen, J. El Khoury and T. K. Means (2013). "The scavenger receptor SCARF1 mediates the clearance of apoptotic cells and prevents autoimmunity." Nat Immunol 14(9): 917-926.

Reddien, P. W. and H. R. Horvitz (2000). "CED-2/CrkII and CED-10/Rac control phagocytosis and cell migration in Caenorhabditis elegans." Nat Cell Biol 2(3): 131-136.

Ren, Y., R. L. Silverstein, J. Allen and J. Savill (1995). "CD36 gene transfer confers capacity for phagocytosis of cells undergoing apoptosis." J Exp Med 181(5): 1857-1862.

Roszer, T., M. P. Menendez-Gutierrez, M. I. Lefterova, D. Alameda, V. Nunez, M. A. Lazar, T. Fischer and M. Ricote (2011). "Autoimmune kidney disease and impaired engulfment of apoptotic cells in mice with macrophage peroxisome proliferator-activated receptor gamma or retinoid X receptor alpha deficiency." J Immunol 186(1): 621-631.

Rovere, P., M. G. Sabbadini, C. Vallinoto, U. Fascio, V. S. Zimmermann, A. Bondanza, P. Ricciardi-Castagnoli and A. A. Manfredi (1999). "Delayed clearance of apoptotic lymphoma cells allows cross-presentation of intracellular antigens by mature dendritic cells." J Leukoc Biol 66(2): 345-349.

Savill, J., I. Dransfield, C. Gregory and C. Haslett (2002). "A blast from the past: clearance of apoptotic cells regulates immune responses." Nat Rev Immunol 2(12): 965-975.

Schlegel, R. A., M. K. Callahan and P. Williamson (2000). "The central role of phosphatidylserine in the phagocytosis of apoptotic thymocytes." Ann N Y Acad Sci 926: 217-225.

Segawa, K., S. Kurata, Y. Yanagihashi, T. R. Brummelkamp, F. Matsuda and S. Nagata (2014). "Caspase-mediated cleavage of phospholipid flippase for apoptotic phosphatidylserine exposure." Science 344(6188): 1164-1168.

Segawa, K., J. Suzuki and S. Nagata (2011). "Constitutive exposure of phosphatidylserine on viable cells." Proc Natl Acad Sci U S A 108(48): 19246-19251.

Shao, W. H. and P. L. Cohen (2011). "Disturbances of apoptotic cell clearance in systemic lupus erythematosus." Arthritis Res Ther 13(1): 202.

Simhadri, V. R., J. F. Andersen, E. Calvo, S. C. Choi, J. E. Coligan and F. Borrego (2012). "Human CD300a binds to phosphatidylethanolamine and phosphatidylserine, and modulates the phagocytosis of dead cells." <u>Blood</u> 119(12): 2799-2809.

Steinman, R. M., S. Turley, I. Mellman and K. Inaba (2000). "The induction of tolerance by dendritic cells that have captured apoptotic cells." <u>J Exp Med</u> 191(3): 411-416.

Stern, M., J. Savill and C. Haslett (1996). "Human monocyte-derived macrophage phagocytosis of senescent eosinophils undergoing apoptosis. Mediation by alpha v beta 3/CD36/thrombospondin recognition mechanism and lack of phlogistic response." <u>Am J Pathol</u> 149(3): 911-921.

Su, H. P., K. Nakada-Tsukui, A. C. Tosello-Trampont, Y. Li, G. Bu, P. M. Henson and K. S. Ravichandran (2002). "Interaction of CED-6/GULP, an adapter protein involved in engulfment of apoptotic cells with CED-1 and CD91/low density lipoprotein receptor-related protein (LRP)." <u>J Biol Chem</u> 277(14): 11772-11779.

Sukkurwala, A. Q., I. Martins, Y. Wang, F. Schlemmer, C. Ruckenstuhl, M. Durchschlag, M. Michaud, L. Senovilla, A. Sistigu, Y. Ma, E. Vacchelli, E. Sulpice, X. Gidrol, L. Zitvogel, F. Madeo, L. Galluzzi, O. Kepp and G. Kroemer (2014). "Immunogenic calreticulin exposure occurs through a phylogenetically conserved stress pathway involving the chemokine CXCL8." <u>Cell Death Differ</u> 21(1): 59-68.

Suzuki, J., D. P. Denning, E. Imanishi, H. R. Horvitz and S. Nagata (2013). "Xk-related protein 8 and CED-8 promote phosphatidylserine exposure in apoptotic cells." <u>Science</u> 341(6144): 403-406.

Suzuki, J., E. Imanishi and S. Nagata (2014). "Exposure of phosphatidylserine by Xk-related protein family members during apoptosis." <u>J Biol Chem</u> 289(44): 30257-30267.

Suzuki, J., M. Umeda, P. J. Sims and S. Nagata (2010). "Calcium-dependent phospholipid scrambling by TMEM16F." <u>Nature</u> 468(7325): 834-838.

Tait, S. W., G. Ichim and D. R. Green (2014). "Die another way--non-apoptotic mechanisms of cell death." <u>J Cell Sci</u> 127(Pt 10): 2135-2144.

Tyurin, V. A., K. Balasubramanian, D. Winnica, Y. Y. Tyurina, A. S. Vikulina, R. R. He, A. A. Kapralov, C. H. Macphee and V. E. Kagan (2014). "Oxidatively modified phosphatidylserines on the surface of apoptotic cells are essential phagocytic 'eat-me' signals: cleavage and inhibition of phagocytosis by Lp-PLA2." <u>Cell Death Differ</u> 21(5): 825-835.

Ucker, D. S., M. R. Jain, G. Pattabiraman, K. Palasiewicz, R. B. Birge and H. Li (2012). "Externalized glycolytic enzymes are novel, conserved, and early biomarkers of apoptosis." <u>J Biol Chem</u> 287(13): 10325-10343.

Vandivier, R. W., C. A. Ogden, V. A. Fadok, P. R. Hoffmann, K. K. Brown, M. Botto, M. J. Walport, J. H. Fisher, P. M. Henson and K. E. Greene (2002). "Role of surfactant proteins A, D, and C1q in the clearance of apoptotic cells in vivo and in vitro: calreticulin and CD91 as a common collectin receptor complex." J Immunol 169(7): 3978-3986.

Voll, R. E., M. Herrmann, E. A. Roth, C. Stach, J. R. Kalden and I. Girkontaite (1997). "Immunosuppressive effects of apoptotic cells." Nature 390(6658): 350-351.

Wu, Y., N. Tibrewal and R. B. Birge (2006). "Phosphatidylserine recognition by phagocytes: a view to a kill." Trends Cell Biol 16(4): 189-197.

Wu, Y. C. and H. R. Horvitz (1998). "The C. elegans cell corpse engulfment gene ced-7 encodes a protein similar to ABC transporters." Cell 93(6): 951-960.

Wu, Y. C. and H. R. Horvitz (1998). "C. elegans phagocytosis and cell-migration protein CED-5 is similar to human DOCK180." Nature 392(6675): 501-504.

Yamaguchi, H., T. Fujimoto, S. Nakamura, K. Ohmura, T. Mimori, F. Matsuda and S. Nagata (2010). "Aberrant splicing of the milk fat globule-EGF factor 8 (MFG-E8) gene in human systemic lupus erythematosus." Eur J Immunol 40(6): 1778-1785.

Yamamoto, N., H. Yamaguchi, K. Ohmura, T. Yokoyama, H. Yoshifuji, N. Yukawa, D. Kawabata, T. Fujii, S. Morita, S. Nagata and T. Mimori (2014). "Serum milk fat globule epidermal growth factor 8 elevation may subdivide systemic lupus erythematosus into two pathophysiologically distinct subsets." Lupus 23(4): 386-394.

Yamazaki, T., D. Hannani, V. Poirier-Colame, S. Ladoire, C. Locher, A. Sistigu, N. Prada, S. Adjemian, J. P. Catani, M. Freudenberg, C. Galanos, F. Andre, G. Kroemer and L. Zitvogel (2014). "Defective immunogenic cell death of HMGB1-deficient tumors: compensatory therapy with TLR4 agonists." Cell Death Differ 21(1): 69-78.

Yang, H., A. Kim, T. David, D. Palmer, T. Jin, J. Tien, F. Huang, T. Cheng, S. R. Coughlin, Y. N. Jan and L. Y. Jan (2012). "TMEM16F forms a Ca2+-activated cation channel required for lipid scrambling in platelets during blood coagulation." Cell 151(1): 111-122.

Yin, Y., X. Huang, K. D. Lynn and P. E. Thorpe (2013). "Phosphatidylserine-targeting antibody induces M1 macrophage polarization and promotes myeloid-derived suppressor cell differentiation." Cancer Immunol Res 1(4): 256-268.

Yuan, J. and H. R. Horvitz (1992). "The Caenorhabditis elegans cell death gene ced-4 encodes a novel protein and is expressed during the period of extensive programmed cell death." Development 116(2): 309-320.

Yuan, J. and H. R. Horvitz (2004). "A first insight into the molecular mechanisms of apoptosis." Cell 116(2 Suppl): S53-56, 51 p following S59.

Yuan, J. Y. and H. R. Horvitz (1990). "The Caenorhabditis elegans genes ced-3 and ced-4 act cell autonomously to cause programmed cell death." Dev Biol 138(1): 33-41.

Zhou, Z., E. Caron, E. Hartwieg, A. Hall and H. R. Horvitz (2001). "The C. elegans PH domain protein CED-12 regulates cytoskeletal reorganization via a Rho/Rac GTPase signaling pathway." Dev Cell 1(4): 477-489.

Zhou, Z., E. Hartwieg and H. R. Horvitz (2001). "CED-1 is a transmembrane receptor that mediates cell corpse engulfment in C. elegans." Cell 104(1): 43-56.

Chapter 9: From caterpillars to clinic: IAP proteins and their antagonists (Vucic)

Domagoj Vucic

Early Discovery Biochemistry, Genentech, South San Francisco, CA, US

Abstract:

Effective regulation of cell death and survival is essential for cellular homeostasis. Inhibitor of apoptosis (IAP) proteins play pivotal roles in cellular survival by blocking cell death, modulating signal transduction, and affecting cellular proliferation. The critical features of these proteins are the Baculovirus IAP Repeat (BIR) domains that are involved in protein-protein interactions and for several IAP proteins also a RING domain, which gives them ubiquitin E3 ligase activity. Through interactions with inducers and effectors of apoptosis IAP proteins can effectively suppress apoptosis triggered by diverse stimuli including death receptor signaling, irradiation, chemotherapeutic agents, or growth factor withdrawal. Evasion of apoptosis, in part due to the action of IAP proteins, enhances resistance of cancer cells to treatment with chemotherapeutic agents and contributes to tumor progression. IAP genes are also subject to amplification, mutation, and chromosomal translocation in human malignancies and autoimmune diseases making them attractive therapeutic targets. Efforts to target IAP proteins in tumors have focused mainly on designing small molecules that mimic the IAP-binding motif of the endogenous IAP antagonist, SMAC, but also include antisense oligonucleotides. This chapter discusses functional roles of IAP proteins and the development of antagonists targeting IAP proteins for cancer treatment.

Discovery of IAPs

In the early 1990's Professor Lois Miller and her team at the University of Georgia, Athens, GA, were interested in the ability of baculoviruses to efficiently replicate their genome without inducing host cell death. By employing genomic deletion analyses, her team discovered *Autographa californica* (a Noctuid moth) nuclear polyhedrosis virus gene *p35*, which encodes a powerful anti-apoptotic protein capable of blocking several caspases (Clem et al., 1991). However, many baculovirus strains do not encode *p35*-like genes. Therefore, Lois' team investigated whether genes from other baculoviruses possess anti-apoptotic capabilities that would allow viral replication.

Two genes, *cp-iap* and *op-iap* from *Cydia pomonella* granulosis virus and *Orgyia pseudotsugata* nuclear polyhedrosis virus (also viruses infecting moths) could rescue deficiency of p35 anti-apoptotic gene (Crook et al., 1993; Birnbaum et al., 1994; Clem and Miller, 1994). Op-IAP and Cp-IAP thus became the founding members of the family of cell death regulators whose functional importance spans viral production, survival of fly embryos and proper regulation of TNF signaling in mammals (Birnbaum et al., 1994; Crook et al., 1993; Silke and Vaux, 2001). Following this initial discovery in baculoviruses, *iap* genes and IAP proteins were identified in a variety of metazoan phylogenetic groups. In Drosophila, DIAP1 and DIAP2 regulate fly survival and innate immune responses, while in *Caenorhabditis elegans* and yeast, IAP homologues regulate cell division (Hay et al., 1995; Silke and Vaux, 2001). Eight IAP proteins are expressed in humans: neuronal AIP (NAIP/BIRC1); cellular IAP1 (c-IAP1/ BIRC2/HIAP2/ MIHB/API2); cellular IAP2 (c-IAP2/ BIRC3/HIAP1/ MIHC/API2); X chromosome-linked IAP (XIAP/BIRC4/MIHA/hILP/ILP-1); survivin (BIRC5/TIAP); BIR-containing ubiquitin conjugating enzyme (BIRC6/BRUCE/Apollon); melanoma IAP (ML-IAP/ BIRC7/KIAP/Livin); and testis-specific IAP (Ts-IAP/BIRC8/hILP2/ILP-2) (reviewed in (Ndubaku et al., 2009a; Salvesen and Duckett, 2002))[Figure 1]. Most of the IAPs were identified through bioinformatic homology searches although a few of them were found in genetic or biochemical experiments. Neuronal apoptosis inhibitory protein (NAIP) was discovered during the search for the gene that causes spinal muscular atrophy (SMA). Although NAIP turned out to be unrelated to SMA, its discovery probably sped the homology searches for related IAPs. Cellular IAP1 and 2 (c-IAP1 and c-IAP2) were identified as components of tumor necrosis factor receptor 2 (TNFR2)-associated complex through their constitutive binding to TNFR-associated factors 1 and 2 (TRAF1 and TRAF2). At the time of this discovery David Goeddel's group at Genentech used over 100 liters of cell culture to identify the c-IAPs by mass spectrometry methods (Rothe et al., 1995). XIAP was independently discovered by several groups, mostly through the BIR domain homology searches (Hunter et al., 2007).

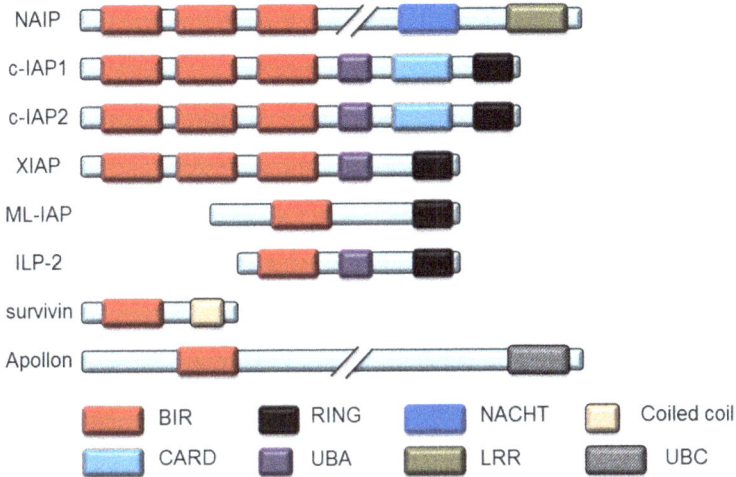

Figure 1. Schematic representation of human IAP proteins. BIR: Baculovirus IAP repeat; CARD: Caspase recruitment domain; LLR: Leucine-rich repeat; NACHT: NAIP, CIITA, HET-E and TP1; RING: Really interesting gene; UBA: Ubiquitin-associated domain; UBC: Ubiquitin conjugating domain.

IAP protein domain organization

IAP proteins are characterized by the presence of the Baculoviral IAP Repeat (BIR) domain – a conserved 70-80 amino acid zinc-binding domain (Hinds et al., 1999; Sun et al., 1999). Most mammalian IAPs contain three copies of the BIR domain, while baculovirus IAPs have two and some IAP proteins that are involved in cell division contain only a single BIR domain [Figure 1]. In addition to BIR domains, several IAP proteins (c-IAP1, c-IAP2, XIAP and ML-IAP) contain a carboxy-terminal RING (really interesting new gene) domain, which gives them E3 ubiquitin ligase activity (Vaux and Silke, 2005; Varfolomeev and Vucic, 2008). From the early days of IAP studies, it was clear that BIR domains might be critical for the anti-apoptotic activity. However, the RING domains were more enigmatic until it became clear that E3 ligase activity of IAP is instrumental for their participation in various signaling pathways and in the regulation of cell death. Actually, with the exception of XIAP, which inhibits caspases through its BIR2 and BIR3 domains, most other IAP proteins use BIR domains as protein-protein interaction modules that allow their recruitment to appropriate signaling complexes where their E3 ligase activity can critically affect cellular outcomes. A few IAP proteins also have a centrally located UBiquitin-Associated (UBA) domain (c-IAPs, XIAP and ILP-2) that enables their association with ubiquitin (Blankenship et al., 2009; Gyrd-Hansen et al., 2008). The UBA domain of IAPs can bind to a variety of ubiquitin linkages (K11, K48, K63, Met1 or linear chains) but the physiological relevance of ubiquitin binding by IAP proteins is not clear yet. There are also several protein domains that are unique to particular IAPs (Figure 1). BRUCE facilitates attachment of ubiquitin moieties to its substrates through the carboxy-terminal ubiquitin-conjugating (UBC) domain (Hauser et al., 1998; Chen et al., 1999) . The coiled-coil region of survivin is required for the proper nuclear localization of the passenger protein complex (PCP), as well as for interactions with other PCP proteins during mitosis (Jeyaprakash et al., 2007). The nucleotide-binding and oligomerization domain (NOD) and leucine-rich repeat (LRR) domains of NAIP are crucial for its function in innate immunity (Liston et al., 1996; Wilmanski et al., 2008). Finally, c-IAP proteins contain a well-conserved Caspase-recruitment domain (CARD) of unknown function. Nature does not often splurge and thus it is really puzzling why c-IAP proteins have this domain with no apparent function. It is possible that interacting partners of c-IAP CARD domains still await discovery, but it is also possible that the CARD domains are simply fillers that allows the proper overall conformation of the c-IAP proteins.

Regulation of programmed cell death by IAPs

Programmed cell death is executed through two evolutionarily conserved signaling pathways: apoptosis and necroptosis [Figure 2]. While apoptotic cell death is solely dependent on the function of caspases, cysteine-dependent aspartyl-specific proteases, necroptosis is initiated once caspase activity is blocked (Salvesen and Abrams, 2004; Vandenabeele et al., 2010). Apoptotic cell death can be initiated through intrinsic and extrinsic signaling [Figure 2]. Cellular stress, developmental cues, or growth factor withdrawal trigger intrinsic cell death leading to disruption of internal cellular integrity including mitochondrial damage (Budihardjo et al., 1999; Kaufmann and Vaux, 2003). Bcl-2-homology 3-only (BH3-only) proteins, such as Bmf, Puma, Noxa and Bim, initiate this signaling (Youle and Strasser, 2008). They activate multi-domain-containing pro-apoptotic Bcl-2 proteins Bax and Bak by neutralizing the inhibitory effects of the Bcl-2 proteins, Bcl-2, Bcl-x$_L$, Mcl-1, Bcl-w and A1. Binding of dATP and cytochrome c released from inner-mitochondrial membranes to Apaf1 leads to formation of the apoptosome complex that recruits and activates initiator caspase-9. Consequently, caspase-9 activates the executioner caspases 3 and 7 (Boatright et al., 2003; Riedl and Salvesen, 2007).

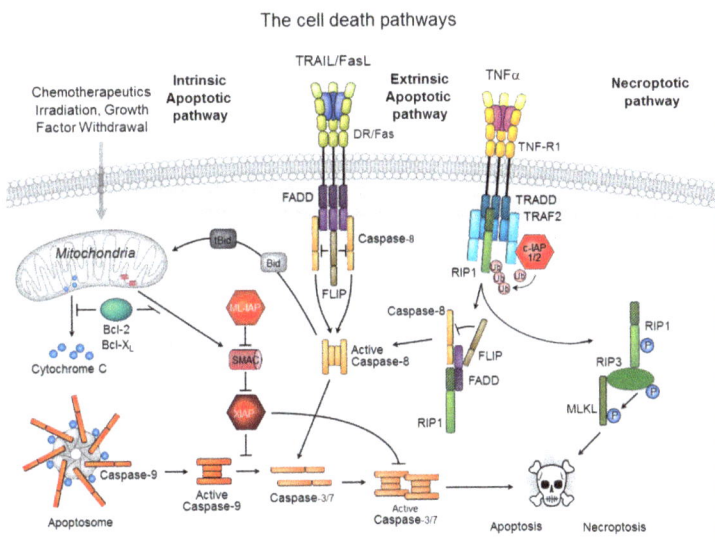

Figure 2. Apoptotic and necroptotic signaling pathways. Programmed cell death is executed through apoptotic (caspase-dependent) and necroptotic (caspase-independent) signaling pathways in response to various stimuli including environmental stress or growth factor deprivation (intrinsic apoptotic pathway), and activation of TNF family receptors (extrinsic apoptotic

or necroptotic pathway). Bcl-2: B-cell lymphoma 2; Bcl-x$_L$: B-cell lymphoma-extra large; BH3: Bcl-2 homology 3; BID: BH3-interacting domain death agonist; c-IAP: cellular IAP; Caspase: Cysteine-aspartic protease; DR: Death receptor; FADD: Fas-associated death domain; FLIP: FLICE inhibitory protein; ML-IAP: Melanoma IAP; RIP: Receptor interacting protein; SMAC: Second mitochondrial activator of caspases; tBID: Truncated BID; TRADD: TNFR-associated death domain; TRAF: TNF receptor-associated factor; TNFα: Tumor necrosis factor α; TNFR1: TNF receptor 1; Ub: Ubiquitin; XIAP: X-chromosome-linked IAP.

The extrinsic apoptotic cell death pathway is typically triggered by ligation of Death Domain (DD)-containing receptors that belong to the Tumor Necrosis Factor Receptor (TNFR) family (Guicciardi and Gores, 2009) [Figure 2]. Trimerization or higher-order aggregation of DR4, DR5 or Fas prompts assembly of the receptor-associated death-inducing signaling complex (DISC). This complex is composed of the receptor, the DD-containing adaptor molecule Fas Associated DD (FADD), initiator caspases 8 and 10, as well as the inhibitory molecule cellular FLICE-inhibitory protein (cFLIP). Oligomerization of apical, initiator caspases causes their self-activation (Boatright et al., 2003), while executioner caspases 3 and 7 are activated when cleaved by the initiator caspases. In some cells, extrinsic apoptosis can be further enhanced through the mitochondrial pathway once the BH3-only molecule Bid is processed by caspase-8 or -10 to its active form tBid (Peter and Krammer, 2003).

TNFR1 induces apoptosis by employing additional adaptor and enzymatic molecules. Additionally, TNFR1 primarily activates canonical NF-κB, JNK and p38 MAP kinase signaling (Karin and Lin, 2002). The TNFR1-signaling complex is initiated upon recruitment of the adaptor molecule TNF Receptor Associated DD (TRADD), which allows engagement of Receptor-Interacting Protein-1 (RIP1) and TNF Receptor-Associated Factor-2 (TRAF2) interactions (Wilson et al., 2009). Cellular IAP proteins are also recruited to this complex through their constitutive association with TRAF2. Inhibition of NF-κB and de novo protein expression or c-IAP deficiency enables apoptosis induction by facilitating translocation of TRADD, RIP1, and TRAF2 to the cytosol and recruitment of FADD and caspases 8 and 10 (Micheau and

Tschopp, 2003; Varfolomeev et al., 2007; Vince et al., 2007; Varfolomeev and Vucic, 2008).

If TNFR1-mediated cell death signaling is halted by inhibition of caspases, an alternative death pathway can be engaged – necroptosis. Necroptosis is a regulated form of necrotic cell death initiated by TNF signaling, and also by pathogen recognition receptors (PRRs) or Toll-like receptor, and mediated by kinases RIP1 and RIP3 (Vandenabeele et al., 2010) [Figure 2]. The presence of RIP3 and caspase inhibition following TNF stimulation enables RIP1-RIP3 interaction via the RIP homologous interaction motif (RHIM) (Sun et al., 2002). Interestingly, while the kinase activity of RIP1 is not required for induction of NF-$\Box$B signaling or apoptosis, it is critical for necroptotic cell death and RIP1-RIP3 association (Newton et al., 2014; Polykratis et al., 2014). Activation of RIP1 and RIP3 leads to their auto-phosphorylation, allowing phosphorylated RIP3 to recruit and phosphorylate its substrate, the pseudokinase MLKL (mixed lineage kinase domain-like) and trigger MLKL translocation to membranes resulting in cell rupture (Sun et al., 2012; Zhao et al., 2012; Cai et al., 2014; Dondelinger et al., 2014; Wang et al., 2014). c-IAP proteins negatively regulate necrotic cell death through their E3 ligase activity, which promotes RIP1 ubiquitination and thereby blocks RIP1 relocation into cytoplasmic cell death-stimulating protein complexes (Vanlangenakker et al., 2011).

Initially, IAP proteins were proposed to block apoptosis by directly binding and inhibiting caspases (Uren et al., 1996). However, later studies clearly demonstrated that XIAP is the only physiologically-relevant direct inhibitor of caspases 3, 7 and 9 (Eckelman et al., 2006). Fully active caspases operate as dimers formed by a pair of interacting large and small catalytic subunits. The peptide-binding groove of the BIR3 domain of XIAP interacts with the conserved amino-terminal four amino-acid IAP-binding motif (IBM) of the p12 small subunit of processed caspase-9. Thus, binding of the XIAP BIR3 domain to caspase-9 prevents homodimerization and, therefore, activation of the caspase (Shiozaki et al., 2003; Srinivasula et al., 2001). In the case of caspases 3 and 7, the linker region between the BIR1 and BIR2 domains of XIAP interacts with the substrate-binding groove of activated caspases, while BIR2 further stabilizes these interactions (Chai et al., 2001; Huang et al., 2001; Riedl et al., 2001; Suzuki et al., 2001b; Scott et al., 2005).

IAP proteins as ubiquitin ligases and signaling regulators

The regulated modification, and in some cases degradation, of cellular proteins by the ubiquitin-proteasome system impacts a wide range of vital cellular processes (Hershko and Ciechanover, 1998). Ubiquitination also plays an important role in the regulation of cell death and signaling pathways. RIP1 ubiquitination in particular has major implications for cell fate (Wertz and Dixit, 2010; Wajant and Scheurich, 2011). The main ubiquitin E3 ligases for RIP1 are the cellular IAP proteins, which promote K11- and K63- linked RIP1 polyubiqutination upon TNF signaling (Bertrand et al., 2008; Mahoney et al., 2008; Varfolomeev et al., 2008; Dynek et al., 2010). RIP1 ubiquitination serves as a docking platform for the assembly of TAK1/TAB1/2/3 and IKK/NEMO protein kinase complexes that mediate canonical NF-κB signaling (Wertz and Dixit, 2008). TNF stimulated autoubiquitnation of c-IAPs, on the other hand, recruits linear ubiquitination assembly complex, LUBAC. Recruitment of IKK and TAK1/TAB complexes leads to their activation, which results in the phosphorylation of IκBα (inhibitor of kappa B) by the kinase IKKβ and its subsequent proteasomal degradation (Wang et al., 2001; Shim et al., 2005). Destruction of IκBα frees the NF-κB transcription factors p50 and RelA/p65 to move into the nucleus and induce the transcription of cell survival genes (Wang et al., 1998; Scheidereit, 2006)[Figure 3].

The alternative or noncanonical NF-κB signaling pathway involves the NF-κB activating kinase, NIK, which activates IKKa through phosphorylation [Figure 3]. NIK protein levels are low in unstimulated cells due to c-IAP1 and c-IAP2-mediated

ubiquitination and subsequent proteasomal degradation of NIK
(Varfolomeev and Vucic, 2008). Some TNF family members like
TWEAK or CD40L induce autoubiquitination and proteasomal
loss of c-IAPs, leading to NIK accumulation, p100 processing and
NF-κB activation (Figure 3) (Varfolomeev et al., 2007;
Matsuzawa et al., 2008; Varfolomeev and Vucic, 2008; Vince et
al., 2008). In addition to NF-κB pathways, IAP proteins have also
been implicated in the activation of several other signaling
pathways including JNK, TGFα, p38 and NOD-dependent
pathways, thus expanding the arsenal of their pro-survival
capabilities (Asselin et al., 2001; Birkey Reffey et al., 2001;
Sanna et al., 2002; Bertrand et al., 2009b).

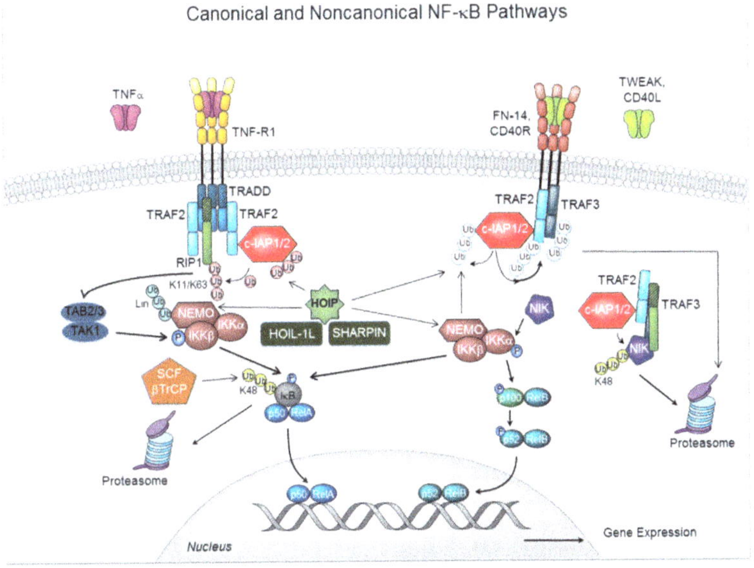

Figure 3. Canonical and noncanonical NF-κB signaling pathways. Cellular IAP
proteins function as positive regulators of canonical and negative regulators of
noncanonical NF-κB signaling pathways. c-IAPs: cellular IAPs; FN14: Fibroblast
growth factor-inducible 14; HOIL-1: Heme-oxidized IRP2 ubiquitin ligase-1;
HOIP: HOIL-1L-interacting protein ; IKK: IκB kinase; NEMO: NF-κB essential
modulator; NIK: NF-κB inducing kinase; RIP: Receptor interacting protein; TAB:
TAK1-binding protein; TAK1: TGF-β activated kinase 1; TRADD: TNFR-
associated death domain; TRAF: TNF receptor-associated factor; TNFα: Tumor
necrosis factor α; TNFR1: TNF receptor 1; TWEAK: TNF-related weak inducer of
apoptosis Ub: Ubiquitin.

The NOD1/2 (nucleotide binding oligomerization domain) proteins are intracellular sensors of bacterial products that regulate expression of genes involved in pathogen-host defense. NOD1/2 oligomerize upon binding to their ligands, and recruit the kinase RIP2 (receptor TNFRSF-interacting serine-threonine kinase 2), XIAP and c-IAPs (Bertrand et al., 2009a; Krieg et al., 2009; Damgaard et al., 2012), which triggers the ubiquitination of RIP2 leading to the NF-κB, MAP kinases p38 and c-Jun N-terminal kinases (http://en.wikipedia.org/wiki/C-Jun_N-terminal_kinases) (JNK) activation (Hasegawa et al., 2008). Recently, mutations affecting the BIR2 and RING domains of XIAP were found in patients with the X-linked lymphoproliferative syndrome type-2 (XLP2). These mutations impaired NOD1/2 signaling, thus confirming the importance of XIAP and RIP2 ubiquitination for NOD1/2 stimulated activation of NF-κB and MAPK signaling pathways (Damgaard et al., 2013).

The ubiquitin E3 ligase activity of IAP proteins also contributes to the regulation of their own stability as ubiquitination of c-IAP2 and XIAP by c-IAP1 promotes their proteasomal degradation (Silke et al., 2005). Other prominent targets of IAP E3 ligase activity are the tumor suppressor protein MAD1 (Zender et al., 2006), and the protein kinase C-RAF whose destabilization by c-IAPs and XIAP potentially modulates the MAPK signaling pathway and cell migration (Dogan et al., 2008).

IAP proteins in human malignancies

Early interest in studying and later targeting IAP proteins stemmed from their functional importance in cell death and signaling pathways but also from the elevated protein and mRNA expression in various malignancies (Hunter et al., 2007; Vucic and Fairbrother, 2007; Vucic, 2008). For example, ML-IAP is highly expressed in melanomas and other cancers where it likely contributes to resistance to chemotherapeutic treatments (Vucic et al., 2000; Tanabe et al., 2004; Crnkovic-Mertens et al., 2006; Crnkovic-Mertens et al., 2007; Wagener et al., 2007; Dynek et al., 2008). Another IAP protein that is present at high levels in the majority of all examined tumor samples is survivin.

Survivin is prominently expressed during embryonic development and in a subset of cells with high renewal potential in adult organisms (placenta, testes and rapidly dividing cells such as CD34+ bone marrow stem cells), but it is almost undetectable in the rest of adult tissues (Adida et al., 1998; Altieri, 2003b; Fukuda and Pelus, 2001). Regulation of survivin expression in cancer cells often displays a mitosis-independent mode, and studies have shown that it may result from augmented survivin promoter activity, increased signaling of MAP kinase pathways, and 17q25 chromosomal amplification of the survivin genetic locus.

In addition to amplification of the survivin locus, several chromosomal abnormalities were also found for the c-IAP1 and c-IAP2 genes (Hunter et al., 2007; Vucic, 2008). The 11q21-q23 chromosomal region that contains the c-IAP1 and c-IAP2 genes is amplified in a number of human malignancies including selected glioblastomas, meduloblastomas, gastric, lung, renal and esophageal squamous-cell carcinomas, resulting in elevated levels of cilAP proteins (Imoto et al., 2001, 2002; Hunter et al., 2007). Amplifications of the syntenic chromosomal region encoding c-IAP1 and c-IAP2 were also found in mouse tumors (Zender et al., 2006). Furthermore, a large number of extranodal non-Hodgkin MALT (mucosa-associated lymphoid tissue protein) lymphomas are caused by the t(11,18)(q21;q21) chromosomal translocation (Isaacson, 2005). The t(11,18)(q21;q21) translocation fuses the BIR and the UBA domains of c-IAP2 with the central and carboxy-terminal portions of paracaspase/MALT1. MALT1 is a critical adaptor and protease in the antigen receptor-stimulated NF-κB signaling pathway. The c-IAP2:MALT1 fusion protein promotes deregulated constitutive NF-κB signaling leading to increased expression of anti-apoptotic genes, which favor cancer progression and survival (Uren et al., 2000; Lucas et al., 2001; Karin et al., 2002).

Natural IAP antagonists

Initial efforts to identify regulators of cell death in Drosophila led to the identification of a genomic region that encoded three cell death genes: reaper, hid (head involution defective) and grim (White et al., 1994; Grether et al., 1995). By binding the BIR domains of Drosophila and baculovirus IAP proteins, RPR, HID and GRIM can promote their ubiquitination and prevent IAP mediated anti-apoptotic activity (Vucic et al., 1997; Vucic et al., 1998; Steller, 2008).

Several years later, two groups independently identified SMAC (second mitochondrial activator of caspases)/DIABLO (Direct IAP Binding protein with Low PI) as mammalian endogenous antagonists of IAP proteins (Du et al., 2000; Verhagen et al., 2000). Upon apoptotic insult, mature SMAC is released from the mitochondria into the cytosol (Burri et al., 2005). Through binding to the peptide-binding grooves of the BIR2 and BIR3 domains of XIAP, the IBM of SMAC blocks XIAP's interactions with caspases (Liu et al., 2000; Wu et al., 2000). While other IAP proteins do not necessarily interact with caspases at physiologically relevant concentrations, the c-IAP proteins, ML-IAP and hILP2, can each bind SMAC with high affinities and block SMAC-mediated inhibition of XIAP (Eckelman and Salvesen, 2006; Shin et al., 2005; Vucic et al., 2005). Apart from SMAC, several other IAP-binding proteins have been identified (Liston et al., 2001; Suzuki et al., 2001a; Arora et al., 2007; Verhagen et al., 2007). However, their physiological relevance for IAP-mediated regulation of cell-death pathways is still not completely understood.

IAP proteins and SMAC also regulate each other's stability through ubiquitination. The RING domain-associated E3 ligase activity of c-IAPs, XIAP and ML-IAP allows them to ubiquitinate and destabilize SMAC (MacFarlane et al., 2002; Hu and Yang, 2003; Morizane et al., 2005; Vaux and Silke, 2005; Ma et al., 2006; Samuel et al., 2006). In addition, XIAP and c-IAPs may promote ubiquitination and proteasomal degradation of caspases 3, 7 and 9 (Suzuki et al., 2001c). SMAC, on the other hand, can stimulate auto-ubiquitination and destabilization of c-IAPs, and block XIAP in a non-degradative ubiquitin-dependent fashion (Yang and Du, 2004; Conze et al., 2005).

Targeting IAP proteins

Antisense oligonucleotides

Several IAP targeting strategies have been investigated so far. One of them is antisense oligonucleotides (AS) to suppress IAP expression during the translational step. This approach is hindered by the poor permeability and cellular uptake of these types of molecules *in vivo* (Tamm, 2006). Antisense oligonucleotides also have rather poor half-lives, making continual dosing necessary in order to achieve efficacy. Nevertheless, this approach has been tested on survivin and XIAP. Targeting survivin mRNA with antisense oligonucleotides can downregulate survivin expression and lead to apoptosis induction *in vivo* (Altieri, 2003a, 2006). Currently, antisense oligonucleotides targeting survivin are in phase II clinical trials being conducted by Eli Lilly. Downregulation of XIAP mRNA levels can effectively induce apoptosis in tumor cells (Hu et al., 2003) and XIAP AS in a xenograft model of lung cancer showed tumor growth suppression together with XIAP protein downregulation *in vivo* (Cao et al., 2004). However, in a randomized Phase II study in patients with refractory AML, XIAP antisense AEG35156 did not provide any therapeutic benefit (Schimmer et al., 2011).

SMAC peptides

SMAC binds with high affinity to the BIR3 domains of XIAP, c-IAP1 and c-IAP2, the single BIR domain of ML-IAP, and with somewhat weaker affinity, the BIR2 domain of XIAP, thereby shutting down their collective ability to directly or indirectly inhibit caspase activity. SMAC - IAP BIR domain interaction occurs through a minimal contact region of SMAC and IAPs, an IBM comprised of four amino acids, AVPI (Chai et al., 2000; Liu et al., 2000; Wu et al., 2000). Initial validation of this targeting approach came from a study that demonstrated that SMAC peptides can abrogate the ability of IAP proteins to inhibit cell death (Vucic et al., 2002). Binding of SMAC peptides to XIAP BIR3, XIAP BIR2, and ML-IAP BIR proteins was further examined using phage-display, which enable identification of amino acid residues with increased binding affinity relative to the SMAC-based peptide (Franklin et al., 2003). The study by Fulda and colleagues using SMAC peptides, and others that used modified SMAC-like peptides designed to increase membrane permeability, established that SMAC-based peptides had the ability to sensitize various solid tumor types to the activity of death ligands and standard chemotherapeutics (Arnt et al., 2002; Fulda et al., 2002; Yang et al., 2003). These early tool molecules had limited application since they lacked favorable physicochemical properties required to elicit anti-tumor responses, but nevertheless, it was clear that further enhancement of their pharmacological properties may render them more drug-like.

SMAC peptidomimetic IAP antagonists

Pioneering work by several groups built upon the established structure-activity relationship (SAR) to generate compounds that were less peptidic in nature and possessed more drug-like properties (Kipp et al., 2002; Oost et al., 2004; Sharma et al., 2006; Zobel et al., 2006). Later efforts characterized a number of IAP antagonists with rigidified scaffold replacements for the peptide backbone and with high affinity towards IAP BIR3 domains, which allowed effective induction of apoptosis in tumor cells (Gaither et al., 2007; Sun et al., 2008). Following the serendipitous discovery of bivalent SMAC mimetic Compound 3 in 2004 (Li et al., 2004), many groups turned their attention to bivalent compounds, primarily because of their increased cellular and in vivo potency. The mechanistic and biophysical understanding of bivalent IAP antagonists was greatly advanced by several studies published in 2007 (Petersen et al., 2007; Sun et al., 2007; Varfolomeev et al., 2007; Vince et al., 2007). One of these studies explored monovalent antagonist MV1 and its bivalent derivative BV6, and found that bivalent antagonist can engage the BIR2 and BIR3 domains of XIAP at the same time and dimerize c-IAP1 via binding to its BIR3 domains (Varfolomeev et al., 2007) [Figure 4]. Other groups confirmed these biophysical properties of bivalent compounds (Nikolovska-Coleska et al., 2008a, b).

IAP antagonists

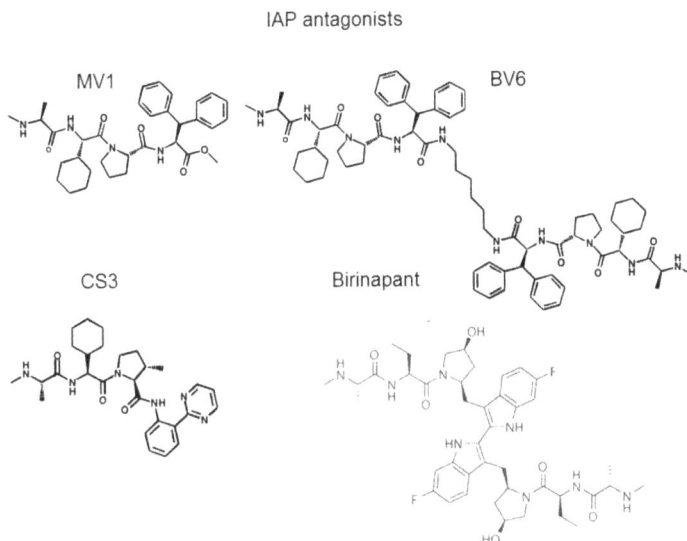

Figure 4. Examples of monovalent (MV1), bivalent (BV6 and Birinapant), and c-IAP selective (CS3) IAP antagonists.

Mechanistic aspects of the IAP antagonism

Discovery of potent bivalent IAP antagonists and their less potent, monovalent counterparts enabled a greater understanding of the biological functions of IAP proteins and the mechanism of action of IAP antagonists. Although IAP antagonists were generally predicted to act as XIAP inhibitors, mechanistic studies revealed several unexpected aspects of IAP antagonism. First, IAP antagonists cause a conformational change in c-IAP1 protein, which rapidly boosts its E3 ligase activity, and results in c-IAP autoubiquitination and subsequent proteasomal degradation (Varfolomeev et al., 2007; Vince et al., 2007; Dueber et al., 2011; Feltham et al., 2011). Second, activation of the ubiquitin ligase activity of c-IAPs is instrumental for the activation of NF-κB signaling pathways. By boosting the E3 ligase activity of c-IAP1 and c-IAP2, IAP antagonists promote RIP1 polyubiquitination and activation of the canonical NF-κB pathway. Ultimately, through autoubiquitination and proteasomal degradation, c-IAP proteins cause their own degradation leading to an accumulation of NIK and activation of the noncanonical NF-κB pathway (Varfolomeev et al., 2007). Both NF-κB pathways stimulate gene expression, especially of TNFα, which binds to TNF receptor I in an autocrine or paracrine fashion, and instigates TNFR-mediated signaling. The presence of c-IAP1 and 2 in the TNFR1-associated signaling complex prevents the induction of apoptosis. However, IAP antagonist-induced absence of the c-IAPs permits formation of the pro-apoptotic signaling complex composed of FADD, caspase-8 and de-ubiquitinated RIP1, and activation of caspase-8 to trigger apoptosis (Petersen et al., 2007; Bertrand et al., 2008). Importantly, TNF-blocking reagents can prevents this apoptotic event thus defining TNF dependence as a critical aspect of IAP antagonism (Gaither et al., 2007; Petersen et al., 2007; Varfolomeev et al., 2007; Vince et al., 2007).

Although monovalent and bivalent SMAC-mimicking IAP antagonists share many functional and mechanistic properties, bivalent compounds are more potent activators of cell death. Probably the major differentiating feature is simultaneous binding to BIR2 and BIR3 domains of XIAP by bivalent antagonists, which promotes stronger and more direct activation of caspases 3 and 7. Together with more prominent degradation of c-IAP1 and 2, robust caspase activation allows bivalent IAP antagonist more potent induction of cell death and inhibition of tumor growth (Gao et al., 2007; Varfolomeev et al., 2007). This is actually quite remarkable given that bivalent IAP antagonists defy established rules on drug-like properties with their bulkiness (over 1000 Da), which often generated skepticism regarding their usability.

IAP protein-selective antagonists

The broad efficacy observed for IAP antagonists prompted the evaluation of compounds with a greater affinity for some family members over the others, in particular, XIAP, c-IAP1 and c-IAP2. The most successful approach for this selective targeting involved structure-guided design and resulted in an IAP antagonist (CS3) with high affinity and selectivity (>2000) for c-IAP1 over XIAP [Figure 4] (Ndubaku et al., 2009b). This c-IAP-selective antagonist stimulated c-IAP1 and c-IAP2 degradation, activation of NF-κB signaling pathways and apoptosis in tumor cells as a single agent. However, compared to a pan-selective antagonist, the c-IAP-selective compound was significantly less potent in inducing cell death, suggesting that antagonism of both XIAP and c-IAP proteins is required for efficient induction of cancer cell death by IAP antagonists (Ndubaku et al., 2009b).

Several groups have reported efforts to target XIAP by specifically disrupting its ability to inhibit caspase-3. One group reported the use of high-throughput biochemical screening of a combinatorial library to find small molecule inhibitors of the XIAP:caspase-3 interaction (TWX-024) that sensitized cancer cells to Apo2L/TRAIL (Wu et al., 2003). Another class of small molecule antagonists of the XIAP:caspase-3 interaction, known as polyphenylureas, could also reverse XIAP-mediated caspase-3 inhibition and exert anti-tumor activity in vivo (Schimmer et al., 2004). Finally, recent publications of compounds with clear selectivity for BIR2 over BIR3 of XIAP have been reported but the in vivo activity of these antagonists is not clear yet (Donnell et al., 2013; Kester et al., 2013; Lukacs et al., 2013).

Clinical development of IAP antagonists and future perspectives

Elucidation of the role of IAP proteins in cancer cell survival and extensive research on IAP antagonism has enabled translation of this research into the clinical setting. The first small-molecule IAP antagonist to enter the clinic was GDC-0152 from Genentech (Call et al., 2008; Flygare and Vucic, 2009). Following GDC-0152, several other IAP targeting compounds have entered into clinical trials. These include AEG40826/HGS-1029 co-developed by Aegera and Human Genome Sciences, LCL-161 developed by Novartis, TL32711/Birinapant developed by Tetralogic Pharmaceuticals, AT-406/DCebio 1143 from Ascenta in partnership with Debiopharm, and GDC-0917/CUDC-427, which was licensed to Curis from Genentech (Gillard, 2008; Zawel, 2009; Allensworth et al., 2013). All of these agents have been well tolerated and a maximum tolerated dose (MTD) has not been reported yet (Wong et al., 2013; Infante et al., 2014). These clinical trials will examine the applicability of IAP antagonists for treatment of human malignancies and pave the way for future clinical investigations of IAP-regulated apoptotic pathways.

The prominent expression of IAP proteins in human malignancies, together with their ability to block apoptosis induced by a variety of extrinsic or intrinsic apoptotic stimuli, and their critical role in the regulation of survival signaling pathways make them attractive targets for the development of novel cancer therapeutics (Fulda and Vucic, 2012). Even though IAP antagonists and other IAP-targeting modalities are effective in tumor growth inhibition on their own, combination with other pro-apoptotic or anti-proliferative agents may synergistically enhance their activity and bring significant benefit to cancer patients. Equally important, recent progress in the mechanistic understanding of IAP targeting in combination with molecular diagnostics (Varfolomeev et al., 2014) should enable identification of cancer patients that will benefit the most from these therapeutic strategies.

Acknowledgments
The author thanks Wayne Fairbrother, Kim Newton and Eugene Varfolomeev for critical reading of the manuscript and help with figures.

Keywords

IAP, Inhibitor of Apoptosis, BIR domain, TNF, SMAC, XIAP, c-IAP, RING domain, apoptosis, NF-kB, cancer, IAP antagonist, SMAC mimetic, necroptosis

References

Adida, C., Berrebi, D., Peuchmaur, M., Reyes-Mugica, M., and Altieri, D.C. (1998). Anti-apoptosis gene, survivin, and prognosis of neuroblastoma. Lancet *351*, 882-883.

Allensworth, J.L., Sauer, S.J., Lyerly, H.K., Morse, M.A., and Devi, G.R. (2013). Smac mimetic Birinapant induces apoptosis and enhances TRAIL potency in inflammatory breast cancer cells in an IAP-dependent and TNF-alpha-independent mechanism. Breast Cancer Res Treat *137*, 359-371.

Altieri, D.C. (2003a). Survivin and apoptosis control. Adv Cancer Res *88*, 31-52.

Altieri, D.C. (2003b). Validating survivin as a cancer therapeutic target. Nat Rev Cancer *3*, 46-54.

Altieri, D.C. (2006). Targeted therapy by disabling crossroad signaling networks: the survivin paradigm. Molecular cancer therapeutics *5*, 478-482.

Arnt, C.R., Chiorean, M.V., Heldebrant, M.P., Gores, G.J., and Kaufmann, S.H. (2002). Synthetic Smac/DIABLO peptides enhance the effects of chemotherapeutic agents by binding XIAP and cIAP1 in situ. J Biol Chem *277*, 44236-44243.

Arora, V., Cheung, H.H., Plenchette, S., Micali, O.C., Liston, P., and Korneluk, R.G. (2007). Degradation of survivin by the X-linked inhibitor of apoptosis (XIAP)-XAF1 complex. J Biol Chem *282*, 26202-26209.

Asselin, E., Mills, G.B., and Tsang, B.K. (2001). XIAP regulates Akt activity and caspase-3-dependent cleavage during cisplatin-induced apoptosis in human ovarian epithelial cancer cells. Cancer Res *61*, 1862-1868.

Bertrand, M.J., Doiron, K., Labbe, K., Korneluk, R.G., Barker, P.A., and Saleh, M. (2009a). Cellular inhibitors of apoptosis cIAP1 and cIAP2 are required for innate immunity signaling by the pattern recognition receptors NOD1 and NOD2. Immunity *30*, 789-801.

Bertrand, M.J., Doiron, K., Labbe, K., Korneluk, R.G., Barker, P.A., and Saleh, M. (2009b). Cellular Inhibitors of Apoptosis cIAP1 and cIAP2 Are Required for Innate Immunity Signaling by the Pattern Recognition Receptors NOD1 and NOD2. Immunity.

Bertrand, M.J., Milutinovic, S., Dickson, K.M., Ho, W.C., Boudreault, A., Durkin, J., Gillard, J.W., Jaquith, J.B., Morris, S.J., and Barker, P.A. (2008). cIAP1 and cIAP2 facilitate cancer cell survival by functioning as E3 ligases that promote RIP1 ubiquitination. Mol Cell *30*, 689-700.

Birkey Reffey, S., Wurthner, J.U., Parks, W.T., Roberts, A.B., and Duckett, C.S. (2001). X-linked inhibitor of apoptosis protein functions as a cofactor in transforming growth factor-beta signaling. J Biol Chem *276*, 26542-26549.

Birnbaum, M.J., Clem, R.J., and Miller, L.K. (1994). An apoptosis-inhibiting gene from a nuclear polyhedrosis virus encoding a polypeptide with Cys/His sequence motifs. J Virol *68*, 2521-2528.

Blankenship, J.W., Varfolomeev, E., Goncharov, T., Fedorova, A.V., Kirkpatrick, D.S., Izrael-Tomasevic, A., Phu, L., Arnott, D., Aghajan, M., Zobel, K., *et al.* (2009). Ubiquitin binding modulates IAP antagonist-stimulated proteasomal degradation of c-IAP1 and c-IAP2. Biochem J *417*, 149-160.

Boatright, K.M., Renatus, M., Scott, F.L., Sperandio, S., Shin, H., Pedersen, I.M., Ricci, J.E., Edris, W.A., Sutherlin, D.P., Green, D.R., *et al.* (2003). A unified model for apical caspase activation. Mol Cell *11*, 529-541.

Budihardjo, I., Oliver, H., Lutter, M., Luo, X., and Wang, X. (1999). Biochemical pathways of caspase activation during apoptosis. Annu Rev Cell Dev Biol *15*, 269-290.

Burri, L., Strahm, Y., Hawkins, C.J., Gentle, I.E., Puryer, M.A., Verhagen, A., Callus, B., Vaux, D., and Lithgow, T. (2005). Mature DIABLO/Smac is produced by the IMP protease complex on the mitochondrial inner membrane. Mol Biol Cell *16*, 2926-2933.

Cai, Z., Jitkaew, S., Zhao, J., Chiang, H.C., Choksi, S., Liu, J., Ward, Y., Wu, L.G., and Liu, Z.G. (2014). Plasma membrane translocation of trimerized MLKL protein is required for TNF-induced necroptosis. Nature cell biology *16*, 55-65.

Call, J.A., Eckhardt, S.G., and Camidge, D.R. (2008). Targeted manipulation of apoptosis in cancer treatment. Lancet Oncol *9*, 1002-1011.

Cao, C., Mu, Y., Hallahan, D.E., and Lu, B. (2004). XIAP and survivin as therapeutic targets for radiation sensitization in preclinical models of lung cancer. Oncogene *23*, 7047-7052.

Chai, J., Du, C., Wu, J.W., Kyin, S., Wang, X., and Shi, Y. (2000). Structural and biochemical basis of apoptotic activation by Smac/DIABLO. Nature *406*, 855-862.

Chai, J., Shiozaki, E., Srinivasula, S.M., Wu, Q., Dataa, P., Alnemri, E.S., and Shi, Y. (2001). Structural basis of caspase-7 inhibition by XIAP. Cell *104*, 769-780.

Chen, Z., Naito, M., Hori, S., Mashima, T., Yamori, T., and Tsuruo, T. (1999). A human IAP-family gene, apollon, expressed in human brain cancer cells. Biochem Biophys Res Commun *264*, 847-854.

Clem, R.J., Fechheimer, M., and Miller, L.K. (1991). Prevention of apoptosis by a baculovirus gene during infection of insect cells. Science *254*, 1388-1390.

Clem, R.J., and Miller, L.K. (1994). Control of programmed cell death by the baculovirus genes p35 and iap. Mol Cell Biol *14*, 5212-5222.

Conze, D.B., Albert, L., Ferrick, D.A., Goeddel, D.V., Yeh, W.C., Mak, T., and Ashwell, J.D. (2005). Posttranscriptional downregulation of c-IAP2 by the ubiquitin protein ligase c-IAP1 in vivo. Mol Cell Biol *25*, 3348-3356.

Crnkovic-Mertens, I., Muley, T., Meister, M., Hartenstein, B., Semzow, J., Butz, K., and Hoppe-Seyler, F. (2006). The anti-apoptotic livin gene is an important determinant for the apoptotic resistance of non-small cell lung cancer cells. Lung Cancer *54*, 135-142.

Crnkovic-Mertens, I., Wagener, N., Semzow, J., Grone, E.F., Haferkamp, A., Hohenfellner, M., Butz, K., and Hoppe-Seyler, F. (2007). Targeted inhibition of Livin resensitizes renal cancer cells towards apoptosis. Cell Mol Life Sci *64*, 1137-1144.

Crook, N.E., Clem, R.J., and Miller, L.K. (1993). An apoptosis-inhibiting baculovirus gene with a zinc finger-like motif. J Virol *67*, 2168-2174.

Damgaard, R.B., Fiil, B.K., Speckmann, C., Yabal, M., zur Stadt, U., Bekker-Jensen, S., Jost, P.J., Ehl, S., Mailand, N., and Gyrd-Hansen, M. (2013). Disease-causing mutations in the XIAP BIR2 domain impair NOD2-dependent immune signalling. EMBO molecular medicine *5*, 1278-1295.

Damgaard, R.B., Nachbur, U., Yabal, M., Wong, W.W., Fiil, B.K., Kastirr, M., Rieser, E., Rickard, J.A., Bankovacki, A., Peschel, C., *et al.* (2012). The ubiquitin ligase XIAP recruits LUBAC for NOD2 signaling in inflammation and innate immunity. Molecular cell *46*, 746-758.

Dogan, T., Harms, G.S., Hekman, M., Karreman, C., Oberoi, T.K., Alnemri, E.S., Rapp, U.R., and Rajalingam, K. (2008). X-linked and cellular IAPs modulate the stability of C-RAF kinase and cell motility. Nat Cell Biol *10*, 1447-1455.

Dondelinger, Y., Declercq, W., Montessuit, S., Roelandt, R., Goncalves, A., Bruggeman, I., Hulpiau, P., Weber, K., Sehon, C.A., Marquis, R.W., *et al.* (2014). MLKL compromises plasma membrane integrity by binding to phosphatidylinositol phosphates. Cell reports *7*, 971-981.

Donnell, A.F., Michoud, C., Rupert, K.C., Han, X., Aguilar, D., Frank, K.B., Fretland, A.J., Gao, L., Goggin, B., Hogg, J.H., *et al.* (2013). Benzazepinones and benzoxazepinones as antagonists of inhibitor of apoptosis proteins (IAPs) selective for the second baculovirus IAP repeat (BIR2) domain. J Med Chem *56*, 7772-7787.

Du, C., Fang, M., Li, Y., Li, L., and Wang, X. (2000). Smac, a mitochondrial protein that promotes cytochrome c-dependent caspase activation by eliminating IAP inhibition. Cell 102, 33-42.

Dueber, E.C., Schoeffler, A.J., Lingel, A., Elliott, J.M., Fedorova, A.V., Giannetti, A.M., Zobel, K., Maurer, B., Varfolomeev, E., Wu, P., et al. (2011). Antagonists induce a conformational change in cIAP1 that promotes autoubiquitination. Science 334, 376-380.

Dynek, J.N., Chan, S.M., Liu, J., Zha, J., Fairbrother, W.J., and Vucic, D. (2008). Microphthalmia-associated transcription factor is a critical transcriptional regulator of melanoma inhibitor of apoptosis in melanomas. Cancer Res 68, 3124-3132.

Dynek, J.N., Goncharov, T., Dueber, E.C., Fedorova, A.V., Izrael-Tomasevic, A., Phu, L., Helgason, E., Fairbrother, W.J., Deshayes, K., Kirkpatrick, D.S., et al. (2010). c-IAP1 and UbcH5 promote K11-linked polyubiquitination of RIP1 in TNF signalling. Embo J 29, 4198-4209.

Eckelman, B.P., and Salvesen, G.S. (2006). The human anti-apoptotic proteins cIAP1 and cIAP2 bind but do not inhibit caspases. J Biol Chem 281, 3254-3260.

Eckelman, B.P., Salvesen, G.S., and Scott, F.L. (2006). Human inhibitor of apoptosis proteins: why XIAP is the black sheep of the family. EMBO Rep 7, 988-994.

Feltham, R., Bettjeman, B., Budhidarmo, R., Mace, P.D., Shirley, S., Condon, S.M., Chunduru, S.K., McKinlay, M.A., Vaux, D.L., Silke, J., et al. (2011). Smac Mimetics Activate the E3 Ligase Activity of cIAP1 Protein by Promoting RING Domain Dimerization. The Journal of biological chemistry 286, 17015-17028.

Flygare, J.A., and Vucic, D. (2009). Development of novel drugs targeting inhibitors of apoptosis. Future Oncol 5, 141-144.

Franklin, M.C., Kadkhodayan, S., Ackerly, H., Alexandru, D., Distefano, M.D., Elliott, L.O., Flygare, J.A., Mausisa, G., Okawa, D.C., Ong, D., et al. (2003). Structure and function analysis of peptide antagonists of melanoma inhibitor of apoptosis (ML-IAP). Biochemistry 42, 8223-8231.

Fukuda, S., and Pelus, L.M. (2001). Regulation of the inhibitor-of-apoptosis family member survivin in normal cord blood and bone marrow CD34(+) cells by hematopoietic growth factors: implication of survivin expression in normal hematopoiesis. Blood 98, 2091-2100.

Fulda, S., and Vucic, D. (2012). Targeting IAP proteins for therapeutic intervention in cancer. Nature reviews Drug discovery 11, 109-124.

Fulda, S., Wick, W., Weller, M., and Debatin, K.M. (2002). Smac agonists sensitize for Apo2L/TRAIL- or anticancer drug-induced apoptosis and induce regression of malignant glioma in vivo. Nat Med 8, 808-815.

Gaither, A., Porter, D., Yao, Y., Borawski, J., Yang, G., Donovan, J., Sage, D., Slisz, J., Tran, M., Straub, C., *et al.* (2007). A Smac mimetic rescue screen reveals roles for inhibitor of apoptosis proteins in tumor necrosis factor-alpha signaling. Cancer Res *67*, 11493-11498.

Gao, Z., Tian, Y., Wang, J., Yin, Q., Wu, H., Li, Y.M., and Jiang, X. (2007). A dimeric Smac/diablo peptide directly relieves caspase-3 inhibition by XIAP. Dynamic and cooperative regulation of XIAP by Smac/Diablo. J Biol Chem *282*, 30718-30727.

Gillard, J.W. (2008). (X)IAP Inhibition and Apoptosis Induction: Explaining the Remarkable Synergy with Death Receptor Agonists. In University of Ulm Collaborative Research Center International Symposium (Ulm, Germany).

Grether, M.E., Abrams, J.M., Agapite, J., White, K., and Steller, H. (1995). The head involution defective gene of Drosophila melanogaster functions in programmed cell death. Genes Dev *9*, 1694-1708.

Guicciardi, M.E., and Gores, G.J. (2009). Life and death by death receptors. Faseb J *23*, 1625-1637.

Gyrd-Hansen, M., Darding, M., Miasari, M., Santoro, M.M., Zender, L., Xue, W., Tenev, T., da Fonseca, P.C., Zvelebil, M., Bujnicki, J.M., *et al.* (2008). IAPs contain an evolutionarily conserved ubiquitin-binding domain that regulates NF-kappaB as well as cell survival and oncogenesis. Nat Cell Biol *10*, 1309-1317.

Hasegawa, M., Fujimoto, Y., Lucas, P.C., Nakano, H., Fukase, K., Nunez, G., and Inohara, N. (2008). A critical role of RICK/RIP2 polyubiquitination in Nod-induced NF-kappaB activation. Embo J *27*, 373-383.

Hauser, H.P., Bardroff, M., Pyrowolakis, G., and Jentsch, S. (1998). A giant ubiquitin-conjugating enzyme related to IAP apoptosis inhibitors. The Journal of cell biology *141*, 1415-1422.

Hay, B.A., Wassarman, D.A., and Rubin, G.M. (1995). Drosophila homologs of baculovirus inhibitor of apoptosis proteins function to block cell death. Cell *83*, 1253-1262.

Hershko, A., and Ciechanover, A. (1998). The ubiquitin system. Annu Rev Biochem *67*, 425-479.

Hinds, M.G., Norton, R.S., Vaux, D.L., and Day, C.L. (1999). Solution structure of a baculoviral inhibitor of apoptosis (IAP) repeat. Nat Struct Biol *6*, 648-651.

Hu, S., and Yang, X. (2003). Cellular inhibitor of apoptosis 1 and 2 are ubiquitin ligases for the apoptosis inducer Smac/DIABLO. J Biol Chem *278*, 10055-10060.

Hu, Y., Cherton-Horvat, G., Dragowska, V., Baird, S., Korneluk, R.G., Durkin, J.P., Mayer, L.D., and LaCasse, E.C. (2003). Antisense oligonucleotides targeting XIAP induce apoptosis and enhance chemotherapeutic activity against human lung cancer cells in vitro and in vivo. Clin Cancer Res 9, 2826-2836.

Huang, Y., Park, Y.C., Rich, R.L., Segal, D., Myszka, D.G., and Wu, H. (2001). Structural basis of caspase inhibition by XIAP: differential roles of the linker versus the BIR domain. Cell 104, 781-790.

Hunter, A.M., LaCasse, E.C., and Korneluk, R.G. (2007). The inhibitors of apoptosis (IAPs) as cancer targets. Apoptosis 12, 1543-1568.

Imoto, I., Tsuda, H., Hirasawa, A., Miura, M., Sakamoto, M., Hirohashi, S., and Inazawa, J. (2002). Expression of cIAP1, a target for 11q22 amplification, correlates with resistance of cervical cancers to radiotherapy. Cancer Res 62, 4860-4866.

Imoto, I., Yang, Z.Q., Pimkhaokham, A., Tsuda, H., Shimada, Y., Imamura, M., Ohki, M., and Inazawa, J. (2001). Identification of cIAP1 As a Candidate Target Gene within an Amplicon at 11q22 in Esophageal Squamous Cell Carcinomas. Cancer Res 61, 6629-6634.

Infante, J.R., Dees, E.C., Olszanski, A.J., Dhuria, S.V., Sen, S., Cameron, S., and Cohen, R.B. (2014). Phase I dose-escalation study of LCL161, an oral inhibitor of apoptosis proteins inhibitor, in patients with advanced solid tumors. J Clin Oncol 32, 3103-3110.

Isaacson, P.G. (2005). Update on MALT lymphomas. Best Pract Res Clin Haematol 18, 57-68.

Jeyaprakash, A.A., Klein, U.R., Lindner, D., Ebert, J., Nigg, E.A., and Conti, E. (2007). Structure of a Survivin-Borealin-INCENP core complex reveals how chromosomal passengers travel together. Cell 131, 271-285.

Karin, M., Cao, Y., Greten, F.R., and Li, Z.W. (2002). NF-kappaB in cancer: from innocent bystander to major culprit. Nat Rev Cancer 2, 301-310.

Karin, M., and Lin, A. (2002). NF-kappaB at the crossroads of life and death. Nat Immunol 3, 221-227.

Kaufmann, S.H., and Vaux, D.L. (2003). Alterations in the apoptotic machinery and their potential role in anticancer drug resistance. Oncogene 22, 7414-7430.

Kester, R.F., Donnell, A.F., Lou, Y., Remiszewski, S.W., Lombardo, L.J., Chen, S., Le, N.T., Lo, J., Moliterni, J.A., Han, X., et al. (2013). Optimization of benzodiazepinones as selective inhibitors of the X-linked inhibitor of apoptosis protein (XIAP) second baculovirus IAP repeat (BIR2) domain. J Med Chem 56, 7788-7803.

Kipp, R.A., Case, M.A., Wist, A.D., Cresson, C.M., Carrell, M., Griner, E., Wiita, A., Albiniak, P.A., Chai, J., Shi, Y., et al. (2002). Molecular targeting of inhibitor of apoptosis proteins based on small molecule mimics of natural binding partners. Biochemistry *41*, 7344-7349.

Krieg, A., Correa, R.G., Garrison, J.B., Le Negrate, G., Welsh, K., Huang, Z., Knoefel, W.T., and Reed, J.C. (2009). XIAP mediates NOD signaling via interaction with RIP2. Proceedings of the National Academy of Sciences of the United States of America *106*, 14524-14529.

Li, L., Thomas, R.M., Suzuki, H., De Brabander, J.K., Wang, X., and Harran, P.G. (2004). A small molecule Smac mimic potentiates TRAIL- and TNFα-mediated cell death. Science *305*, 1471-1474.

Liston, P., Fong, W.G., Kelly, N.L., Toji, S., Miyazaki, T., Conte, D., Tamai, K., Craig, C.G., McBurney, M.W., and Korneluk, R.G. (2001). Identification of XAF1 as an antagonist of XIAP anti-Caspase activity. Nat Cell Biol *3*, 128-133.

Liston, P., Roy, N., Tamai, K., Lefebvre, C., Baird, S., Cherton-Horvat, G., Farahani, R., McLean, M., Ikeda, J.E., MacKenzie, A., *et al.* (1996). Suppression of apoptosis in mammalian cells by NAIP and a related family of IAP genes. Nature *379*, 349-353.

Liu, Z., Sun, C., Olejniczak, E.T., Meadows, R.P., Betz, S.F., Oost, T., Herrmann, J., Wu, J.C., and Fesik, S.W. (2000). Structural basis for binding of Smac/DIABLO to the XIAP BIR3 domain. Nature *408*, 1004-1008.

Lucas, P.C., Yonezumi, M., Inohara, N., McAllister-Lucas, L.M., Abazeed, M.E., Chen, F.F., Yamaoka, S., Seto, M., and Nunez, G. (2001). Bcl10 and MALT1, independent targets of chromosomal translocation in malt lymphoma, cooperate in a novel NF-kappa B signaling pathway. J Biol Chem *276*, 19012-19019.

Lukacs, C., Belunis, C., Crowther, R., Danho, W., Gao, L., Goggin, B., Janson, C.A., Li, S., Remiszewski, S., Schutt, A., *et al.* (2013). The structure of XIAP BIR2: understanding the selectivity of the BIR domains. Acta crystallographica Section D, Biological crystallography *69*, 1717-1725.

Ma, L., Huang, Y., Song, Z., Feng, S., Tian, X., Du, W., Qiu, X., Heese, K., and Wu, M. (2006). Livin promotes Smac/DIABLO degradation by ubiquitin-proteasome pathway. Cell Death Differ *13*, 2079-2088.

MacFarlane, M., Merrison, W., Bratton, S.B., and Cohen, G.M. (2002). Proteasome-mediated degradation of Smac during apoptosis: XIAP promotes Smac ubiquitination in vitro. J Biol Chem *277*, 36611-36616.

Mahoney, D.J., Cheung, H.H., Mrad, R.L., Plenchette, S., Simard, C., Enwere, E., Arora, V., Mak, T.W., Lacasse, E.C., Waring, J., *et al.* (2008). Both cIAP1 and cIAP2 regulate TNFalpha-mediated NF-kappaB activation. Proc Natl Acad Sci U S A *105*, 11778-11783.

Matsuzawa, A., Tseng, P.H., Vallabhapurapu, S., Luo, J.L., Zhang, W., Wang, H., Vignali, D.A., Gallagher, E., and Karin, M. (2008). Essential cytoplasmic translocation of a cytokine receptor-assembled signaling complex. Science *321*, 663-668.

Micheau, O., and Tschopp, J. (2003). Induction of TNF receptor I-mediated apoptosis via two sequential signaling complexes. Cell *114*, 181-190.

Morizane, Y., Honda, R., Fukami, K., and Yasuda, H. (2005). X-linked inhibitor of apoptosis functions as ubiquitin ligase toward mature caspase-9 and cytosolic Smac/DIABLO. J Biochem (Tokyo) *137*, 125-132.

Ndubaku, C., Cohen, F., Varfolomeev, E., and Vucic, D. (2009a). Targeting inhibitor of apoptosis (IAP) proteins for therapeutic intervention. Future Med Chem *1*, 1509-1525.

Ndubaku, C., Varfolomeev, E., Wang, L., Zobel, K., Lau, K., Elliott, L.O., Maurer, B., Fedorova, A.V., Dynek, J.N., Koehler, M., *et al.* (2009b). Antagonism of c-IAP and XIAP proteins is required for efficient induction of cell death by small-molecule IAP antagonists. ACS Chem Biol *In press.*

Newton, K., Dugger, D.L., Wickliffe, K.E., Kapoor, N., de Almagro, M.C., Vucic, D., Komuves, L., Ferrando, R.E., French, D.M., Webster, J., *et al.* (2014). Activity of protein kinase RIPK3 determines whether cells die by necroptosis or apoptosis. Science *343*, 1357-1360.

Nikolovska-Coleska, Z., Meagher, J.L., Jiang, S., Kawamoto, S.A., Gao, W., Yi, H., Qin, D., Roller, P.P., Stuckey, J.A., and Wang, S. (2008a). Design and characterization of bivalent Smac-based peptides as antagonists of XIAP and development and validation of a fluorescence polarization assay for XIAP containing both BIR2 and BIR3 domains. Anal Biochem *374*, 87-98.

Nikolovska-Coleska, Z., Meagher, J.L., Jiang, S., Yang, C.Y., Qiu, S., Roller, P.P., Stuckey, J.A., and Wang, S. (2008b). Interaction of a cyclic, bivalent smac mimetic with the x-linked inhibitor of apoptosis protein. Biochemistry *47*, 9811-9824.

Oost, T.K., Sun, C., Armstrong, R.C., Al-Assaad, A.S., Betz, S.F., Deckwerth, T.L., Ding, H., Elmore, S.W., Meadows, R.P., Olejniczak, E.T., *et al.* (2004). Discovery of potent antagonists of the antiapoptotic protein XIAP for the treatment of cancer. J Med Chem *47*, 4417-4426.

Peter, M.E., and Krammer, P.H. (2003). The CD95(APO-1/Fas) DISC and beyond. Cell Death Differ *10*, 26-35.

Petersen, S.L., Wang, L., Yalcin-Chin, A., Li, L., Peyton, M., Minna, J., Harran, P., and Wang, X. (2007). Autocrine TNFalpha signaling renders human cancer cells susceptible to smac-mimetic-induced apoptosis. Cancer cell *12*, 445-456.

Polykratis, A., Hermance, N., Zelic, M., Roderick, J., Kim, C., Van, T.M., Lee, T.H., Chan, F.K., Pasparakis, M., and Kelliher, M.A. (2014). Cutting edge: RIPK1 Kinase inactive mice are viable and protected from TNF-induced necroptosis in vivo. J Immunol *193*, 1539-1543.

Riedl, S.J., Renatus, M., Schwarzenbacher, R., Zhou, Q., Sun, C., Fesik, S.W., Liddington, R.C., and Salvesen, G.S. (2001). Structural basis for the inhibition of caspase-3 by XIAP. Cell *104*, 791-800.

Riedl, S.J., and Salvesen, G.S. (2007). The apoptosome: signalling platform of cell death. Nat Rev Mol Cell Biol *8*, 405-413.

Rothe, M., Pan, M.G., Henzel, W.J., Ayres, T.M., and Goeddel, D.V. (1995). The TNFR2-TRAF signaling complex contains two novel proteins related to baculoviral inhibitor of apoptosis proteins. Cell *83*, 1243-1252.

Salvesen, G.S., and Abrams, J.M. (2004). Caspase activation - stepping on the gas or releasing the brakes? Lessons from humans and flies. Oncogene *23*, 2774-2784.

Salvesen, G.S., and Duckett, C.S. (2002). IAP proteins: blocking the road to death's door. Nat Rev Mol Cell Biol *3*, 401-410.

Samuel, T., Welsh, K., Lober, T., Togo, S.H., Zapata, J.M., and Reed, J.C. (2006). Distinct BIR domains of cIAP1 mediate binding to and ubiquitination of tumor necrosis factor receptor-associated factor 2 and second mitochondrial activator of caspases. J Biol Chem *281*, 1080-1090.

Sanna, M.G., da Silva Correia, J., Ducrey, O., Lee, J., Nomoto, K., Schrantz, N., Deveraux, Q.L., and Ulevitch, R.J. (2002). IAP suppression of apoptosis involves distinct mechanisms: the TAK1/JNK1 signaling cascade and caspase inhibition. Mol Cell Biol *22*, 1754-1766.

Scheidereit, C. (2006). IkappaB kinase complexes: gateways to NF-kappaB activation and transcription. Oncogene *25*, 6685-6705.

Schimmer, A.D., Herr, W., Hanel, M., Borthakur, G., Frankel, A., Horst, H.A., Martin, S., Kassis, J., Desjardins, P., Seiter, K., *et al.* (2011). Addition of AEG35156 XIAP antisense oligonucleotide in reinduction chemotherapy does not improve remission rates in patients with primary refractory acute myeloid leukemia in a randomized phase II study. Clin Lymphoma Myeloma Leuk *11*, 433-438.

Schimmer, A.D., Welsh, K., Pinilla, C., Wang, Z., Krajewska, M., Bonneau, M.J., Pedersen, I.M., Kitada, S., Scott, F.L., Bailly-Maitre, B., *et al.* (2004). Small-molecule antagonists of apoptosis suppressor XIAP exhibit broad antitumor activity. Cancer cell *5*, 25-35.

Scott, F.L., Denault, J.B., Riedl, S.J., Shin, H., Renatus, M., and Salvesen, G.S. (2005). XIAP inhibits caspase-3 and -7 using two binding sites: evolutionarily conserved mechanism of IAPs. Embo J *24*, 645-655.

Sharma, S.K., Straub, C., and Zawel, L. (2006). Development of peptidomimetics targeting IAPs. International Journal of Peptide Research and Therapeutics *12*, 21-32.

Shim, J.H., Xiao, C., Paschal, A.E., Bailey, S.T., Rao, P., Hayden, M.S., Lee, K.Y., Bussey, C., Steckel, M., Tanaka, N., *et al.* (2005). TAK1, but not TAB1 or TAB2, plays an essential role in multiple signaling pathways in vivo. Genes Dev *19*, 2668-2681.

Shin, H., Renatus, M., Eckelman, B.P., Nunes, V.A., Sampaio, C.A., and Salvesen, G.S. (2005). The BIR domain of IAP-like protein 2 is conformationally unstable: implications for caspase inhibition. Biochem J *385*, 1-10.

Shiozaki, E.N., Chai, J., Rigotti, D.J., Riedl, S.J., Li, P., Srinivasula, S.M., Alnemri, E.S., Fairman, R., and Shi, Y. (2003). Mechanism of XIAP-mediated inhibition of caspase-9. Mol Cell *11*, 519-527.

Silke, J., Kratina, T., Chu, D., Ekert, P.G., Day, C.L., Pakusch, M., Huang, D.C., and Vaux, D.L. (2005). Determination of cell survival by RING-mediated regulation of inhibitor of apoptosis (IAP) protein abundance. Proc Natl Acad Sci U S A *102*, 16182-16187.

Silke, J., and Vaux, D.L. (2001). Two kinds of BIR-containing protein - inhibitors of apoptosis, or required for mitosis. Journal of cell science *114*, 1821-1827.

Srinivasula, S.M., Hegde, R., Saleh, A., Datta, P., Shiozaki, E., Chai, J., Lee, R.A., Robbins, P.D., Fernandes-Alnemri, T., Shi, Y., *et al.* (2001). A conserved XIAP-interaction motif in caspase-9 and Smac/DIABLO regulates caspase activity and apoptosis. Nature *410*, 112-116.

Steller, H. (2008). Regulation of apoptosis in Drosophila. Cell Death Differ *15*, 1132-1138.

Sun, C., Cai, M., Gunasekera, A.H., Meadows, R.P., Wang, H., Chen, J., Zhang, H., Wu, W., Xu, N., Ng, S.C., *et al.* (1999). NMR structure and mutagenesis of the inhibitor-of-apoptosis protein XIAP. Nature *401*, 818-822.

Sun, H., Nikolovska-Coleska, Z., Lu, J., Meagher, J.L., Yang, C.Y., Qiu, S., Tomita, Y., Ueda, Y., Jiang, S., Krajewski, K., *et al.* (2007). Design, synthesis, and characterization of a potent, nonpeptide, cell-permeable, bivalent Smac mimetic that concurrently targets both the BIR2 and BIR3 domains in XIAP. J Am Chem Soc *129*, 15279-15294.

Sun, H., Nikolovska-Coleska, Z., Yang, C.Y., Qian, D., Lu, J., Qiu, S., Bai, L., Peng, Y., Cai, Q., and Wang, S. (2008). Design of small-molecule peptidic and nonpeptidic Smac mimetics. Acc Chem Res *41*, 1264-1277.

Sun, L., Wang, H., Wang, Z., He, S., Chen, S., Liao, D., Wang, L., Yan, J., Liu, W., Lei, X., *et al.* (2012). Mixed Lineage Kinase Domain-like Protein Mediates Necrosis Signaling Downstream of RIP3 Kinase. Cell *148*, 213-227.

Sun, X., Yin, J., Starovasnik, M.A., Fairbrother, W.J., and Dixit, V.M. (2002). Identification of a novel homotypic interaction motif required for the phosphorylation of receptor-interacting protein (RIP) by RIP3. J Biol Chem *277*, 9505-9511.

Suzuki, Y., Imai, Y., Nakayama, H., Takahashi, K., Takio, K., and Takahashi, R. (2001a). A serine protease, htra2, is released from the mitochondria and interacts with xiap, inducing cell death. Mol Cell *8*, 613-621.

Suzuki, Y., Nakabayashi, Y., Nakata, K., Reed, J.C., and Takahashi, R. (2001b). X-linked inhibitor of apoptosis protein (XIAP) inhibits caspase-3 and -7 in distinct modes. J Biol Chem *276*, 27058-27063.

Suzuki, Y., Nakabayashi, Y., and Takahashi, R. (2001c). Ubiquitin-protein ligase activity of X-linked inhibitor of apoptosis protein promotes proteasomal degradation of caspase-3 and enhances its anti-apoptotic effect in Fas-induced cell death. Proc Natl Acad Sci U S A *98*, 8662-8667.

Tamm, I. (2006). Antisense therapy in malignant diseases: status quo and quo vadis? Clin Sci (Lond) *110*, 427-442.

Tanabe, H., Yagihashi, A., Tsuji, N., Shijubo, Y., Abe, S., and Watanabe, N. (2004). Expression of survivin mRNA and livin mRNA in non-small-cell lung cancer. Lung Cancer *46*, 299-304.

Uren, A.G., O'Rourke, K., Aravind, L.A., Pisabarro, M.T., Seshagiri, S., Koonin, E.V., and Dixit, V.M. (2000). Identification of paracaspases and metacaspases: two ancient families of caspase-like proteins, one of which plays a key role in MALT lymphoma. Mol Cell *6*, 961-967.

Uren, A.G., Pakusch, M., Hawkins, C.J., Puls, K.L., and Vaux, D.L. (1996). Cloning and expression of apoptosis inhibitory protein homologs that function to inhibit apoptosis and/or bind tumor necrosis factor receptor-associated factors. Proc Natl Acad Sci USA *93*, 4974-4978.

Vandenabeele, P., Galluzzi, L., Vanden Berghe, T., and Kroemer, G. (2010). Molecular mechanisms of necroptosis: an ordered cellular explosion. Nat Rev Mol Cell Biol *11*, 700-714.

Vanlangenakker, N., Vanden Berghe, T., Bogaert, P., Laukens, B., Zobel, K., Deshayes, K., Vucic, D., Fulda, S., Vandenabeele, P., and Bertrand, M.J. (2011). cIAP1 and TAK1 protect cells from TNF-induced necrosis by preventing RIP1/RIP3-dependent reactive oxygen species production. Cell death and differentiation *18*, 656-665.

Varfolomeev, E., Blankenship, J.W., Wayson, S.M., Fedorova, A.V., Kayagaki, N., Garg, P., Zobel, K., Dynek, J.N., Elliott, L.O., Wallweber, H.J., et al. (2007). IAP antagonists induce autoubiquitination of c-IAPs, NF-κB activation, and TNFα-dependent apoptosis. Cell 131, 669-681.

Varfolomeev, E., Goncharov, T., Fedorova, A.V., Dynek, J.N., Zobel, K., Deshayes, K., Fairbrother, W.J., and Vucic, D. (2008). c-IAP1 and c-IAP2 Are Critical Mediators of Tumor Necrosis Factor alpha (TNFα)-induced NF-κB Activation. J Biol Chem 283, 24295-24299.

Varfolomeev, E., Izrael-Tomasevic, A., Yu, K., Bustos, D., Goncharov, T., Belmont, L.D., Masselot, A., Bakalarski, C.E., Kirkpatrick, D.S., and Vucic, D. (2014). Ubiquitination profiling identifies sensitivity factors for IAP antagonist treatment. Biochem J.

Varfolomeev, E., and Vucic, D. (2008). (Un)expected roles of c-IAPs in apoptotic and NF-κB signaling pathways. Cell Cycle 7, 1511-1521.

Vaux, D.L., and Silke, J. (2005). IAPs, RINGs and ubiquitylation. Nat Rev Mol Cell Biol 6, 287-297.

Verhagen, A.M., Ekert, P.G., Pakusch, M., Silke, J., Connolly, L.M., Reid, G.E., Moritz, R.L., Simpson, R.J., and Vaux, D.L. (2000). Identification of DIABLO, a mammalian protein that promotes apoptosis by binding to and antagonizing IAP proteins. Cell 102, 43-53.

Verhagen, A.M., Kratina, T.K., Hawkins, C.J., Silke, J., Ekert, P.G., and Vaux, D.L. (2007). Identification of mammalian mitochondrial proteins that interact with IAPs via N-terminal IAP binding motifs. Cell Death Differ 14, 348-357.

Vince, J.E., Chau, D., Callus, B., Wong, W.W., Hawkins, C.J., Schneider, P., McKinlay, M., Benetatos, C.A., Condon, S.M., Chunduru, S.K., et al. (2008). TWEAK-FN14 signaling induces lysosomal degradation of a cIAP1-TRAF2 complex to sensitize tumor cells to TNFalpha. The Journal of cell biology 182, 171-184.

Vince, J.E., Wong, W.W., Khan, N., Feltham, R., Chau, D., Ahmed, A.U., Benetatos, C.A., Chunduru, S.K., Condon, S.M., McKinlay, M., et al. (2007). IAP antagonists target cIAP1 to induce TNFalpha-dependent apoptosis. Cell 131, 682-693.

Vucic, D. (2008). Targeting IAP (inhibitor of apoptosis) proteins for therapeutic intervention in tumors. Current cancer drug targets 8, 110-117.

Vucic, D., Deshayes, K., Ackerly, H., Pisabarro, M.T., Kadkhodayan, S., Fairbrother, W.J., and Dixit, V.M. (2002). SMAC Negatively Regulates the Anti-apoptotic Activity of Melanoma Inhibitor of Apoptosis (ML-IAP). J Biol Chem 277, 12275-12279.

Vucic, D., and Fairbrother, W.J. (2007). The inhibitor of apoptosis proteins as therapeutic targets in cancer. Clin Cancer Res 13, 5995-6000.

Vucic, D., Franklin, M.C., Wallweber, H.J., Das, K., Eckelman, B.P., Shin, H., Elliott, L.O., Kadkhodayan, S., Deshayes, K., Salvesen, G.S., et al. (2005). Engineering ML-IAP to produce an extraordinarily potent caspase 9 inhibitor: implications for Smac-dependent anti-apoptotic activity of ML-IAP. Biochem J 385, 11-20.

Vucic, D., Kaiser, W.J., Harvey, A.J., and Miller, L.K. (1997). Inhibition of Reaper-induced apoptosis by interaction with inhibitor of apoptosis proteins (IAPs). Proc Natl Acad Sci U S A 94, 10183-10188.

Vucic, D., Kaiser, W.J., and Miller, L.K. (1998). Inhibitor of apoptosis proteins physically interact with and block apoptosis induced by Drosophila proteins HID and GRIM. Mol Cell Biol 18, 3300-3309.

Vucic, D., Stennicke, H.R., Pisabarro, M.T., Salvesen, G.S., and Dixit, V.M. (2000). ML-IAP, a novel inhibitor of apoptosis that is preferentially expressed in human melanomas. Curr Biol 10, 1359-1366.

Wagener, N., Crnkovic-Mertens, I., Vetter, C., Macher-Goppinger, S., Bedke, J., Grone, E.F., Zentgraf, H., Pritsch, M., Hoppe-Seyler, K., Buse, S., et al. (2007). Expression of inhibitor of apoptosis protein Livin in renal cell carcinoma and non-tumorous adult kidney. Br J Cancer 97, 1271-1276.

Wajant, H., and Scheurich, P. (2011). TNFR1-induced activation of the classical NF-kappaB pathway. Febs J.

Wang, C., Deng, L., Hong, M., Akkaraju, G.R., Inoue, J., and Chen, Z.J. (2001). TAK1 is a ubiquitin-dependent kinase of MKK and IKK. Nature 412, 346-351.

Wang, C.Y., Mayo, M.W., Korneluk, R.G., Goeddel, D.V., and Baldwin, A.S., Jr. (1998). NF-κB antiapoptosis: induction of TRAF1 and TRAF2 and c-IAP1 and c-IAP2 to suppress caspase-8 activation. Science 281, 1680-1683.

Wang, H., Sun, L., Su, L., Rizo, J., Liu, L., Wang, L.F., Wang, F.S., and Wang, X. (2014). Mixed lineage kinase domain-like protein MLKL causes necrotic membrane disruption upon phosphorylation by RIP3. Molecular cell 54, 133-146.

Wertz, I.E., and Dixit, V.M. (2008). Ubiquitin-mediated regulation of TNFR1 signaling. Cytokine Growth Factor Rev 19, 313-324.

Wertz, I.E., and Dixit, V.M. (2010). Regulation of death receptor signaling by the ubiquitin system. Cell Death Differ 17, 14-24.

White, K., Grether, M.E., Abrams, J.M., Young, L., Farrell, K., and Steller, H. (1994). Genetic control of programmed cell death in Drosophila. Science 264, 677-683.

Wilmanski, J.M., Petnicki-Ocwieja, T., and Kobayashi, K.S. (2008). NLR proteins: integral members of innate immunity and mediators of inflammatory diseases. J Leukoc Biol *83*, 13-30.

Wilson, N.S., Dixit, V., and Ashkenazi, A. (2009). Death receptor signal transducers: nodes of coordination in immune signaling networks. Nat Immunol *10*, 348-355.

Wong, H., Gould, S.E., Budha, N., Darbonne, W.C., Kadel, E.E., 3rd, La, H., Alicke, B., Halladay, J.S., Erickson, R., Portera, C., *et al.* (2013). Learning and confirming with preclinical studies: modeling and simulation in the discovery of GDC-0917, an inhibitor of apoptosis proteins antagonist. Drug metabolism and disposition: the biological fate of chemicals *41*, 2104-2113.

Wu, G., Chai, J., Suber, T.L., Wu, J.W., Du, C., Wang, X., and Shi, Y. (2000). Structural basis of IAP recognition by Smac/DIABLO. Nature *408*, 1008-1012.

Wu, T.Y., Wagner, K.W., Bursulaya, B., Schultz, P.G., and Deveraux, Q.L. (2003). Development and characterization of nonpeptidic small molecule inhibitors of the XIAP/caspase-3 interaction. Chem Biol *10*, 759-767.

Yang, L., Mashima, T., Sato, S., Mochizuki, M., Sakamoto, H., Yamori, T., Oh-Hara, T., and Tsuruo, T. (2003). Predominant suppression of apoptosome by inhibitor of apoptosis protein in non-small cell lung cancer H460 cells: therapeutic effect of a novel polyarginine-conjugated Smac peptide. Cancer Res *63*, 831-837.

Yang, Q.H., and Du, C. (2004). Smac/DIABLO selectively reduces the levels of c-IAP1 and c-IAP2 but not that of XIAP and livin in HeLa cells. J Biol Chem *279*, 16963-16970.

Youle, R.J., and Strasser, A. (2008). The BCL-2 protein family: opposing activities that mediate cell death. Nat Rev Mol Cell Biol *9*, 47-59.

Zawel, L. (2009). LCL161. In Targeted Therapies of the Treatment of Lung Cancer, (Santa Monica, CA).

Zender, L., Spector, M.S., Xue, W., Flemming, P., Cordon-Cardo, C., Silke, J., Fan, S.T., Luk, J.M., Wigler, M., Hannon, G.J., *et al.* (2006). Identification and validation of oncogenes in liver cancer using an integrative oncogenomic approach. Cell *125*, 1253-1267.

Zhao, J., Jitkaew, S., Cai, Z., Choksi, S., Li, Q., Luo, J., and Liu, Z.G. (2012). Mixed lineage kinase domain-like is a key receptor interacting protein 3 downstream component of TNF-induced necrosis. Proceedings of the National Academy of Sciences of the United States of America *109*, 5322-5327.

Zobel, K., Wang, L., Varfolomeev, E., Franklin, M.C., Elliott, L.O., Wallweber, H.J., Okawa, D.C., Flygare, J.A., Vucic, D., Fairbrother, W.J., *et al.* (2006). Design, synthesis, and biological activity of a potent Smac mimetic that sensitizes cancer cells to apoptosis by antagonizing IAPs. ACS Chem Biol *1*, 525-533.

Chapter 10: How do cells stay alive? (Green, Llambi, and Fienberg)

Douglas R. Green[1], Fabien Llambi[1], and Harris Fienberg[2]

[1]Department of Immunology, St. Jude Children's Research Hospital, Memphis, TN., [2]Department of Molecular Pharmacology, Stanford University Medical Center, Palo Alto, CA.

Abstract

The mitochondrial pathway of apoptosis is probably the major way in which cells in an animal body die. Studies at the single cell level have shown that variations at the level of protein expression dictate the timing and extent of engagement of the mitochondrial pathway when a pro-apoptotic signal is given to a homogeneous population of cells. But cells in a metazoan body must persist to provide integrity to the multicellular state, and therefore the signaling states of surviving versus dying cells must be stable. Generally, such stability requires negative feedback mechanisms, yet no such feedback mechanisms are known for the key event in this pathway, the process of mitochondrial outer membrane permeabilization (MOMP), mediated and controlled by BCL2 family protein interactions. Here, we discuss this problem of how cells survive from the perspective of the "random walks" cells take in their variations in the levels of proteins expressed in any single cells, and how such walks can move cells into the state of cell death upon stress or other pro-apoptotic signals. Some of our conclusions, based on both theory and observation, are counter-intuitive.

Cells on the random walk of life

Cells die, of course. Some die within hours of their "birth" while others live for decades. Cell death can be accidental, the simple accrual of more damage than can be successfully repaired, but for the most part, cells die as a consequence of regulated, molecular events. As a reader of this book, you are cognizant that most regulated cell death is by apoptosis, but there are other forms of regulated cell death as well. But that is not what this chapter is about: Instead, this is about how cells stay alive.

When we discuss apoptosis and other forms of regulated cell death, we consider the pathways responsible for the death of the cell and the inhibitory interactions that block them. Therefore, we are usually quick to consider that cells stay alive because they actively block cell death pathways. And indeed, genetic studies inform us that these inhibitors are often essential not only for the life of a cell, but also for the life of the multicellular organism.

Clearly, though, the survival of a cell is more than the inhibition of cell death pathways. Cells require energy in the form of metabolites to sustain homeostasis, especially of the plasma membrane that separates the cell from the outside world. Cellular processes must be intimately coordinated to ensure that incompatible events do not compromise cellular integrity (e.g., endoplasmic reticulum genesis without sufficient lipid; mitosis without chromosome duplication; generation of reactive oxygen from an inappropriately powered electron transport chain). There are many ways that a cell can die, even active ways that a cell can die, that do not involve the cell death pathways generally considered in this volume. While potentially relevant to the concepts that will be discussed herein, our focus will be on apoptosis, and in particular, the mitochondrial pathway of apoptosis, and how cells manage not to engage it more than they do.

The mitochondrial pathway of apoptosis is likely to be the most common form of active cell death that occurs in animals. Although some phyla, such as the Nematoda and probably the Arthropoda, do not display a canonical mitochondrial pathway of apoptosis (but, instead, variants that may not involve the core mitochondrial involvement), this pathway has been rigorously demonstrated not only in the Chordates, but also in the Echinodermata (both Deuterostomes), and in the Protostome phylum Lophotrochazoa (flatworms) (Oberst, et al., 2008; Bender, et al., 2012).

First principles
Life and death can be viewed as two stable states—a cell is either alive or it is not. A living cell can transition to a dead cell, but the "arrow" goes in only one direction; dead cells cannot transition to living cells. Therefore, at some point in the process of apoptosis, for example, there must be a point of no return—a

point from which the cell inexorably commits to die. For some time, it was thought that the point of no return in the mitochondrial pathway of apoptosis occurred at mitochondrial outer membrane permeabilization (MOMP), when the BCL2 family protein interactions resulted in outer membrane disruption and release of the proteins of the intermembrane space (Chipuk, et al., 2010; Green, 2011). These include cytochrome c and IAP-antagonists (such as Smac and Omi), which contribute to activation of caspases downsteam of MOMP (Tait and Green, 2010). However, even if caspase activation is blocked or disrupted, cells generally die following MOMP, either as a consequence of the release of other "toxic" molecules (AIF, endonuclease G) (Chipuk and Green, 2005) or due to bioenergetic catastrophe (Lartique, et al., 2009). A hypothetical scheme for the relationships between stress, the extent of MOMP, and cell survival is shown in Figure 1a. In this model, a threshold level of stress is required to engage MOMP, leading to apoptosis and/or caspase-independent cell death. More recent studies, however, have shown that cells can survive MOMP when caspase activity is limited (Colell, et al., 2007; Tait, et al., 2010). Such survival appears to depend on the presence of intact mitochondria (i.e., that did not undergo MOMP) (Tait, et al., 2010), and we can envision a continuum wherein the likelihood of death following MOMP depends on how many mitochondria were engaged in the process (Figure 1b). It should be noted, however, that while the relationship between stress and apoptosis is well established, any role for differential levels for MOMP are purely speculative (Figure 1b). The alternative, that MOMP follows more of an all-or-nothing response (Goldstein, et al., 2000) remains the favored concept (Figure 1a).

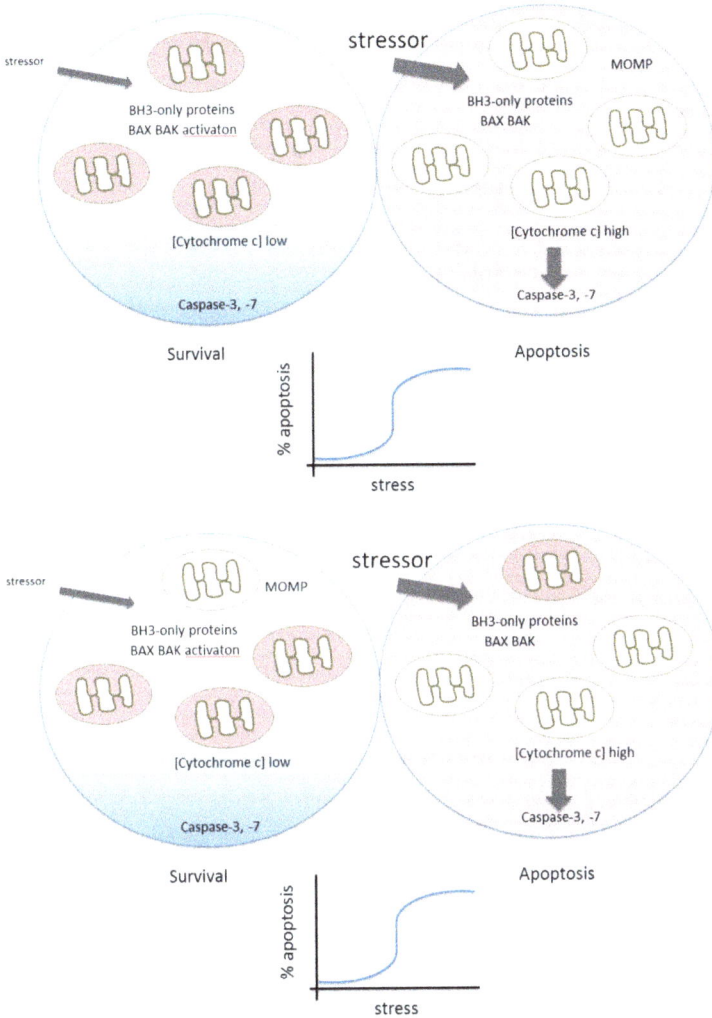

Figure 1. Relationships between cell stress and apoptosis via the mitochondrial pathway. A. Conventional view that MOMP occurs at a threshold level of stress, leading to caspase activation and apoptosis. This model requires that negative feedback mechanisms function prior to MOMP to enforce cell survival. B. Alternative view that the extent of MOMP corresponds to engagement of apoptosis. This model suggests that feedback mechanisms may function post-MOMP to limit caspase activation (or caspase-independent death) until a threshold level of MOMP occurs.

Either scheme likely relates to other mechanisms of apoptosis as well, such as that induced by ligation of death receptors, where the extent of caspase activation can dictate the likelihood that a cell will survive. This is, in at least some cases, dictated by the expression of caspase-inhibiting IAPs and the release of IAP antagonists (such as Smac and Omi) upon MOMP.

For cells to stay alive, cell survival must be stable. For cells to die, it must be possible to move between this stable state to the stable state of death. The transition (such as illustrated in Figure 1) is formally a transcritical bifurcation (Strogatz, 2001). Generally, systems displaying the property of transcritical bifurcation (sometimes referred to as "bistability") contain negative feedback loops to ensure the stability of each state (Pomerening, 2008). Such feedback is readily apparent, for example, in the function of transcriptional activators and repressors that function in the transcritical bifurcation operating in cell differentiation. For the two models shown in Figure 1, such feedback will function to sustain cell survival either upstream (Figure 1a) or downstream (Figure 1b) of MOMP.

An example of such negative feedback is seen in the case of p53 in cell death (and its other functions). The expression of p53 is constitutive, but it drives the expression of MDM2, an E3-ligase that functions to ubiquitinate p53 for its degradation (Puszynski, et al., 2008). Only when the MDM2-p53 interaction is disrupted, for example by the expression of ARF, phosphorylation of p53 in response to DNA damage, or by ribosomal proteins that become available under conditions of ribosomal dysfunction does p53 stabilize to perform its functions (Vousden and Prives, 2007). Once active, p53 can drive apoptosis via its induction of expression of the BH3-only proteins, PUMA and NOXA, and by direct interaction of cytosolic p53 with BAX to induce MOMP and apoptosis (Green and Kroemer, 2009). Thus, through feedback regulation the effects of p53 are stably controlled.

It is noteworthy, then, that the function of the BCL2 protein family in controlling MOMP does not seem to contain any such negative feedback regulation. There is no known process, for example, wherein an increase in a pro-apoptotic BCL2 family protein, such as BIM, triggers increased expression or functions of an anti-apoptotic BCL2 family protein, such as BCL-2, BCL-xl, or MCL-1. While such homeostatic mechanisms may yet be uncovered, the fact that they have not so far been described is intriguing, given the considerations that follow. Thus, while there are positive and negative effects of different BCL2 proteins on MOMP and apoptosis, there does not appear to be a signaling feedback mechanism that ensures that this system remains in check to preserve cell survival.

Whenever cells are induced to die by apoptosis, the process occurs in a manner that is stochastic, that is, one dying cell does not influence whether or not a neighboring cell will die. Indeed, this was one of the "hallmarks" of apoptosis as originally described (Kerr, et al., 1972). Even when a clonal population of cells is examined, irrespective of cell cycle synchronization or other manipulations, the probability that a given cell will die at a particular time is described by a bell-shaped distribution for the population (Goldstein, et al., 2000; Spencer, et al., 2009).

Further insights into this variation were obtained using live cell imaging techniques, observing MOMP, caspase activation, and cell death following an insult, in this case, exposure to TRAIL (Spencer, et al., 2009). While the time between MOMP and cell death was fairly constant, the time from initial exposure to MOMP varied considerably. Similar results had previously been obtained using other stimuli to induce MOMP and apoptosis (Goldstein, et al., 2000). Strikingly, however, if a cell divided, the two daughter cells showed a strong tendency to die at about the same time, that is, if a given daughter cell died shortly after cell division, there was a strong probability that its sister cell would also die within a given time window (Figure 2). This probability of coordinated cell death declined a few hours after cell division, until the probability that the sister cell would die was no different from that of any other cell in the culture (Spencer, et al., 2009). The effect is not related to cell cycle, per se, that is, synchronizing the cell cycle does not synchronize cell death in the population. Instead, this is an effect that can be observed when a cell divides, and the two daughters die in (relative) synchrony. This finding tells us that upon cell division, conditions in each cell are sufficiently similar that the response to the stimulus to induce apoptosis will be mirrored in each cell, but in time this tendency diffuses. The most likely explanation for this is that levels of proteins that dictate the response fluctuate such that within a few hours, the response to the apoptotic stimulus becomes again stochastic (that is, the timing of cell death for the two daughters is no more similar than that of any other cell in the population).

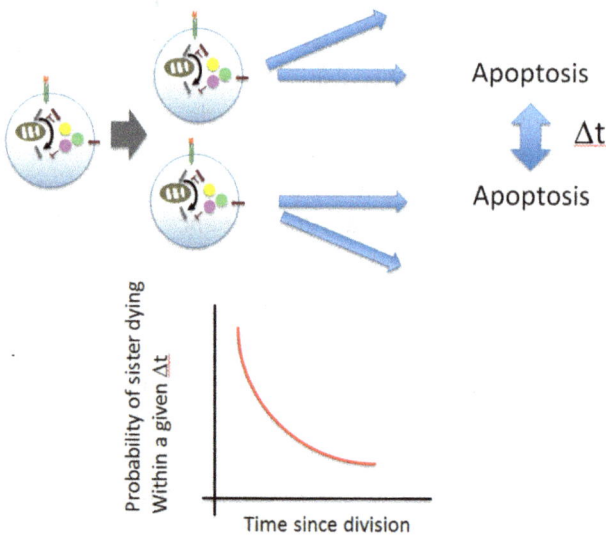

Figure 2. Sensitivity to TRAIL-induced death rapidly diverges in daughter cells post-division.

Cells exposed to TRAIL die at various rates. If a cell divides prior to death, the likelihood that the two daughters will die synchronously diverges with the time since division. From Spencer, et al., 2009.

This effect is non-genetic, that is, if we transiently treat cells with an agent that induces apoptosis and recover the surviving cells ("persisters"), expanding this population will yield the same response to the stimulus as was seen in the original treatment. Of course, long term treatment of a genetically variant population can select for mutant, resistant cells, but here we are concerned with the non-genetic variation that exists in any population of cells, even those that are genetically identical. As we will see, there are concepts that emerge from this simple idea.

Clearly, individual cells can vary in a number of ways. In the above experiments, the time frame of the effects strongly suggests that variation in protein levels and protein interactions/modifications, are the most relevant for determination of cell death events under these conditions. Of course, variations in gene expression controlled by epigenetics, or mutational variation in the genomes of cells in the population can certainly contribute to susceptibility or resistance to cell death, but this is unlikely to be relevant in short-term effects. Nevertheless, short term effects to stochastically preserve survival of some cells provides an opportunity for longer term influences (epigenetic changes, mutation) to occur, and be selected. Understanding how cells can persist in the face of an apoptotic signal may be important.

Random walks, Gambler's ruin, and cell survival
Clearly then, cells can exist in a range of signaling states by varying the levels of proteins. External signals perturb the overall signaling of the cells, but again, there is variability with respect to the average state, and the state of each cell changes fairly rapidly (within hours in the above experiment (Spencer, et al., 2009)). This variation is stochastic, and thus cells in a population perform a "random walk" from one signaling state to another, within constraints dictated by genetics and epigenetics. We can represent this as in Figure 3, where cell variation is random but constrained, but there exists a state where cell death can occur. Only those cells that "cross the line"

into this state die. But there is a problem with this random walk idea, a problem that is sometimes referred to as "Gambler's Ruin." If a gambler (the cell) is playing a fair game against a bank with infinite resources, the gambler will always lose, that is, the random walk will eventually bankrupt the gambler. In our scheme, given enough time, every cell must eventually leave the "alley of survival" and drop off the "cliff of death." Survival factors can tend to push cells deeper into the "alley", while pro-apoptotic factors (such as cell stress) can tend to push cells closer to the "cliff," but given that the range of variation includes cell death, all cells must eventually die. We can suppose that this is true (all cells do, eventually, either divide or die), and even cells that live a very long time without dividing eventually die. But how can this be structured so that survival is the standard condition, while the probability of cell death is kept extremely low until conditions favor such cell death? If we simply make the "alley of survival" extremely long, cell death will be rare (Figure 3).

Signaling states

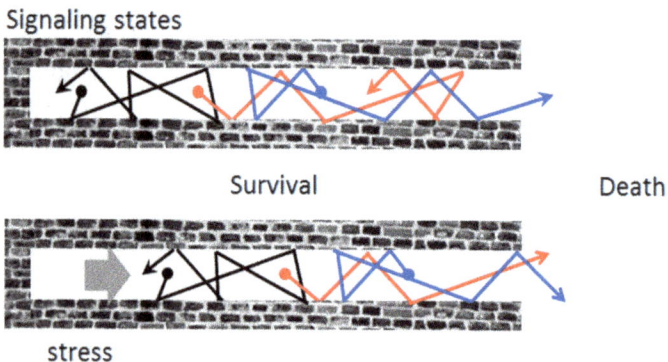

Survival   Death

stress

Figure 3. Random walks and cell death. Cells randomly change their signaling space over time (i.e., levels of proteins and interactions vary), in agreement with the relationships shown in Figure 2. The cells progress in time from the "ball" end to the "arrow" end of each individual track.

Therefore, without feedback in the system, changing the limits can sustain cell survival. As a concrete example, a cell population expressing high levels of BCL-xL persists despite changes in the levels of Bim. If a stress signal (DNA damage) engages a regulated process (stabilization of p53 to induce PUMA), the susceptibility to these changes shifts, and more (but not necessarily all) cells die. Similar effects may relate (at a non-transcriptional level) to activation of BID, changes in MCL-1 stability, and similar events.

Dangerous Adventures and Safe Havens

Rather than thinking of a one dimensional random walk, we can move this idea into multidimensional signaling space, imaging a "cloud" of possible states for individual cells in a population. Representing this in two dimensions (for convenience) the above considerations translate into the model shown in Figure 4. A pro-apoptotic signal moves the "cloud" so that some cells will be in the state representing cell death (from which they cannot return). If this is correct, then anything that increases the range of variation among individual cells will then lead to more cells dying in response to the pro-apoptotic signal. For example, if resting cells are driven to proliferate, it may be that variation will increase (as a consequence of increased translation, in general, e.g., by increasing ribosome biogenesis). This could lead to the idea that a cell in cycle is "more sensitive" to cell death; perhaps it may be more accurate to state that greater variability in a population caused by entry into cell cycle increases the probability that a cell will occupy the "cell death space" when the population is shifted towards cell death. In our scheme, resting and proliferating cells are well away from the cell death "cliff." The pro-apoptotic signal, then, launches the cells into a "dangerous adventure" where the population moves closer to the cell death "cliff," and some cells perish.

There is, however, at least one alternative to this view. Rather than consider the state of cell death as crossing a threshold (as in the Figures discussed above), we can consider cell death as a constrained set of signaling states (e.g., leading directly to the activation of BAX and/or BAK). In this scheme, cells are free to sample a wide variety of signaling states (prescribed by genetics, epigenetics, and environmental inputs) with only a very few representing that of cell death. Pro-death stimuli, in this setting, might "nudge" the cell in the direction of the death state (Figure 5). In this two-dimensional representation, it is easy to envision that any cell is likely to "wander" into the cell death state, and more will do so when "nudged" in that direction. But, of course, the signaling state of a cell is described by more than two variables. And as the number of relevant variables (e.g. levels of a particular protein or signaling event) increases, the probability that a cell can move into a particular state decreases. At three variables (dimensions) the probability of reaching a specific destination during a random walk falls from near certainty in two dimensions to about 34% (Weistein, undated). With 7 or 8 relevant variables/dimensions this falls to about 8% and 7%, respectively. This can increase as pro-apoptotic signals move cells toward the cell death state, but most cells will survive despite a random walk that can by chance enter the signaling space associated with cell death.

For the purposes of this discussion, the model outlined in Figure 4, involving a threshold beyond which cell death is engaged, will be referred to as the "dangerous adventure." The model in Figure 5, in which the cell death state in multidimensional space is more specific and therefore spontaneously engaged only rarely (provided the relevant variables are more than a few) is the "safe haven."

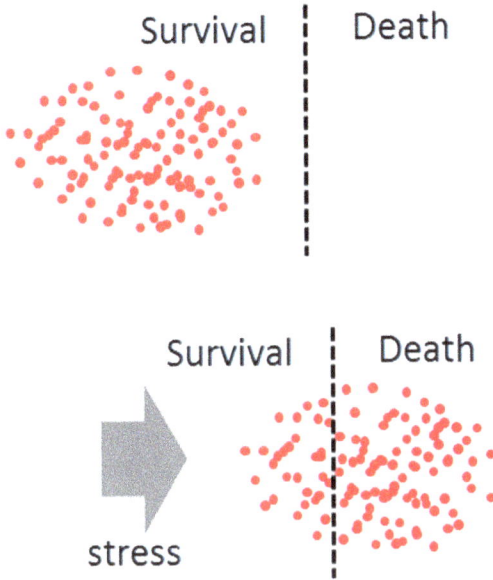

Figure 4. The Dangerous Adventure Model. The signaling space of an individual cell is represented in n-dimensional space. Stress drives the population towards the space representing cell death. This simple model (derived from Figure 3) predicts that all cells will change their signaling relationships upon exposure to a pro-apoptotic signal, such that some die and some survive. While intuitively consistent with the relationships between cell stress and cell death, examinations of signaling relationships within single cells in a population have not supported this model to date.

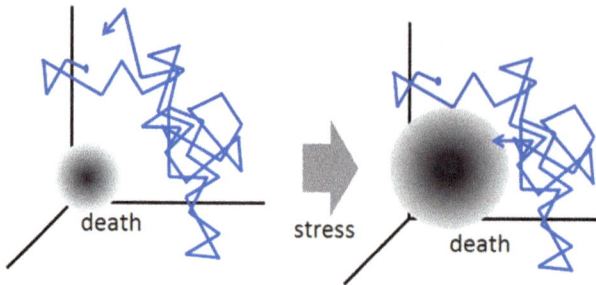

Figure 5. The Safe Haven Model. Individual cells in a population move randomly among different signaling states in n-dimensional space. If n is large (i.e., many independent signaling variables contribute to survival versus death), and if the state space describing cell death is constrained (left) then the probability that a cell will enter the signaling state associated with cell death is low. Pro-apoptotic signals increase the death-associated signaling space, increasing the probability that cells will enter that space and die. While counter-intuitive, this model may be consistent with observations made on signaling relationships in a population of cells exposed to a pro-apoptotic signal.

Can we distinguish between these models? One way to do so would be to determine the relative values of a number of (hopefully relevant) variables simultaneously in each individual cell in a population, and then determine how these change in response to a pro-death stimulus. Until fairly recently, this would be technically impossible (or at best, extremely difficult), but advances in single cell analysis using mass cytometry now allow the interrogation of multiple signaling events using metal-labeled antibodies, detected by time of flight mass spectroscopy of single cells (Bandura, et al., 2009; Bendall, et al., 2011). Indeed, we have embarked on such an exploration, using a panel of antibodies that detect phosphorylation events in a number of pathways, including MAP kinases, NF-κB, the PI3K-TORC1 pathway, BCL2 protein levels, apoptotic events (e.g., cleavage of executioner caspases and their substrates) and others. We then examined the effects of addition of a pro-apoptotic signal (TRAIL) and examined the effects on these signals.

While a detailed description of our unpublished results is not suitable here, one of our findings is particularly relevant to this discussion. We examined the signaling space occupied by those cells that did *not* engage apoptosis (the "persisters"), that is, the relationships between each variable and every other variable in our set, and found that *these did not change at any time during the pro-apoptotic treatment*, including time 0. In three distinct cell lines, the signaling space describing persistence, while different in each line (and highly reproducible for each cell line), did not "move" during the process of engagement of apoptosis.

These results, provided we have interrogated relevant variables (which we think is likely), help us discriminate between our models. If the "dangerous adventure" model were correct, we should have seen the persisters move towards the signaling space occupied by the dying cells (Figure 4). In contrast, our results are more consistent with our "safe haven" model.

If our analysis is correct, then, cells wander a landscape that includes cell death as a possibility, but because there are many relevant variables, the vast majority of cells do not enter the cell death state simply because of the high dimensionality of the signaling states. When a pro-apoptotic signal is present, the number of states describing persistence decreases, but those in that state continue to survive. Cell persistence is selected, not induced.

From Persistance to Resistance

When a population of proliferating cells is genomically unstable, such as in many cancers, the induction of apoptosis can select cells that are genetically resistant to cell death. There are two ways that this can happen. In one scenario, cells mutate and radiate from a founder cell, competing for suitable niches, and resistant mutants are selected from this diverse collection. This idea is supported by the genotypes of pediatric ALL before and after relapse; the relapsed cancer was often genetically distinct from that of the initial cancer (Notta, et al., 2011). Alternatively, a genetically resistant clone can emerge from the original cancer, as determined by the genotypes (Notta, et al., 2011). In this scenario, any cells that persist despite being genetically "susceptible" to this treatment can subsequently mutate, even as a consequence of the treatment itself (which is often mutagenic). In such settings, the principles underlying non-genetic persistence are clearly relevant (Brock, et al, 2009; Pisco, et al., 2013; Sharma, et al., 2010). If so, then the considerations here are important.

From the perspective we have developed, then, a goal of cancer therapy (for example) can be to sufficiently constrain the signaling state of the cells in a population to the point that no cells occupy the state that defines persistence. Thus, an agent (or, more likely, a set of agents) that might not kill cells itself, but only perturb the signaling state away from that of persistence, would then ensure that no cell survives the primary, pro-apoptotic insult. If we can identify all (or many) relevant variables that define the persistent state, it follows that we could ultimately determine, for each individual cancer, the set of perturbations (and the agents that will effect it) necessary to achieve this outcome.

Is this feasible? Using a completely different approach, cells can be analyzed for how "close" they are to the cell death state, that is, how "primed for death" they are. This is achieved by determining the effects of different BH3 peptides (or, probably, BH3 mimetic drugs) on mitochondrial outer membrane permeabilization in a cancer versus primary cells from the same individual. This "primed for death" status is remarkably predictive of the response to conventional therapies in a number of cancers (Ni Chonghaile, et al., 2011; Vo, et al., 2012). It follows, then, that if we can determine the signaling state, using mass cytometry, that corresponds to the "primed for death" status, we can refine the approach. Currently, no simple determination of levels of BCL2 proteins is predictive of whether a cancer is more "primed for death" than primary cells

(Vo, et al., 2012). But it is certainly likely that a deeper interrogation of the many variables dictating cell survival and death in cells will yield such a correlation, to that point that we can predict the state. Combining this with knowledge of persistence and how it can be constrained would then optimize treatments.

We are far from achieving this goal in practice, but the principles are becoming clear, and as our understanding improves, we can envision a day where interrogation of a cancer in terms of both its genetic and non-genetic variation will dictate effective therapy. It's worth thinking about.

Nearly every study of cell death focuses on the cells that die and how they do so. Perhaps something more can be learned from the cells that live. We think so.

Acknowledgements

The authors gratefully acknowledge the wealth of knowledge provided over the past 25 years to the community by the pioneering studies a many investigators studying the process of apoptosis. In our referencing of our statements made in this essay, we have failed to cite the vast majority of these, for which we apologize. Our studies are supported by the US National Institutes of Health, and by the American Lebanese Syrian Associated Charities.

References

Bandura, D.R., Baranov, V.I., Ornatsky, O.I., Antonov, A., Kinach, R., Lou, X., Pavlov, S., Vorobiev, S., Dick, J.E., and Tanner, S.D. (2009). Mass cytometry: technique for real time single cell multitarget immunoassay based on inductively coupled plasma time-of-flight mass spectrometry. Analytical chemistry 81, 6813-6822.

Bendall, S.C., Simonds, E.F., Qiu, P., Amir el, A.D., Krutzik, P.O., Finck, R., Bruggner, R.V., Melamed, R., Trejo, A., Ornatsky, O.I., et al. (2011). Single-cell mass cytometry of differential immune and drug responses across a human hematopoietic continuum. Science 332, 687-696.

Bender, C.E., Fitzgerald, P., Tait, S.W., Llambi, F., McStay, G.P., Tupper, D.O.,

Pellettieri, J., Sanchez Alvarado, A., Salvesen, G.S., and Green, D.R. (2012). Mitochondrial pathway of apoptosis is ancestral in metazoans. Proceedings of the National Academy of Sciences of the United States of America 109, 4904-4909.

Brock, A., Chang, H., and Huang, S. (2009). Non-genetic heterogeneity —a mutation- independent driving force for the somatic evolution of tumours. Nature Reviews Genetics 10, 336–342.

Chipuk, J.E., and Green, D.R. (2005). Do inducers of apoptosis trigger caspase-independent cell death? Nature reviews. Molecular cell biology 6, 268-275.

Chipuk, J.E., Moldoveanu, T., Llambi, F., Parsons, M.J., and Green, D.R. (2010). The BCL-2 family reunion. Molecular cell 37, 299-310.

Colell, A., Ricci, J.E., Tait, S., Milasta, S., Maurer, U., Bouchier-Hayes, L., Fitzgerald, P., Guio-Carrion, A., Waterhouse, N.J., Li, C.W., et al. (2007). GAPDH and autophagy preserve survival after apoptotic cytochrome c release in the absence of caspase activation. Cell 129, 983-997.

Goldstein, J.C., Kluck, R.M., and Green, D.R. (2000). A single cell analysis of apoptosis. Ordering the apoptotic phenotype. Annals of the New York Academy of Sciences 926, 132-141.

Green, D.R. (2011) *Means to an end: Apoptosis and other cell death mechanisms.* Cold Spring Harbor Press.

Green, D.R., and Kroemer, G. (2009). Cytoplasmic functions of the tumour suppressor p53. Nature 458, 1127-1130.

Kerr, J.F., Wyllie, A.H., and Currie, A.R. (1972). Apoptosis: a basic biological phenomenon with wide-ranging implications in tissue kinetics. British journal of cancer 26, 239-257.

Lartigue, L., Kushnareva, Y., Seong, Y., Lin, H., Faustin, B., and Newmeyer, D.D. (2009). Caspase-independent mitochondrial cell death results from loss of respiration, not cytotoxic protein release. Molecular biology of the cell 20, 4871-4884.

Ni Chonghaile, T., Sarosiek, K.A., Vo, T.T., Ryan, J.A., Tammareddi, A., Moore Vdel, G., Deng, J., Anderson, K.C., Richardson, P., Tai, Y.T., et al. (2011). Pretreatment mitochondrial priming correlates with clinical response to cytotoxic chemotherapy. Science 334, 1129-1133.

Nightingale, D.F. (1998). *Games, Gods, and Gambling: A History of Probability and Statistical Ideas.* Courier Dover Publications.

Notta, F., Mullighan, C.G., Wang, J.C., Poeppl, A., Doulatov, S., Phillips, L.A., Ma, J., Minden, M.D., Downing, J.R., and Dick, J.E. (2011). Evolution of human BCR-ABL1 lymphoblastic leukaemia-initiating cells. Nature 469, 362-367.

Oberst, A., Bender, C., and Green, D.R. (2008). Living with death: the evolution of the mitochondrial pathway of apoptosis in animals. Cell death and differentiation 15, 1139-1146.

Pearson, K. (1905) The problem of the random walk. Nature, 72: 294.

Pisco, A.O., Brock, A., Zhou, J., Moor, A., Mojtahedi, M., Jackson, D., and Huang, S. (2013). Non-Darwinian dynamics in therapy-induced cancer drug resistance. Nature Communications 4, 1–11.

Pomerening, J.R. (2008). Uncovering mechanisms of bistability in biological systems. Current opinion in biotechnology 19, 381-388.

Puszynski, K., Hat, B., and Lipniacki, T. (2008). Oscillations and bistability in the stochastic model of p53 regulation. Journal of theoretical biology 254, 452-465.

Sharma, S.V., Lee, D.Y., Li, B., Quinlan, M.P., Takahashi, F., Maheswaran, S., McDermott, U., Azizian, N., Zou, L., Fischbach, M.A., et al. (2010). A chromatin-mediated reversible drug-tolerant state in cancer cell subpopulations. Cell 141, 69–80.

Spencer, S.L., Gaudet, S., Albeck, J.G., Burke, J.M., and Sorger, P.K. (2009). Non-genetic origins of cell-to-cell variability in TRAIL-induced apoptosis. Nature 459, 428-432.

Strogatz, Steven (2001). Nonlinear dynamics and chaos: with applications to physics, biology, chemistry, and engineering. Boulder: Westview Press.

Tait, S.W., and Green, D.R. (2010). Mitochondria and cell death: outer membrane permeabilization and beyond. Nature reviews. Molecular cell biology 11, 621-632.

Tait, S.W., Parsons, M.J., Llambi, F., Bouchier-Hayes, L., Connell, S., Munoz-Pinedo, C., and Green, D.R. (2010). Resistance to caspase-independent cell death requires persistence of intact mitochondria. Developmental cell 18, 802-813.

Vo, T.T., Ryan, J., Carrasco, R., Neuberg, D., Rossi, D.J., Stone, R.M., Deangelo, D.J., Frattini, M.G., and Letai, A. (2012). Relative mitochondrial priming of myeloblasts and normal HSCs determines chemotherapeutic success in AML. Cell 151, 344-355.

Vousden, K.H., and Prives, C. (2009). Blinded by the Light: The Growing Complexity of p53. Cell 137, 413-431.

Weisstein, Eric W. (undated) "Pólya's Random Walk Constants." From MathWorld--A Wolfram Web Resource. http://mathworld.wolfram.com/PolyasRandomWalkConstants.html

PART 3: AUTOPHAGY AND NECROPTOSIS

Chapter 11. Apoptosis and Autophagy *face to face*: Apaf1 and Ambra1 as a paradigm (De Zio and Cecconi)

Daniela De Zio[1,2] and Francesco Cecconi[1,2]

[1] Unit of Cell Stress and Survival, Danish Cancer Society Research Center, Strandboulevarden 49, Copenhagen DK-2100, Denmark

[2]Department of Biology, University of Rome "Tor Vergata", Via della Ricerca Scientifica, Rome 00133, Italy

Abstract

During the 1990s, knock-out mice models in biomedical research finally took off. Knock-out genes made it possible to understand their functions simply by looking at the phenotype of the targeted knock-out animals. In the last 20 years, these models, along with the more sophisticated *conditional* knock-out mice, have been playing a valuable and indispensable role in unravelling the main players and characteristics of two crucial cellular processes: "apoptosis" and "autophagy". Apart from providing general information, in this chapter we aim to retell the story of how everything has started and how we have contributed to characterizing the intricate signaling network underlying both processes, providing an overview and a personal interpretation of how apoptosis and autophagy are intertwined.

Introduction

Apoptosis, a striking name coined to identify the programmed suicide of the cell, is a form of cellular death that can occur in all tissues in order to eliminate non-necessary cells, during both development and adulthood (Fuchs and Steller, 2011). Since this represents a lifeline for the entire organism, apoptosis can be considered a "social" feature of the cells, since its role concurs with the maintenance of tissue homeostasis by mediating the removal of the unwanted cells without affecting the surrounding healthy ones.

While apoptosis is certainly a mechanism of cell death, autophagy has been proposed as a mechanism of both survival and death (namely, type II cell death), although autophagy's survival role remains the most remarkable (Rubinstein and Kimchi, 2012). Autophagy is a lysosome-dependent degradative process activated under conditions of nutrient withdrawal (Parzych and Klionsky, 2014). It promotes the degradation and the recycling of damaged organelles and macromolecules and, in doing so, sustains the cell in overcoming stressful conditions. From this point of view, autophagy is a more "individualistic" process the main role of which is to maintain homeostasis of each single cell. The multiple interconnections among the diverse molecular pathways culminating in apoptosis and autophagy support the idea that autophagy is usually activated as an early response in order to overcome stress conditions, with any failure leading to cell death. In this chapter, we will attempt to describe how apoptosis and autophagy crosstalk to dictate cell fate, focusing on two key proteins, Apaf1 and Ambra1, previously found to be crucial molecular players in orchestrating both these pathways.

Apaf1 knock-out mouse: a milestone in the history of Apoptosis

The advent of the gene targeting and gene trap technologies developed in the early '90s led to an insight into the factors involved in the regulation of apoptosis. The multimolecular machinery "apoptosome", which was described for the first time in 1997 (Li et al., 1997; Zou et al., 1997), is a case in point. It is now commonly accepted that the apoptosome is a huge complex composed by the Apoptotic protease activating factor 1 (Apaf1), cytochrome c and pro-Caspase-9 (Bratton et al., 2001; Li et al., 1997; Zou et al., 1999). From a mechanistic point of view, the full activation of the apoptosome takes place when seven Apaf1 monomers stoichiometrically bind to cytochrome c released from mitochondria and, then, interact with adenine nucleotides (dATP or ATP) and pro-Caspase 9 through the Nucleotide Binding Domain (NBD) and the Caspase Recruitment Domain (CARD), respectively. The resulting multi-molecular machinery is then composed by a ring-like platform, this being required for the activation of the effector caspases (Caspase-3, -6 and -7) to execute cell death (Yuan et al., 2013).

In those years, we were dealing with the gene trap approach to generate a library of mutant mice displaying defects in normal development. The gene trap strategy is based on the integration of specific vectors (containing β-galactosidase and neomycin genes) into a genomic locus downstream of a functional promoter (Cecconi and Meyer, 2000). The tagged gene is, therefore, entrapped by this vector and turns out to be no longer active. As integration occurs randomly, the tagged gene can be recognized by the use of the anchored PCR procedure. During the screening of gene trapped mice, we isolated a null allele of the murine Apaf1 gene, the same Apaf1 that had been just discovered to be involved in apoptosis regulation by forming the apoptosome in vitro (Li et al., 1997). We were amazed to observe that deficiency of this gene led to embryonic lethality around embryonic stage 16.5 (e16.5); even more surprising was the phenotype affecting many aspects of developmental apoptosis. The most striking alterations observed were brain hyperplasia in the diencephalon and midbrain; neural tube closure defects (e.g. spina bifida); severe craniofacial malformations; alterations of the retina, lens and vascular system; persistence of interdigital webs (Cecconi et al., 1998) [partly shown in Fig. 11.1, panels b and e, (Cecconi et al., 2008)]. We realized that Apaf1 was involved in: i) histogenetic cell death because of cell number control in the developing retina and brain; ii) morphogenetic cell death in the neural tube, lens, skull, face, and limbs; iii) phylogenetic cell death as it affected the elimination of the hyaloid artery system in the developing eye. Moreover, we provided the first in vivo evidence that Apaf1 was required for activating Caspase-3, confirming that Apaf1 and Caspase-3 were components of the same apoptotic pathway during brain development. This finding was consistent with the results that Zou and colleagues obtained one year before demonstrating, at a molecular level, the functional link between Apaf1 and Caspase-3 (Zou et al., 1997).

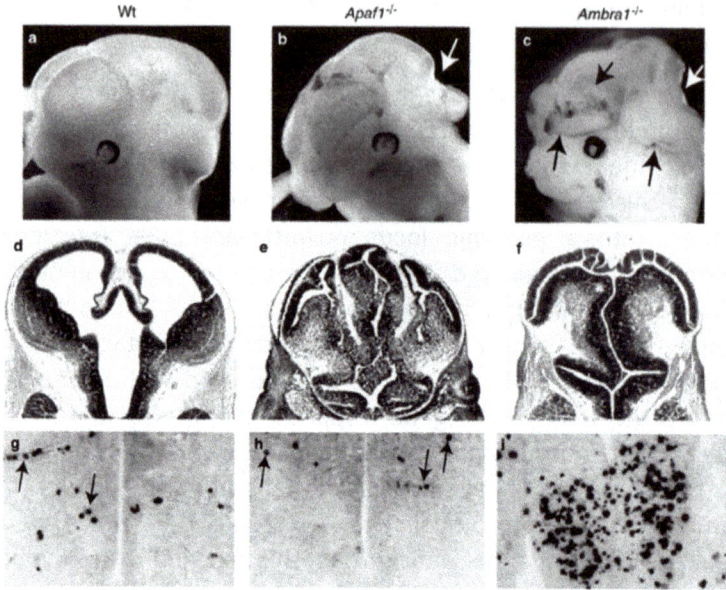

Figure 11.1. *Apaf1* and *Ambra1* knock-out mice *face to face*.

A comparison between the phenotypes of *Apaf1* and *Ambra1* knock-out mice is shown [excerpts from (Cecconi et al., 2008)]. In panels **a-c**, e12 embryo heads of *Apaf1* and *Ambra1* knock-outs (*Apaf1* $^{-/-}$ and *Ambra1* $^{-/-}$, **b** and **c**) are displayed *versus* wild-type (Wt) counterpart (**a**). Haematoxylin-eosin sections from the same heads are shown in panels **d-f** and highlight neuroepithelium over-proliferation due to *Apaf1* and *Ambra1* deficiency (**e** and **f**). TUNEL assay shows different level of apoptotic cells in the region of the rostral spinal cord of *Apaf1* (**h**) and *Ambra1* (**i**) knock-out e12 embryos compared with Wt counterpart (**g**). For all panels reprinted we acknowledge Nature Publishing Group: Cell Death & Differentiation, Cecconi et al., *15*, 1170-1177, copyright 2008.

The pieces of the puzzle were all fitting into place: Apaf1-ablated mice recapitulated the neuronal phenotype already observed in Caspase-9 and, less severely, in Caspase-3 knock-out animals, where the decreased neuronal apoptosis led to neurodevelopmental abnormalities including an expanded ventricular zone, ectopic and duplicated neuronal structures, and gross brain malformations (Hakem et al., 1998; Kuida et al., 1996; Kuida et al., 1998). The evidence that Apaf1 gene trap phenocopied Caspase-9/-3 knock-out mice was the final proof that the three genes lie on the same molecular pathway, with the common phenotype being the result of a disproportionate number of neural precursor cells (NPCs), which fail to correctly die, but keep their normal rate of growth and differentiation. Our findings were published in 1998 and were soon supported by the results obtained studying another Apaf1 mutant mouse generated through gene targeting technology and published few days later by Yoshida and colleagues (Yoshida et al., 1998). In this model, the targeted disruption of the Apaf1 locus was mediated by the insertion of the neomycin cassette within the NBD region of Apaf1 with the same resulting inactivation of Apaf1 gene. Brain defects and craniofacial abnormalities observedin Apaf1 knock-out embryos were analogous to those observed in our mouse, reinforcing our findings and the biological role of Apaf1. The similar phenotypes displayed by the two Apaf1 knock-out mouse models and the similarities with Caspase-9 and -3 knock-outs in terms of brain development represented an important proof of the role of these molecules in the regulation of the intrinsic (mitochondrial) apoptotic pathway during mammalian development.

Elucidation seemed to be complete. The central role of the apoptosome in tuning apoptosis, as well as its implication in embryonic development, had at last been unveiled. From that time on, most studies focused on the comprehension of Apaf1 as an etiopathogenetic determinant of several diseases (namely those in which apoptosis was excessively activated, such as neurodegeneration) and, in turn, on the possible manipulation of its pro-apoptotic activity. In this context, we generated an immortalized NPCs line, coming from the e14.5 telencephalon of $Apaf1^{-/-}$ mice, which we called ETNA (Embryonic Telencephalic Naïve Apaf1). As expected, $Apaf1^{-/-}$ ETNA cells failed to undergo apoptosis and were resistant to various neurodegenerative stimuli, such as the administration of α-amyloid or the overexpression of the G93A mutant isoform of SOD1 (Cozzolino et al., 2004), used as *in vitro* models of Alzeheimer's disease and amyotrophic lateral sclerosis, respectively. This was a crucial finding, as it provided proof of concept for exploring potential novel therapies targeting Apaf1 for the treatment of human neurodegeneration; at the same time, it confirmed that the intrinsic pathway of apoptosis was the elective process underlying neuronal cell death observed in several neurological disorders. This assumption was strengthened and expanded few years later, when we discovered that Caspase-3 was activated very early on, exclusively at the level of dendritic spines in still asymptomatic young mice developing Alzheimer's disease. This finding was a milestone in neurological research. Not only did it demonstrate that a functional apoptosome assembly was not only required for the complete loss of specific neuronal populations *via* apoptosis; it also provided evidence that its "local" and "sub-apoptotic" induction finely contributed to all those synaptic degenerations responsible for the loss of function of neuronal transmission, and preparatory to the pathology's clinical onset (D'Amelio et al., 2011).

In those years, there was significant scientific progress concerning possible clinical applications of Apaf1-targeting therapies. However, the same trend was not supported by basic research. As an exercise, the investigation of the molecular factors/signals downstream of Apaf1 and responsible for the neurodevelopmental aberrations observed in *Apaf1* deficient mice fell into neglect. Only 15 years later did two different papers study these issues in depth, by revisiting and elucidating Apaf1's in regulating apoptosis. It was in 2013 that Long and colleagues discovered that a novel loss-of-function allele of *Apaf1* (named *yautja*), which disrupts nervous system and craniofacial development, thus resembling the previously characterized *Apaf1* deletion alleles (Long et al., 2013). Such a similarity depended on the fact that, although the *yautja* mutant *Apaf1* (harboring Leu375Pro point mutation) was produced at normal levels during embryonic development, it was, substantially, a functionally-null protein. Interestingly, the absence of any apoptotic function linked to the *yautja* mutant *Apaf1* was found to be closely associated with the persistent activation of the Sonic hedgehog (Shh) signaling pathway. Shh is one of the best characterized morphogens regulating vertebrate organogenesis. Therefore, craniofacial defects described in *yautja Apaf1* mice underlined the crucial role of Apaf1 in morphogenetic cell death. Along these lines, a few months later, Nonomura and colleagues elegantly demonstrated that Apaf1-mediated apoptosis was required to selectively eliminate FGF8 morphogen-expressing cells in the anterior neural ridge (ANR), which acts as an organizing center of the forebrain during early embryonic development (Nonomura et al., 2013). Using a highly sophisticated live-imaging technique, they clearly showed that apoptosis deficiency, due to *Apaf1* or *Caspase-9* gene inactivation, led to the survival of undead and non-proliferative cells in the ANR, which propagated the FGF8 signaling and determined the improper forebrain development. This work finally closed the circle, so suggesting that apoptosis temporally controls the proliferation of neural progenitors in the ANR area and, in turn, is able to modulate early mammalian brain development. However, an important question still remains unsolved: Are the over-proliferating NPCs in *Apaf1 gene trap* embryos the same cells that escaped death because of the lack of Apaf1? Or are the cells escaping death the ones that stimulate the proliferation of neural progenitors? Although the difference could seem somewhat specious at a glance, the underlying biological significance is pivotal to understanding whether the Apaf1-mediated apoptosis is, respectively, cell-autonomous or cell-non-autonomous. Currently, we have, as yet, no satisfactory answers. However, we have accumulated evidence arguing that, in $Apaf1^{-/-}$ embryos, NPCs escape programmed cell death, proliferate and retain their potential to differentiate (Cozzolino et al., 2004). Moreover, the hypomorphic Apaf1 mutant *fog*(forebrain over-growth), in which Apaf1 expression is strongly reduced but not fully compromised, clearly indicates that Apaf1 is involved in brain-mass control in a dose-dependent manner (Moreno et al., 2002). Despite this increasing amount of evidence, we have not yet been able to come to a unequivocal conclusion as regards the cell-dependence of the Apaf-1-mediated neurodevelopment. Therefore, over those years, we took the line that the best solution for unravelling this issue was to selectively inactivate *Apaf1* in different neuronal populations (neuronal progenitors and post-mitotic neurons), according to the various stages of brain development. Accordingly, we sought to generate an *Apaf1* conditional knock-out mouse which displayed *Apaf1* inactivation in a time and tissue-dependent manner by the integration of Cre-recombination sites within *Apaf1* gene. Indeed, Cre-recombinase expression allows the selective inactivation of *Apaf1* gene only in the specific Cre-bearing neuronal population. Despite the efforts made, the high recombinogenic nature of *Apaf1 locus* has not hitherto allowed the generation of any conditional model. At least with the techniques for genetic manipulation currently in use, the fine dissection of apoptosis in neurodevelopment remains elusive, so posing a challenge for coming generations.

A proverb has it that "For everything that comes to an end, there is always something new which begins". Indeed, while attempting to clarify the molecular mechanisms allowing *Apaf1$^{-/-}$* ETNA cells to survive under stressful conditions, we came upon evidence that apoptosome-deficient cells kept generating ATP by glycolysis and activated a Beclin1-dependent autophagy pathway to sustain ATP production (Ferraro et al., 2008). Thus, the capability of Apaf1-deficient cells to proliferate seemed to be guaranteed by an autophagy-dependent mechanism. This discovery occurred exactly when we were embarking on the autophagy field and adjusting our interests towards a new direction.

Ambra1 knock-out mouse: an advance in the world of Autophagy

Among *gene trap* mutants we collected in the late '90s, we were particularly attracted to one showing neuronal phenotypes surprisingly similar to the *Apaf1* knock-out: neuroepithelium over-proliferation, mid and hindbrain exencephaly, neural tube closure defects and embryonic lethality around e16.5 [partly shown in Fig. x.1, panels a-f, (Fimia et al., 2007; Cecconi et al., 2008)]. In detail, we had found that, at early stages during development, the expression of this unknown gene was restricted to the nervous system, becoming, at later developmental stages, abundant throughout the developing nervous system, as well as in other tissues. The striking difference with $Apaf1^{-/-}$ was the high level of apoptosis detectable by TUNEL assay in embryo sections, a finding that was completely unexpected [Fig. 11.1, panels g-i, (Fimia et al., 2007; Cecconi et al., 2008)]. What was this mysterious gene which, once mutated, concomitantly displayed enhanced cell proliferation and death?

In fact, after a long period of brainstorming to decipher the gene's role, the breakthrough came through studying its possible interactors by the yeast two-hybrid approach. We were at once surprised and enthusiastic to find among its molecular partners Beclin 1, which had been previously described as a master regulator of autophagy in mammals (Liang et al., 1999). Indeed, Beclin 1 had been discovered to be a component of the class III phosphatidylinositol-3-kinase (also known as Vps34) complex, regulating the formation of the autophagosome, a double-membraned vesicle required for the occurrence of autophagy (Kihara et al., 2001; Liang et al., 1999). Moreover, a fundamental role for Beclin 1 had been previously shown in the regulation of tumor growth; also, the hemizygosity of the *BECN1* allele had been associated with human breast and ovarian cancers (Liang et al., 1999). We realized that autophagy was the keyword, and that we were on the right path, albeit many questions were still awaiting an answer.

For example, what was the functional significance of this crosstalk? And what was exactly the role of autophagy in neurodevelopment? We had long been studying the programmed death of the cell as a well-defined process regulating embryo development or involved in cell response to stress. Therefore, we focused our attention on another process, autophagy, which was also emerging as being involved in both phenomena. In those years, autophagy had been already described as a lysosome-dependent degradative process underlying cell response induced by the absence of nutrients (Dunn, 1994; Klionsky and Emr, 2000). Also, there was already evidence arguing for its involvement in human diseases, such as Parkinson's disease (Anglade et al., 1997). Interestingly, autophagy had been discovered to be implicated in survival after the early neonatal starvation period: Atg5 (Autophagy related gene 5) knock-out mice, although they survived embryogenesis, died soon after birth because of an incapacity to feed and overcome the critical starvation perinatal period (Kuma et al., 2004). Autophagy had been suggested as also playing a role in the clearance of apoptotic cells during embryonic development (Qu et al., 2007). Two different groups in 2006 found that basal autophagy was a homeostatic process needed for constitutive removal of normal soluble proteins from neural cells (Hara et al., 2006; Komatsu et al., 2006). In fact, the evidence for defective autophagy observed in Atg5 and Atg7 conditional knock-out mice (Hara et al., 2006; Komatsu et al., 2006), starting from e15.5, had been associated with the progressive and lethal accumulation of ubiquitylated proteins (hallmark of defective *basal* autophagy), and characterized as an early symptom of neurodegeneration. This suggested that autophagy played a key role in maintaining the survival of neural cells.

Once we found that knock-out embryos for this mysterious gene showed impaired autophagy and accumulation of ubiquitylated proteins, we went on to confirm that the interaction with Beclin 1 was functional and provided the first evidence about this gene's role of in regulation of autophagy

(Fimia et al., 2007). Eventually, we decided to name the new gene Ambra1, which stood for Activating molecule in beclin1 regulated autophagy 1 (Fimia et al., 2007). We were convinced that we had identified a new and essential element regulating the autophagy program and deeply involved in correct brain development. Indeed, we demonstrated that the abnormal cell proliferation and the subsequent accumulation of apoptotic cells observed in *Ambra1* knock-out embryos were both caused by the dysregulation of Beclin 1-dependent autophagy. Therefore, the absence of Ambra1 and its interaction with Beclin 1 resulted in impaired formation of the autophagosome and, thus, in a lack of basal autophagy in neurodevelopment. The evidence that this phenomenon was associated with an excess of apoptotic cells argued for the existence of a strict relationship between autophagy, apoptosis and cell proliferation (Cecconi et al., 2007).

Subsequent characterization studies have indicated that Ambra1, through its binding to Beclin 1, favors Beclin 1-VPS34 interaction. In particular, we discovered that the Ambra1/Beclin 1 complex is recruited under normal conditions at the cytoskeleton through direct binding to the dynein light chains 1/2. By contrast, when autophagy is induced by environmental stimuli, Ulk1 (UNC51-like kinase 1) promotes the dissociation of Ambra1/Beclin 1 from the dynein complex by catalyzing Ambra1 phosphorylation, thus enabling the execution of the autophagic program (Di Bartolomeo et al., 2010). By studying in depth the molecular mechanisms governing Ambra1's role in autophagy, we gradually realized that, besides its primary role as a Beclin 1 interactor, Ambra1 is a multi-functional player in autophagy, able to interact with several other proteins and to coordinate their activity in different ways. For instance, we have recently demonstrated that Ambra1 is not only a target of Ulk1, but can also regulate Ulk1 stability by mediating its regulative ubiquitylation through TRAF6 (tumor necrosis factor receptor–associated factor 6) (Nazio et al., 2013). This suggests that Ambra1 can regulate its activation state by means of positive feedback mechanisms. Coherently, under nutrient-rich conditions, mTORC1 (mammalian target of rapamycin complex 1), which is the major nutrient sensor inhibiting autophagy, specifically phosphorylates Ambra1 and blocks autophagy, this being a further means through which mTORC1 exerts its anti-autophagic function (Nazio et al., 2013). Altogether, these findings have contributed to providing important molecular details clarifying part of the cellular signalling network that finely and redundantly controls autophagy.

Although our studies were unveiling, step by step, Ambra1's role in autophagy, from the time of our very first studies on Ambra1 mice brains we had been unable to explain how Ambra 1 could directly affect cell proliferation. The answer to this question emerged very recently, when we discovered evidence that, through the physical interaction with the protein phosphatase PP2A, Ambra1 promotes c-Myc de-phosphorylation and inhibits cell cycle (Cianfanelli et al., 2014). Such a functional relationship between Ambra1 and c-Myc also explains the spontaneous insurgence of tumors found in *Ambra1* heterozygous mice. *Ambra1* mutations associated to some kind of human cancers finally confirmed Ambra1's crucial role in regulating cell proliferation, the implication being that this new protein is an intriguing scaffold protein able to finely modulate and integrate many cellular processes, *e.g.* autophagy *versus* cell cycle. On the basis of this assumption, Ambra1 could likely represent a key target for therapies aimed at counteracting pathologies associated to impaired autophagy or defective apoptosis, such as neurodegeneration and cancer.

Autophagy and Apoptosis crosstalk: could Ambra1 be a linking player?

Although Apaf1 and Ambra1 represent two crucial proteins in two opposite cellular processes, autophagy (survival) *versus* apoptosis (death), comparison of *Apaf1* and *Ambra1* knock-out phenotypes displays numerous similarities (Fig. 11.1, panels a-f) (Cecconi et al., 2008). Apoptosis serves to eliminate excess cells by means of a negative selection of proliferating NPCs, which are required for the correct brain development. As a result,

Apaf1$^{-/-}$ mice show exencephaly and brain hyperplasia. By contrast, autophagy is basically needed for cell renewal during neural differentiation and cell cycle control. Therefore, in *Ambra1*$^{-/-}$ mice, the lack of autophagy directly affects NPCs proliferation, and cell death is over-induced as a compensatory mechanism. Nevertheless, the final result is, however surprisingly, the same. Actually, under stressful conditions, autophagy and apoptosis can be contextually activated, with the resulting outcome changing on the basis of the cellular context or the stimulus applied, but primarily depending on the complex crosstalk between these pathways.

How do apoptosis and autophagy interplay in neurodevelopment or in other physiological contexts? Which are the key molecule(s) bridging these two processes? In this regard, we have recently accumulated compelling evidence arguing for Ambra1 being a molecular player integrating autophagic and apoptotic response in cells. In particular, we discovered that Ambra1 specifically interacts with Bcl2, the anti-apoptotic protein which negatively controls the mitochondrial release of cytochrome *c*, thereby interfering with the early activation steps of the intrinsic pathway of apoptosis (Youle and Strasser, 2008; Strappazzon et al., 2011). Intriguingly, we discovered that this interaction results in the inhibition of autophagy, since Ambra1/Bcl2 binding at the mitochondria sequesters Ambra1 and impairs its autophagic activity in normal conditions. Conversely, upon autophagic stimuli, Ambra1 dissociates from Bcl2, while it increases its affinity for Beclin1; this supports the hypothesis that Ambra1 integrates autophagic and apoptotic signaling by shuttling between Bcl2 and Beclin1. However, Ambra1 and Bcl2 interaction is also disrupted upon apoptotic stimuli; in support of t his argument, it has been found that Ambra1 is digested by caspases and calpains and fully degraded during apoptosis's early phases (Pagliarini et al., 2012), thus indirectly contributing to inhibition of autophagy. This integration of opposite signals (apoptosis *versus* autophagy) regulated by Ambra1 is fundamental to harmonizing cell response. In other words, Ambra1 works to modulate cell survival and death in a way that ensures a functional cellular outcome, whereby degradation of Ambra1 by caspases and calpains becomes crucial for the suppression of the cytoprotective autophagy activated as a defensive response following stress-induced cell death. Although many "side", but crucial, aspects of Ambra1's regulatory role in cellular processes dealing with cell death and survival have emerged in recent years, whether Ambra1 protein degradation is functional in triggering death of neuronal cells during neurodevelopment is a question that we are still seeking to answer. What we can state categorically is that Ambra1 represents an important link between autophagy and apoptosis, it being an example of how the cell can take advantage of utilizing an autophagy protein to regulate apoptosis and *vice-versa*.

References

Anglade, P., Vyas, S., Javoy-Agid, F., Herrero, M.T., Michel, P.P., Marquez, J., Mouatt-Prigent, A., Ruberg, M., Hirsch, E.C., and Agid, Y. (1997). Apoptosis and autophagy in nigral neurons of patients with Parkinson's disease. Histology and histopathology *12*, 25-31.

Bratton, S.B., Walker, G., Srinivasula, S.M., Sun, X.M., Butterworth, M., Alnemri, E.S., and Cohen, G.M. (2001). Recruitment, activation and retention of caspases-9 and -3 by Apaf-1 apoptosome and associated XIAP complexes. The EMBO journal *20*, 998-1009.

Cecconi, F., Alvarez-Bolado, G., Meyer, B.I., Roth, K.A., and Gruss, P. (1998). Apaf1 (CED-4 homolog) regulates programmed cell death in mammalian development. Cell *94*, 727-737.

Cecconi, F., Di Bartolomeo, S., Nardacci, R., Fuoco, C., Corazzari, M., Giunta, L., Romagnoli, A., Stoykova, A., Chowdhury, K., Fimia, G.M., *et al.* (2007). A novel role for autophagy in neurodevelopment. Autophagy *3*, 506-508.

Cecconi, F., and Meyer, B.I. (2000). Gene trap: a way to identify novel genes and unravel their biological function. FEBS letters *480*, 63-71.

Cecconi, F., Piacentini, M., and Fimia, G.M. (2008). The involvement of cell death and survival in neural tube defects: a distinct role for apoptosis and autophagy? Cell death and differentiation *15*, 1170-1177.

Cianfanelli, V., Fuoco, C., Lorente, M., Salazar, M., Quondamatteo, F., Gherardini, P.F., De Zio, D., Nazio, F., Antonioli, M., D'Orazio, M., *et al.* (2014).

AMBRA1 links autophagy to cell proliferation and tumorigenesis by promoting c-MYC dephosphorylation and degradation. Nature cell biology *in press*.

Cozzolino, M., Ferraro, E., Ferri, A., Rigamonti, D., Quondamatteo, F., Ding, H., Xu, Z.S., Ferrari, F., Angelini, D.F., Rotilio, G., *et al.* (2004). Apoptosome inactivation rescues proneural and neural cells from neurodegeneration. Cell death and differentiation *11*, 1179-1191.

D'Amelio, M., Cavallucci, V., Middei, S., Marchetti, C., Pacioni, S., Ferri, A., Diamantini, A., De Zio, D., Carrara, P., Battistini, L., *et al.* (2011). Caspase-3 triggers early synaptic dysfunction in a mouse model of Alzheimer's disease. Nature neuroscience *14*, 69-76.

Di Bartolomeo, S., Corazzari, M., Nazio, F., Oliverio, S., Lisi, G., Antonioli, M., Pagliarini, V., Matteoni, S., Fuoco, C., Giunta, L., *et al.* (2010). The dynamic interaction of AMBRA1 with the dynein motor complex regulates mammalian autophagy. The Journal of cell biology *191*, 155-168.

Dunn, W.A., Jr. (1994). Autophagy and related mechanisms of lysosome-mediated protein degradation. Trends in cell biology *4*, 139-143.

Ferraro, E., Pulicati, A., Cencioni, M.T., Cozzolino, M., Navoni, F., di Martino, S., Nardacci, R., Carri, M.T., and Cecconi, F. (2008). Apoptosome-deficient cells lose cytochrome c through proteasomal degradation but survive by autophagy-dependent glycolysis. Molecular biology of the cell *19*, 3576-3588.

Fimia, G.M., Stoykova, A., Romagnoli, A., Giunta, L., Di Bartolomeo, S., Nardacci, R., Corazzari, M., Fuoco, C., Ucar, A., Schwartz, P., *et al.* (2007). Ambra1 regulates autophagy and development of the nervous system. Nature *447*, 1121-1125.

Fuchs, Y., and Steller, H. (2011). Programmed cell death in animal development and disease. Cell *147*, 742-758.

Hakem, R., Hakem, A., Duncan, G.S., Henderson, J.T., Woo, M., Soengas, M.S., Elia, A., de la Pompa, J.L., Kagi, D., Khoo, W., *et al.* (1998). Differential requirement for caspase 9 in apoptotic pathways in vivo. Cell *94*, 339-352.

Hara, T., Nakamura, K., Matsui, M., Yamamoto, A., Nakahara, Y., Suzuki-Migishima, R., Yokoyama, M., Mishima, K., Saito, I., Okano, H., *et al.* (2006). Suppression of basal autophagy in neural cells causes neurodegenerative disease in mice. Nature *441*, 885-889.

Kihara, A., Kabeya, Y., Ohsumi, Y., and Yoshimori, T. (2001). Beclin-phosphatidylinositol 3-kinase complex functions at the trans-Golgi network. EMBO reports *2*, 330-335.

Klionsky, D.J., and Emr, S.D. (2000). Autophagy as a regulated pathway of cellular degradation. Science *290*, 1717-1721.

Komatsu, M., Waguri, S., Chiba, T., Murata, S., Iwata, J., Tanida, I., Ueno, T., Koike, M., Uchiyama, Y., Kominami, E., et al. (2006). Loss of autophagy in the central nervous system causes neurodegeneration in mice. Nature 441, 880-884.

Kuida, K., Haydar, T.F., Kuan, C.Y., Gu, Y., Taya, C., Karasuyama, H., Su, M.S., Rakic, P., and Flavell, R.A. (1998). Reduced apoptosis and cytochrome c-mediated caspase activation in mice lacking caspase 9. Cell 94, 325-337.

Kuida, K., Zheng, T.S., Na, S., Kuan, C., Yang, D., Karasuyama, H., Rakic, P., and Flavell, R.A. (1996). Decreased apoptosis in the brain and premature lethality in CPP32-deficient mice. Nature 384, 368-372.

Kuma, A., Hatano, M., Matsui, M., Yamamoto, A., Nakaya, H., Yoshimori, T., Ohsumi, Y., Tokuhisa, T., and Mizushima, N. (2004). The role of autophagy during the early neonatal starvation period. Nature 432, 1032-1036.

Li, P., Nijhawan, D., Budihardjo, I., Srinivasula, S.M., Ahmad, M., Alnemri, E.S., and Wang, X. (1997). Cytochrome c and dATP-dependent formation of Apaf-1/caspase-9 complex initiates an apoptotic protease cascade. Cell 91, 479-489.

Liang, X.H., Jackson, S., Seaman, M., Brown, K., Kempkes, B., Hibshoosh, H., and Levine, B. (1999). Induction of autophagy and inhibition of tumorigenesis by beclin 1. Nature 402, 672-676.

Long, A.B., Kaiser, W.J., Mocarski, E.S., and Caspary, T. (2013). Apaf1 apoptotic function critically limits Sonic hedgehog signaling during craniofacial development. Cell death and differentiation 20, 1510-1520.

Moreno, S., Ferraro, E., Eckert, S., and Cecconi, F. (2002). Apaf1 reduced expression levels generate a mutant phenotype in adult brain and skeleton. Cell death and differentiation 9, 340-342.

Nazio, F., Strappazzon, F., Antonioli, M., Bielli, P., Cianfanelli, V., Bordi, M., Gretzmeier, C., Dengjel, J., Piacentini, M., Fimia, G.M., et al. (2013). mTOR inhibits autophagy by controlling ULK1 ubiquitylation, self-association and function through AMBRA1 and TRAF6. Nature cell biology 15, 406-416.

Nonomura, K., Yamaguchi, Y., Hamachi, M., Koike, M., Uchiyama, Y., Nakazato, K., Mochizuki, A., Sakaue-Sawano, A., Miyawaki, A., Yoshida, H., et al. (2013). Local apoptosis modulates early mammalian brain development through the elimination of morphogen-producing cells. Developmental cell 27, 621-634.

Pagliarini, V., Wirawan, E., Romagnoli, A., Ciccosanti, F., Lisi, G., Lippens, S., Cecconi, F., Fimia, G.M., Vandenabeele, P., Corazzari, M., et al. (2012). Proteolysis of Ambra1 during apoptosis has a role in the inhibition of the autophagic pro-survival response. Cell death and differentiation 19, 1495-1504.

Parzych, K.R., and Klionsky, D.J. (2014). An overview of autophagy: morphology, mechanism, and regulation. Antioxidants & redox signaling *20*, 460-473.

Qu, X., Zou, Z., Sun, Q., Luby-Phelps, K., Cheng, P., Hogan, R.N., Gilpin, C., and Levine, B. (2007). Autophagy gene-dependent clearance of apoptotic cells during embryonic development. Cell *128*, 931-946.

Rubinstein, A.D., and Kimchi, A. (2012). Life in the balance - a mechanistic view of the crosstalk between autophagy and apoptosis. Journal of cell science *125*, 5259-5268.

Strappazzon, F., Vietri-Rudan, M., Campello, S., Nazio, F., Florenzano, F., Fimia, G.M., Piacentini, M., Levine, B., and Cecconi, F. (2011). Mitochondrial BCL-2 inhibits AMBRA1-induced autophagy. The EMBO journal *30*, 1195-1208.

Yoshida, H., Kong, Y.Y., Yoshida, R., Elia, A.J., Hakem, A., Hakem, R., Penninger, J.M., and Mak, T.W. (1998). Apaf1 is required for mitochondrial pathways of apoptosis and brain development. Cell *94*, 739-750.

Youle, R.J., and Strasser, A. (2008). The BCL-2 protein family: opposing activities that mediate cell death. Nature reviews Molecular cell biology *9*, 47-59.

Yuan, S., Topf, M., Reubold, T.F., Eschenburg, S., and Akey, C.W. (2013). Changes in Apaf-1 conformation that drive apoptosome assembly. Biochemistry *52*, 2319-2327.

Zou, H., Henzel, W.J., Liu, X., Lutschg, A., and Wang, X. (1997). Apaf-1, a human protein homologous to C. elegans CED-4, participates in cytochrome c-dependent activation of caspase-3. Cell *90*, 405-413.

Zou, H., Li, Y., Liu, X., and Wang, X. (1999). An APAF-1.cytochrome c multimeric complex is a functional apoptosome that activates procaspase-9. The Journal of biological chemistry *274*, 11549-11556.

Chapter 12. Cell Death and Autophagy: A Historical Perspective (Gozuacik and Kig)

Devrim Gozuacik* and Cenk Kig

SABANCI University

Faculty of Engineering and Natural Sciences, Molecular Biology Genetics and Bioengineering Program, Orhanli-Tuzla 34956, Istanbul, TURKEY

*Corresponding author

Abstract

The term "autophagy" defines an evolutionarily conserved mechanism by which eukaryotic cells recycle or degrade internal constituents in a membrane-trafficking pathway. Thus, autophagy helps providing the cells with a constant supply of biomolecules and energy for maintenance of homeostasis under stressful conditions. From this point of view autophagy seems to come into play as a survival mechanism. However, accumulating evidence also suggests that autophagy may contribute to cellular demise, and intricate connections between autophagy and cell death were reported. Autophagy dysregulations were observed in a wide spectrum of diseases, including cancer, neurodegenerative diseases, metabolic diseases and infections. Therefore, how autophagy contributes to cell death, and how autophagy-related cell death switches are controlled are important issues to be addressed from both a basic scientific and a clinical point of view. In this chapter, we summarized the history of authophagy and its relation to cell death, and discussed the evolution of the much debated "autophagic cell death" concept.

Introduction

Cells have developed several mechanisms to survive stressful conditions such as nutrient deprivation, energy shortage, hypoxia, accumulation of non-functional proteins or toxic molecules. One of these stress responses is a catabolic process termed "autophagy" (Greek for "self eating"), an evolutionarily conserved mechanism from baker's yeast to man. Basically, macroautophagy (autophagy hereinafter) defines the process by which eukaryotic cells recycle or degrade internal constituents in a membrane-trafficking pathway. Thus, autophagy helps providing the cells with a constant supply of biomolecules and energy for maintenance of homeostasis via recycling of macromolecules under stressful conditions. Intricate connections between autophagy and cell death were reported. In this chapter, we will first summarize landmark findings that led to the appreciation of autophagy as a key cellular mechanism, and then elaborate on the evolution of research on autophagy and cell death.

Early discoveries

The first published observation about autophagy can be traced back to as early as (Effront J, 1905). In the book "The Yeasts (1920)", Alexandre Guilliermonde defined autophagy as the "autodigestion or autolysis" of yeast glycogen and a cellular protein degradation mechanism activated by glucose starvation (Guilliermond AaT, 1920). Although in mammalian cells, the process was described by Clark in 1957 (Clark, 1957), the term "autophagy" was first coined by Christian de Duve and Wattiaux at the CIBA Foundation Symposium on Lysosomes in 1963. De Duve's pioneering work on lysosomes won him the Nobel Prize in Physiology or Medicine in 1974 (De Duve et al., 1955; De Duve and Wattiaux, 1966).

Research conducted on lysosomes paved the way for studies in the field of autophagy (Shi et al., 2011). Christian de Duve assayed activity of enzymes in mitochondrial fractions (heavy and light) of rat liver homogenates using consecutive centrifugation, a technique which was developed by Albert Claude (Claude, 1946). De Duve observed that the light fraction showed very high lytic enzyme activities, such as acid phosphatase, ribonuclease, deoxyribonuclease, cathepsin and β-glucuronidase activities with an optimal activity at acidic pH (De Duve and Wattiaux, 1966). Based on these observations, de Duve and colleagues decided to call the organelle in this fraction a "lysosome" ('digestive body' for Greek) (de Duve, 2005; Ohsumi, 2014). Soon after, Novikoff and colleagues identified lysosomes within various cell types *in situ* using electron microscopy techniques (Novikoff et al., 1956).

During the same period, the process termed "endocytosis" was under focus. Electron microscopy (EM) analysis revealed that foreign materials outside the cell were first engulfed into membrane-bound structures called phagosomes, which then fused with lysosomes to yield a digestive body known as the "phagolysosome" (De Duve and Wattiaux, 1966). Although, these observations provided an insight to the path leading to lysosomes, it was not yet clear how the intracellular components were degraded. Meanwhile, Novikoff et al. (1956) and Clark (1957) reported the presence of stress-induced and irregular shaped vacuoles containing amorphous material and mitochondria within cells in addition to lysosomes (Novikoff et al., 1956; Clark, 1957). Soon after, Ashfold and Porter (1962) showed that glucagon treatment induced an increase in the number of of lysosomes containing cytoplasmic components in hepatic cells (Ashford and Porter, 1962). De Duve introduced the term "autophagy" the following year (de Duve, 1963). Morphological appearance of glucagon-induced autophagic vacuoles in hepatic homogenates was then published in 1967 (Deter et al., 1967).

EM was an ideal technique for the study of morphology and contents of autophagosomes in mammalian cells. Using EM, Arstila and Trump showed formation of autophagic vacuoles as double membrane-bound structures containing portions of cytoplasm and organelles, but without hydrolytic enzymes (Arstila and Trump, 1968). But electron micrographs failed to reveal the details of the intricate process of membrane formation. Seglen's studies using radioactively labeled probes in late 1980's, led to the identification of the phagophore (origin and precursor of autophagosomal membranes) and the amphisome (a non-lysosomal vesicle formed by the fusion of autophagosomes and endosomes) (Gordon and Seglen, 1988). In yeast, a perinuclear compartment called "preautophagosomal structure" (PAS) was described in 2001 as the origin of autophagosomes (Suzuki et al., 2001). Recent studies showed that in mammals, autophagosomes may form on various membranes, inlcuding the outer membrane of the endoplasmic reticulum (Hayashi-Nishino et al., 2009; Ylä-Anttila et al., 2009).

Physiological role of autophagy

Due to the lack of our understanding of the molecular mechanisms underlying autophagic regulation in cells, early research on autophagy mainly relied on the ultrastructural and morphological observations until the first biochemical findings were published.

Following the discovery that glucagon could induce autophagy in rat liver (Deter et al., 1967), role of autophagy in cellular and organismal homeostasis started to gain interest. Mortimore found that protein degradation rates were inversely correlated with the level of amino acids in a perfusion solution, suggesting that protein degradation via autophagy could be controlled by amino-acid levels (Mortimore and Ward, 1976; Mortimore et al., 1983). Seglen and collegues (Seglen et al., 1980) as well as Pfeifer's group also showed that autophagy was tightly regulated by nutrient conditions and majority of liver proteins were degraded upon starvation (Pfeifer and Strauss, 1981)(Pfeifer and Warmuth-Metz, 1983). In contrast to the stimulatory effect of glucagon, Pfeifer's group reported in the late 1970s the inhibitory effect of insulin on autophagy (Pfeifer, 1977).

Molecular revolution: Discovery of autophagy-related genes

a) Discovery of autophagy-related genes in lower eukaryotes

The 1992 paper by Ohsumi's group first demonstrated that the morphology of autophagy in yeast was similar to that described in mammals (Takeshige et al., 1992). Subsequent analyses in yeast led to a breakthrough in our understanding of how autophagy is controlled.

The Ohsumi group observed under phase contrast microscopy significant morphological changes in the vacuoles of yeasts in response to starvation (Takeshige et al., 1992). Autophagic bodies gradually accumulated in the vacuole after 30 min of starvation. Therefore, progression of autophagy could easily be monitored in the vacuole using light microscopy. Further EM studies revealed that membrane dynamics were highly similar to those described in mammals (Baba et al., 1995). Using this simple approach, which was based on light microscopic selection, the Ohsumi group carried out the first genetic screen for yeast mutants that failed to accumulate autophagic bodies under nitrogen-starvation conditions. In this manner, they isolated first autophagy-defective mutant, the atg1 mutant (Tsukada and Ohsumi, 1993). atg1 mutant cells died when subjected to long periods of nitrogen starvation (Tsukada and Ohsumi, 1993). By exploiting this loss of viability phenotype in a primary screen, they then performed a secondary microscopic screen to select further mutants. This selection technique led to the identification of about one hundred apg mutants that fell into 15 complementation groups (Tsukada and Ohsumi, 1993).

Klionsky and collegues were on the other hand focusing on the transport of cytoplasm-to vacuole transport pathway (Cvt pathway), a constitutive pathway allowing the transport of enzymes to the vacuole, including vacuolar enzyme α-aminopeptidase I (Apel), in yeast. By searching mutants defective in apel transport, they discovered several genes that were necessary for the Cvt pathway (Harding et al., 1995). Electron microscopic studies of the Cvt pathway revealed striking similarities with macroautophagy (Baba et al., 1997). Although it is a constitutive and biosynthetic pathway of vacuolar enzymes rather than a degradative pathway, the Cvt pathway served as a model system for selective autophagy. Later research revealed that some proteins that played a role in the Cvt pathway were also necessary for autophagy (Ohsumi, 2014).

Thumm and collegues also performed independent yeast screens to isolate autophagy-related genes. They isolated several mutants of Saccharomyces cerevisiae that they called aut mutants. Indeed, *aut1*, *aut2* and *aut3* mutant diploids exhibited reduced sporulation frequency and viability under nitrogen-deficient growth conditions (Thumm et al., 1994). The same group later found the *AUT2* and *AUT7* genes, which are also essential for autophagocytosis in the yeast *S. cerevisiae* (Lang et al., 1998).

In addition to *S. cerevisiae*, other yeast species were also used as model systems in autophagy research. For example, *Pichia pastoris* (*P. pastoris*) was used to study the selective autophagy of peroxisomes. Subramani and his colleagues labeled the vacuolar membrane and peroxisomal matrix in living *P. pastoris* cells, and studied vacuolar degradation of peroxisomes through microautophagy and macroautophagy. Deployment of this technique led to isolation of *P. pastoris* genes, mutation of which resulted in a defect in peroxisome microautophagy (*PAG* genes) (Sakai et al., 1998).

In 1998, Mizushima et al. discovered the Atg5-Atg12 conjugation system in yeast (Mizushima et al., 1998). Discovery of the role of the Atg16 protein in Apg12-Apg5-Apg16 oligomerization followed (Kuma et al., 2002). Soon after, similar systems were analyzed and dissected in various organisms, including human (see below). Further screens for mutants that affected selective mitochondrial degradation (mitophagy), led to the identification of more than 30 *ATG* genes (Ohsumi, 2014). Availability of complete genome sequences accelerated the process of identification of *ATG* genes. *ATG* orthologues were soon identified in other model organisms, including *Dictyostelium*, *Arabidopsis*, *C. elegans* and *Drosophila* (Doelling et al., 2002; Hanaoka et al., 2002; Li and Vierstra, 2014), mice and man.

b) Autophagy systems in mammals

In addition to seminal findings in yeast, Ohsumi's laboratory reported the identification of the first mammalian autophagy genes, *ATG5* and *ATG12*. While studies in the yeast were pursued, mechanistic studies on autophagy regulation were already conducted in mammalian systems. Mizushima from the same laboratory first demostrated that the Atg12–Atg5 conjugation system was also conserved in man (Mizushima et al., 1998). Two years later, Yoshimori and collegues identified a microtubule associated protein MAP1LC3 (or shortly LC3) as the mammalian Atg8 orthologue, leading to the development of various LC3-based assays for monitoring autophagy in mammals and other organisms (Kabeya, 2000). p62/SQSTM1, a commonly used autophagy marker and a selective autophagy receptor was discovered in 2005 by Johansen group (Bjørkøy et al., 2005; Pankiv et al., 2007).

Another breakthrough came from the study of BCL2 interacting proteins in the mammalian system. In a yeast two-hybrid screen using Bcl-2 as the bait, Levine group identified the mammalian ortholog of yeast Atg6/Vps30, and they called the protein Beclin 1 (Liang et al., 1998). Subsequent studies by the same group revealed that Beclin 1-Bcl-2 complex was a key regulator of autophagy and autophagy-dependent cell death (Pattingre et al., 2005).

Mizushima et al. successfully generated a GFP-LC3 expressing transgenic mice line, which offered an elegant technique to visualize autophagosome formation *in vivo* (Mizushima et al., 2004). In 2004, the Mizushima group generated the first Atg5 knockout mice model (Kuma et al., 2004). Although Atg5 deficient mice appeared normal at birth they died shortly after delivery. Beclin 1 knockout mice died at early developmental stages (Yue et al., 2003) and the heterozygous mice were tumor-prone (Qu et al., 2003). GFP-LC3 transgenesis and *ATG* gene knockouts were performed in several other model organisms confirming and broadening the *in vivo* relevance of experimental findings (Ichimura and Komatsu, 2011; Ohsumi, 2014).

Drugs and kinases

In 1995, Meijer's group showed that rapamycin, a drug that blocked TOR kinase activity, was able to induce autophagy in rat hepatocytes and overcome the inhibitory effect of amino acids on autophagy (Blommaart et al., 1995). Three years later, Ohsumi's laboratory reported that rapamycin induced autophagy in yeast and that TOR kinase controlled autophagy in yeast (Noda and Ohsumi, 1998). Of note, rapamycin bound to FKBP12 and affected TOR activity, most probably through blocking substrate recruitment to the TORC1 complex (Yang et al., 2013). Since then, rapamycin and later introduced torin 1 (a potent selective mTOR kinase inhibitor) are used as a classical inducers of autophagy in various experimental settings.

Another landmark in autophagy studies was characterization of the pharmacological reagent 3-methyladenine (3-MA) as an inhibitor of autophagy (Seglen and Gordon, 1982). Further work showed that 3-MA in fact inhibited the type III phosphatidylinositol 3-kinase (PI-3K), which was shown to be essential for autophagosome formation (Blommaart et al., 1997). Codogno et al., in collaboration with Meijer's laboratory, showed that the class III PtdIns3K product, phosphatidylinositol 3-phosphate (PtdIns(3)P), was essential for autophagy, whereas the class I PtdIns3K products, phosphatidylinositol (3,4)-bisphosphate (PtdIns(3,4)P2) and phosphatidylinositol (3,4,5)-trisphosphate (PtdIns(3,4,5)P3), exerted inhibitory effects (Petiot et al., 2000). In fact, Class I PI3K enzymes result in the activation of Akt/PKB pathway, an upstream signal that stimulates mTOR (Esclatine et al., 2009) and inhibits autophagy (Arico et al., 2001; Meijer and Codogno, 2004). Accordingly, overexpression of PTEN, which hydrolyzed PtdIns(3,4)P2 and PtdIns(3,4,5)P3, and reversed the effects of class I PtdIns3K, was shown to stimulate autophagy (Arico et al., 2001). Yet, more general inhibitors of phosphatidylinositol 3-kinases, wortmannin and LY294002 were also found to block autophagy in the absence of amino acids in rat hepatocytes (Blommaart et al., 1997).

The mTORC1 complex is also inhibited by AMP-activated kinase (AMPK). In fact, AMPK is a sensor of cellular ATP levels. During starvation the AMP/ATP ratio increases and binding of AMP to

AMPK promotes its activation through LKB1 (Shaw et al., 2004). Although initial studies suggested the opposite (Samari and Seglen, 1998), AMPK activation was found to inhibit TOR-dependent signaling mainly by interfering with the activity of Rheb GTPase (Meijer and Codogno, 2004).

Nomeclature issues

Isolation of autophagy-related genes by independent groups and in screens focusing on a number of autophagy-related phenomena resulted in a nomenclature where a gene had more than one name as (e.g. *APG, AUT, CVT, GSA, PAG, PAZ, PDD* and *PDG*). As the number of autophagy-related genes increased, lack of unification of the gene nomenclature led to confusion among researhcers and in the litterature. A consensus was reached in a Gordon Conference that was held in 2003. A new and unified system of nomenclature was voted and accepted, and the *ATG* (*AuTophaGy* related genes) abbreviation was adopted to name genes controlling autophagy (Klionsky et al., 2003).

Nomenclature on specific types of autophagy was revisited as well. Depending on the type of cargo destined for autophagic degradation, various terms have been proposed. Some of these include: the ER (reticulophagy), peroxisomes (pexophagy), mitochondria (mitophagy), lipid droplets (lipophagy), secretory granules (zymophagy), nucleus (nucleophagy), pathogens (xenophagy), ribosomes (ribophagy) and protein aggregates (aggrephagy) (Klionsky et al., 2007).

Autophagy as a cell death mechanism

The story of autophagic or autophagy-related cell death is a long one with several ups and downs. The possible contribution of lysosomes to cell death was already suggested in the 50's and 60's in the works of de Duve, Berthet and Wattiaux. The cell death field was transformed with Saunders' work on chick limbs and Lockshin and Williams' work on insect development, leading them to suggest that cell death might be programmed. These publications were followed by studies by Kerr and Wyllie, coining the term "apoptosis" as an important type of cell death

(Kerr J. F. R., Wyllie A. H., 1972). Yet, apoptosis was not the only morphological process that was observed in dying cells. Ultrastructural analyses of mice and rat embryos by Schweichel and Merker (1973) showed that there were more than one pattern associated with cell death (Schweichel and Merker, 1973). This classification was revisited and refined by Clarke et al. (1990), who described cell death as Type I, apoptotic cell death; Type II cell death or autophagic cell death, Type III nonlysosomal vesiculate degradation cell death (Clarke, 1990).

Discovery of simple techniques of DNA fragmentation analysis, and later on, molecular pathways, resulted in a gain of momentum in apoptosis research. In 1977, Sulston and Horvitz demonstrated that a portion of somatic cells in the *C. elegans* embryo died in an organized way, and study of mutants with cell death defects revealed basic molecular pathways of apoptosis (Sulston and Horvitz, 1977). Discovery of caspases, Bcl-2 proteins and death receptors-ligands, provided a toolbox to study enzymatic and cellular outcomes of apoptosis and allowed the appreciation of its clinical implications (Trauth et al., 1989; Vaux et al., 1988; Yonehara et al., 1989). Discovery of the cell death-related functions of p53 constituted another landmark (Yonish-Rouach et al., 1991; Lowe et al., 1993). As a result, other types of cell death mechanisms were neglected, and the term "apoptosis" started to be used as a generic name for "programmed cell death". Other cell death types were simply referred to as "necrosis".

Introduction of 3-methyl adenine as an autophagy inhibitor was a turning point in the study of autophagy-related events in a cell death context. For example, following blockage of autophagy by the addition of 3-MA, Schwarze and Seglen were able to prevent death of 24-h amino-acid starved primary hepatocytes (Schwarze and Seglen, 1985). Meanwhile, observation of autophagic vesicles in dying cells was reported in several independent publications. In some systems, total area of autophagic vacuoles was reported to exceed that of cytosol and organelles outside the autophagic vesicles (Clarke, 1990).

Therefore, an autophagic activity at this level was suspected to lead to the total collapse of cellular functions and result in cell death.

An autophagic cell death literature was soon accumulating, and it has not stopped growing ever since. A current PubMed search using "autophagic cell death" as a key word revealed 825 publications. Of note, in dozens of additional publications, authors carefully used "autophagy-dependent cell death" or "autophagy-related cell death" terms as alternatives to define the phenomenon. The list includes well-controlled studies demonstrating attenuation of cell death through the use of 3-MA and/or genetic downregulation of key autophagy genes, including *ATG5*, *ATG7* and Beclin 1 [for examples see (Eberhart et al., 2014; Gozuacik and Kimchi, 2007)], as well as reports relying on mere accumulation of autophagic vesicles in cells dying in a non-apoptotic, caspase-independent manner (Gozuacik and Kimchi, 2004, 2007).

Conversely, several other studies focused on the prosurvival aspects of autophagy, and argued that inhibition of autophagy in the context of stress, including growth factor deprivation, starvation or exposure to toxins, rather potentiated cellular demise [for example see (Levine and Yuan, 2005)]. Moreover in yeasts and other organisms where autophagy mechanisms were analyzed in detail, basic and survival-related role of autophagy was prominent and undebatable. Yet, in some studies arguing against the autophagic cell death concept, lysosomal inhibitors (e.g. hydroxychloroquine) were used in order to block autophagy (Boya et al., 2005). But in addition to their negative effect on autolysosomal degradation, these substances permeabilized lysosomes, an event that would potentiate cell death in an autophagy-independent manner. The above-mentioned discrepancies and others resulted in a vivid discussion among scientists in the field about the prosurvival versus death-inducing role of autophagy in cells (Gozuacik and Kimchi, 2007; Kroemer and Levine, 2008; Levine and Yuan, 2005; Maiuri et al., 2007).

Therefore, a careful definition and description of autophagic cell death was required. In their review, Shen and Codogno categorized autophagic cell death as a form of programmed cell death in which autophagy serves as a cell death mechanism (Shen and Codogno, 2011). They proposed the following criteria for the definition of autophagic cell death: (i) Cell death occurs without the involvement of the apoptosis machinery, such as caspase activation. (ii) There is an increase of autophagic flux, and not just an increase of the autophagic markers, in the dying cells. (iii) Suppression of autophagy via both pharmacological inhibitors and genetic approaches (either *ATG* siRNA knock-down, knockout or overexpression of dominant negative *ATG* genes) is able to rescue or prevent cell death. These rules and definitions currently set the standard for an observation to be defined as autophagic cell death (Galluzzi et al., 2012).

Yet even in similar systems, different groups reported divergent results about the role of autophagy in cell death and survival, underlining the complexity and context dependency of autophagy-cell death connections. For example, apoptosis-defective cells derived from Bax/Bak double knockout mice responded to autophagy inhibition in contradictory ways. Bax/Bak knockout fibroblasts were rescued from etoposide, staurosporine, tunicamycin, obatoclax, rapamycin or photodynamic therapy-induced death, following inhibition of autophagy (Shimizu et al., 2004; Buytaert et al., 2006; Gozuacik et al., 2008; Bonapace et al., 2010). But, bone marrow-derived cells from the same mice, could survive for weeks autophagy activated by deprivation of growth factor (IL-3) that was imposed on them. In this case, inhibition of autophagy rather than the autophagic activity was lethal to cells (Lum et al., 2005).

On the other hand, expression of various death-related proteins, including BNIP3 (Vande Velde et al., 2000), death-associated protein kinase (DAPk) and DRP-1/DAPk2 (Inbal et al., 2002), ZIPk/DAPk3 (Shani et al., 2004) and smARF (Reef et al., 2006), and a mutant Beclin 1 (Pattingre et al., 2005) activated autophagic cell death. Moreoever, autophagy was suppressed by proteins protecting cells from death, such as Bcl-2 family proteins (Saeki et al., 2000; Vande Velde et al., 2000; Xue et al., 2001; Cárdenas-Aguayo et al., 2003; Yanagisawa et al., 2003; Pattingre et al., 2005)

Studies in model organisms

Contribution of autophagy to cell death in model organisms was reported as well. In *Dictyostelium*, autophagy and autophagy genes were shown to be important for development, differentiation and fruiting body formation (Otto et al., 2003; Roisin-Bouffay et al., 2004). In *C. elegans*, developmental signals as well as necrosis-inducing signals activated autophagic activity that resulted in cellular destruction (Kang et al., 2007; Samara et al., 2008). In *Drosophila*, autophagy was stimulated during developmental programs requiring elimination of cells and tissues. Indeed, Baehrecke's group showed that autophagy was required for developmental steroid hormone-induced degradation of *Drosophila* salivary gland cells (Lee and Baehrecke, 2001). Another developmental event, midgut destruction in *Drosophila* larvae, was reported to proceed through activation of autophagy and autophagy-related genes that were required for this event (Lee et al., 2002). Autophagy-related cell death was observed in plants as well. Autophagy contributed to hypersensitive cell death in *Arabidopsis* (Hofius et al., 2009). Cell death observed during tracheary element differentiation in *Arabidopsis* also involved autophagy (Kwon et al., 2010).

Different from above cited model organisms, autophagy did not seem to affect physiological and developmental cell death in mammals. Autophagy-deficient mice developed to term and showed no obvious morphological or anatomical abnormalities (Kuma et al., 2004). But with age, these mice accumulated protein aggregates, especially in neurons, and they suffered from neurodegeneration (Hara et al., 2006; Komatsu et al., 2006). On the other hand, Beclin 1-deficiency was lethal as early as in the preimplantation period, but the fact that a clear lethal phenotype was not observed in other Atg-deficient mice suggested that the effects were related to an autophagy-independent or specific developmental function of the protein (Qu et al., 2003; Yue et al., 2003). Moreover, tissue-specific deletion of Atg genes resulted in heart failure, muscle atrophy, anemia or lymphopenia etc. depending on the targeted tissues (Ichimura and Komatsu, 2011). But, most of these phenotypes were attributed to a defect in the clearance of abnormal mitochondria or accumulation of protein aggregates containing the SQSTM1/p62 protein. So, there was no obvious proof that these phenotypes were secondary to a defect in autophagic cell death.

But later studies showed that autophagy-related phenomena were rate-limiting for cellular demise in animal models of stress. In mammals, two newly discovered necrotic phenomena, namely necroptosis and autosis, involved autophagy activation and cell death. Both necroptosis and autosis cell death pathways were shown to be activated under stress conditions, including brain ischemia-reperfusion in rodents, and their inhibition prevented cell death both in *in vitro* and *in vivo* set-ups (Degterev et al., 2005; Liu et al., 2013).

Necroptosis was described by Yuan's group as a caspase-independent cell death type that required the activity of RIPK1/3, and novel chemical inhibitors called necrostatins prevented this type of cell death (Degterev et al., 2005). In fact, contribution of RIPK proteins to a caspase-independent type of cell death and autophagy activation was already demostrated by others (Holler et al., 2000; Yu et al., 2004), but introduction of necrostatins as tools to study necroptosis increased the interest to the field, and facilitated further studies.

While contribution of autophagy to cellular demise by necroptosis seemed to be context-dependent (Degterev et al., 2005; Bell et al., 2008; Bonapace et al., 2010; Khan et al., 2012), autosis clearly depended on autophagy and autophagy-related genes (Liu et al., 2013). Levine and collegues discovered autosis as a cell death type-induced following introduction of a Tat-

Beclin peptide into cells. Further analyses showed that starvation and ischemia also activated autosis. This type of cell death was mostly observed in non-apoptotic and adherent cells, and it was caspase- and RIPK-independent. Autophagy activation and perinuclear swelling were described as the hallmarks of the phenotype. Cardiac glycosides (Na$^+$, K$^+$-ATPase antagonists) or Na, K-α1–subunit knockdown were able to inhibit autosis type of cell death. Further analyses are required to dissect how autophagy contributes to cell death observed during necroptosis and autosis.

Autophagy-apoptosis connections

Despite the fact that apoptosis and autophagy are regulated through distinct signaling mechanisms (caspase cascades and atg protein systems, respectively) crosstalk among components of the two pathways exists (Gozuacik and Kimchi, 2004). In fact, autophagy and apoptosis connection may reveal itself in various ways [see (Gozuacik and Kimchi, 2004) for some early descriptive studies in the literature].

From a molecular point of view, one of the first clues about the presence of crosstalk between autophagy and apoptosis pathways came from the discovery that Beclin-1 was a BCL2-interacting protein (Liang et al., 1998, 1999). Soon after, Saeki et al. reported that Bcl-2 downregulation by antisense messangers activated autophagy and killed leukemia cells by a caspase-independent manner (Saeki et al., 2000). Later, Levine and colleagues provided molecular evidence that interaction of Beclin 1 with BCL-2 family anti-apoptotic proteins inhibited autophagy, and a binding-defective Beclin 1 was toxic to cells (Pattingre et al., 2005).

Cell death-related proteases were reported to play a determining role in apoptosis-autophagy crosstalk. A study by Lenardo and collegues suggested that caspase-8-dependent degradation of RIPK blocked autophagy and zVAD-induced cell death in L929 mouse fibroblasts (Yu et al., 2004). But later studies questioned the role of autophagy in cell death that was observed in this system (Wu et al., 2008), whereas a similar RIPK-dependent pathway was reported to be related to a hyperactive autophagy and cell death in T cells (Bell et al., 2008). Another report on the crosstalk came from Simon's lab, showing that calpain-mediated cleavage of Atg5 protein blocked autophagy (Yousefi et al., 2006). Moreover, Atg5 cleavage resulted in the appearance of a truncated form of the protein that translocated from cytosol to mitochondria, and triggered cytochrome c release and caspase activation. In fact, further studies demostrated that several key autophagy proteins were targets of caspase-mediated cleavage. Lane and collegues showed that Atg4D was a caspase-3 target (Betin and Lane, 2009). Interestingly, caspase-cleaved form of Atg4D had increased GABARAP-L1 priming and delipidation activity; it showed an inhibitory effect on autophagy and sensitized cells to starvation and staurosporine-induced cell death. Similarly, Luo and Rubinsztein showed that caspase-3 cleaved Beclin 1 and this event inhibited autophagy (Luo and Rubinsztein, 2010). In line with above-mentioned studies, our group reported that another autophagy protein, Atg3 was directly cleaved by caspase-8, and this event blocked autophagy that was activated by TNF-α and TRAIL treatment (Oral et al., 2012). In this case, autophagy supported cell survival and its activation under these conditions protected cells from death receptor-activated apoptosis.

Atg5 protein was reported to contribute to apoptosis-autophagy coordination and cell death regulation by several independent studies. Pyo et al. demonstrated that Atg5 played a role in IFN-gamma-induced cell death through its interaction with the proapoptotic protein FADD (Pyo et al., 2005). Walsh and collegues observed that T cells lacking FADD function or caspase-8 had hyperactive autophagy that could lead to cell death (Bell et al., 2008). Interestingly, they proposed that Atg5-Atg12-Atg16 formed a new complex with RIPK, FADD and caspase-8, and that this new platform was instrumental in a switch controlling caspase-8 activation or RIPK1- and autophagy-dependent cell death. The idea that Atg5-positive compartments served as a platform for caspase-8 activation was supported by the findings of others (Young et al., 2012).

Furthermore, autophagy-independent and cell death-related functions of a number of autophagy proteins were described. Kimchi's group showed that Atg12 in its free, Atg5 unconjugated form, associated with antiapoptotic BCL2 members and promoted mitochondrial apoptosis (Rubinstein et al., 2011). Simon and collegues demostrated that Atg5 played an autophagy-independent role during DNA damage-induced mitotic catastrophe and cell death (Maskey et al., 2013). UVRAG, a human homolog of yeast Vps38 which plays a role in autophagosome formation and endocytosis was shown to exhibit anti-apoptotic activity through its interaction with the pro-apoptotic BAX protein (Yin et al., 2011).

Tumor suppressor p53protein was redefined as a molecule regulating both apoptosis and autophagy. The nuclear form of p53 was responsible for the expression of DRAM protein, which could stimulate both autophagy and apoptosis (Crighton et al., 2006). On the other hand, genetic or pharmacological inhibition of p53 could activate autophagy in some systems. These observations led to the identification of a cytoplasmic form of p53 that acted as an inhibitor of autophagy (Tasdemir et al., 2008). More recently, the p53-induced glycolysis and apoptosis regulator (TIGAR) was reported to inhibit both apoptosis and autophagy (Xie et al., 2014).

Death-associated protein kinase (DAPK) family members constitute other examples of proteins involved in the regulation of both apoptosis and autophagy (Gozuacik and Kimchi, 2006). DAPK family members were shown to activate apoptosis (Raveh et al., 2001; Jang et al., 2002) or autophagy (Inbal et al., 2002; Shani et al., 2004) in various systems. Moreover, analysis of DAPK1 knockout mice revealed that the protein was necessary for both caspase activation and autophagosome formation, thereby, linking these two distinct pathways and introducing the protein as a co-regulator of apoptosis and autophagy (Gozuacik et al., 2008). To activate autophagy, DAPK1 directly phosphorylated Beclin 1 and promoted its dissociation from Bcl-X_L (Zalckvar et al., 2009). On the other hand, the other family member DAPK2 phosphorylated mTORC1 components and downregulated mTOR activity (Ber et al., 2014).

Autophagy-apoptosis interplay was reported to take place during developmental programmed cell of *Drosophila melanogaster*. Baehrecke's group first reported that caspases as well as core cell death machinery played a role in autophagy-related cell death of *Drosophila* salivary glands and during midgut destruction (Lee and Baehrecke, 2001; Lee et al., 2002). In line with these findings, Gorski's group performed a gene expression profiling study using steroid-triggered development in *Drosophila* as a model, and discovered that several apoptosis genes were upregulated together with autophagy genes in this system (Gorski et al., 2003). In fact, caspase-3 activity was found to be increased in *Drosophila* salivary glands undergoing autophagy (Martin and Baehrecke, 2004). Neufeld et al. demonstrated that overexpression of Atg1 was able to induce apoptotic cell death in *Drosophila* (Scott et al., 2007). Moreover, effector caspase Dcp-1 was found to control starvation-induced autophagy during oogenesis of *Drosophila* (Hou et al., 2009, 2008). Although these studies underline presence of a connection and interaction between apoptosis and autophagy components during developmental cell death, Baehrecke et al confirmed the need for autophagy for the degradation of salivary gland cells in *Drosophila* (Berry and Baehrecke, 2007).

Interestingly, a number of studies suggested that cell survival- or cell death-related proteins were direct targets of autophagic protein degradation. For example, Xiao et al. showed that IκB kinase (IKK) protein was degraded by autophagy following Hsp90 inhibition (Qing et al., 2006). Lenardo et al. reported that an antioxidant prosurvival protein, catalase, was a target of autophagy-dependent destruction (Yu et al., 2006). Another interesting finding was reported by Rabinowich and colleagues. They showed that active caspase-8, which was generated during TRAIL-stimulated apoptosis, was sequestered in autophagosomes and degraded in autolysosomes (Hou et al., 2010). In T cells, autophagy blockade by Beclin 1 deficiency resulted in the accumulation of caspase-3, caspase-8 and Bim, while autophagy induction by rapamycin led to a decrease in caspase levels, indicating that autophagy in this context could operate to block apoptosis (Kovacs et al., 2012). Recently, Thorburn and colleagues revealed that, to regulate mitochondrial apoptosis, p62/SQSTM1-dependent selective autophagy kept PUMA levels low through an indirect mechanism (Thorburn et al., 2014).

The above-mentioned studies show that a complex and context-dependent interaction exists between apoptosis and autophagy pathways. The crosstalk seems to be a determining factor in cell death/survival decisions in many cell types.

Discussion and Conclusions

Characterization of *ATG* genes, almost 30 years after the discovery of lysosomes, opened a new era in autophagy research. Since then, our understanding of the molecular mechanisms participating in autophagy regulation was greatly improved. The importance of autophagy under both physiological and disease-related contexts started to be appreciated only in the last couple of years. Knowledge and tools provided by basic autophagy research revealed the significance of autophagy dysregulation in a wide spectrum of diseases, including cancer, neurodegenerative diseases, metabolic diseases and infections. Consequently, modulation of autophagy with drugs is now considered as a potent and novel approach that could help treat a variety of health problems.

Although autophagy is basically regarded as a mechanism associated with survival or stress adaptation, as discussed above, under certain contexts autophagy may lead to cellular demise. But the contribution of autophagy to cell death is still a debated issue in some communities. In our view, ambiguity of some researchers is partly related to the natural evolution of the field: Availability of molecular tools other than cumbersome electron microscopy techniques and radioactive protein degradation assays, including inhibitors (e.g. 3-MA) and autophagy protein-related techniques (e.g. LC3-based tests, p62 degradation etc) and wide-spread use of RNAi techniques against autophagy genes, allowed documentation of autophagy in stressed and/or dying cells by researchers from relevant fields all around the world. Absence of a general definition and consensus in the field led to a heterogeneous literature where terms such as "autophagic cell death", "autophagy-related", "autophagy-dependent" or "autophagy gene-dependent cell death" were used without a clear consensus on their definition. We have previously pointed out to the need to use genetic approaches of targeting autophagy genes, and emphasized the importance of revisiting cases where autophagic activity was claimed to contribute to the cell death solely based on morphological observations or on the use of chemical inhibitors (Gozuacik and Kimchi, 2007). Lack of commonly accepted standards and an inflation of descriptive and uncontrolled studies in the literature resulted in publications categorically denying a possible role of autophagy in cell death (Levine and Yuan, 2005; Kroemer and Levine, 2008; Galluzzi et al., 2009). A clear and commonly accepted definition of autophagic cell death was published only in 2011 (Shen and Codogno, 2011), that later was adopted by the cell death community (Galluzzi et al., 2012). Accordingly, in order to classify cell death as autophagic, usage of genetic interventions targeting at least two autophagy components become the gold standard of the the well-controlled publications in the literature. But one should emphasize the fact that contribution of autophagic degradative activity is equally important, since unproductive autophagy leading to the accumulation of supernumerary autophagic vesicles was also shown to be toxic to cells in many cases. Moreover, caution should be exerced when using lysosomal inhibitors such as hydroxychloroquine since the substance was shown to lead to lysosomal rupture and cathepsin release to cytosol, potentiating non-autophagic cell death.

Recent advances in the field allowed a better appreciation of the relation between autophagy and cell death. It was revealed that autophagy might contribute to cell death and survival in a variety of ways: In addition to the classical autophagic type II cell death, which now becomes a generic term that describes autophagic activity- and autophagy gene-dependent, but caspase-independent, cell death type, accumulating data indicate a direct contribution of autophagy to different necrotic-like cell death types, including necroptosis and autosis. In fact, autosis, and in some situations necroptosis, fit into the current definition of "autophagic cell death", although other molecular components such as Na^+,K^+-ATPase activity (autosis) and RIPK1/3 function (necroptosis) were also described.

Current literature indeed points to a more complex role for autophagy in cell survival and death. In addition to dynamic molecular interactions with apoptotic pathways, the role of autophagy in a variety of cellular events was documented. Maintenance of mitochondria, limitation of reactive oxygen damage (ROS), preservation of ATP levels and targeted or non-targeted recycling of cellular components by autophagy possibly contribute to cell death-related outcomes of the phenomenon. Hyperactivation of autophagy, which was observed in some contexts, could possibly lead to an excessive degradation of vital organelles or molecules that are key to survival of cells, and affect intarcellular energy and ROS levels, tipping the survival/death balance toward cell death. In addition to nuclear changes that are observed during cell death, cytoplasmic destruction by autophagy might significantly contribute to the rapid elimination of large cells such as muscle cells or neurons, and collaborate with other cell death types.

Therefore, considering its relevance in human health and disease, autophagy-cell death connection will certainly continue to be one of the important fields of research in the coming years. Further studies will reveal in which other contexts autophagy contributes to cell death, and how autophagy-related cell death switches are controlled, opening way to their modulation for therapeutic purposes, especially in cancer and neurodegenerative diseases.

Abbreviations:

3-MA: 3-methyladenine

AMP: Adenosine monophosphate

AMPK : AMP-activated kinase

ATG: autophagy-related

ATP: Adenosine triphosphate

Bak: Bcl-2 homologous antagonist killer

Bax: BCL2-associated X protein

Bcl-2 : B-cell lymphoma 2 protein

Bcl-XL: B-cell lymphoma-extra large

BNIP3: BCL2/adenovirus E1B 19kDa interacting protein 3

Cvt pathway: cytoplasm-to vaculole transport pathway

DAPk: Death-associated protein kinase

DAPk: Death-associated protein kinase

Dcp-1: death caspase-1

DRAM: damage-regulated autophagy modulator

DRP-1: dynamin-related protein

EM: Electron microscopy

ER: endoplasmic reticulum

FADD: Fas-Associated protein with Death Domain

FKBP12: FK506/rapamycin-binding protein

GABARAP-L1: GABA(A) receptor-associated protein-like 1

GFP: green fluorescent protein

GTP: Guanosine-5'-triphosphate

Hsp90: heat shock protein 90

IKK : IκB kinase

IL-3: interleukin-3

LC3: (microtubule associated protein 1) light chain 3B

Lkb1: liver kinase B1

MAP1LC3: microtubule associated protein 1 light chain 3B

mTOR : mammalian target of rapamycin

mTORC1: mammalian target of rapamycin complex 1

PI-3K : phosphatidylinositol 3-kinase

PtdIns(3)P: phosphatidylinositol 3-phosphate

PtdIns(3,4)P2: phosphatidylinositol (3,4)-bisphosphate

PtdIns(3,4,5)P3: phosphatidylinositol (3,4,5)-trisphosphate

PTEN: Phosphatase and tensin homolog

PUMA: p53 upregulated modulator of apoptosis

Rheb: Ras homolog enriched in brain

RIPK: receptor-interacting protein kinase

ROS: reactive oxygen species

TIGAR: TP53-inducible glycolysis and apoptosis regulator

TNF-α : Tumor necrosis factor alpha

TRAIL: TNF-related apoptosis-inducing ligand

UVRAG: UV radiation resistance-associated gene protein

Vps30: Vacuolar protein sorting-associated protein 30

ZIPk: Zipper-Interacting Protein Kinase

zVAD: pan caspase inhibitor (carbobenzoxy-valyl-alanyl-aspartyl-[O-methyl]-fluoromethylketone)

References

Arico, S., Petiot, A., Bauvy, C., Dubbelhuis, P.F., Meijer, A.J., Codogno, P., and Ogier-Denis, E. (2001). The tumor suppressor PTEN positively regulates macroautophagy by inhibiting the phosphatidylinositol 3-kinase/protein kinase B pathway. J. Biol. Chem. *276*, 35243–35246.

Arstila, A.U., and Trump, B.F. (1968). Studies on cellular autophagocytosis. The formation of autophagic vacuoles in the liver after glucagon administration. Am. J. Pathol. *53*, 687–733.

Ashford, T.P., and Porter, K.R. (1962). Cytoplasmic components in hepatic cell lysosomes. J. Cell Biol. *12*, 198–202.

Baba, M., Osumi, M., and Ohsumi, Y. (1995). Analysis of the membrane structures involved in autophagy in yeast by freeze-replica method. Cell Struct. Funct. *20*, 465–471.

Baba, M., Osumi, M., Scott, S. V, Klionsky, D.J., and Ohsumi, Y. (1997). Two distinct pathways for targeting proteins from the cytoplasm to the vacuole/lysosome. J. Cell Biol. *139*, 1687–1695.

Bell, B.D., Leverrier, S., Weist, B.M., Newton, R.H., Arechiga, A.F., Luhrs, K.A., Morrissette, N.S., and Walsh, C.M. (2008). FADD and caspase-8 control the outcome of autophagic signaling in proliferating T cells. Proc. Natl. Acad. Sci. U. S. A. *105*, 16677–16682.

Ber, Y., Shiloh, R., Gilad, Y., Degani, N., Bialik, S., and Kimchi, A. (2014). DAPK2 is a novel regulator of mTORC1 activity and autophagy. Cell Death Differ.

Berry, D.L., and Baehrecke, E.H. (2007). Growth arrest and autophagy are required for salivary gland cell degradation in Drosophila. Cell *131*, 1137–1148.

Betin, V.M.S., and Lane, J.D. (2009). Caspase cleavage of Atg4D stimulates GABARAP-L1 processing and triggers mitochondrial targeting and apoptosis. J. Cell Sci. *122*, 2554–2566.

Bjørkøy, G., Lamark, T., Brech, A., Outzen, H., Perander, M., Overvatn, A., Stenmark, H., and Johansen, T. (2005). p62/SQSTM1 forms protein aggregates degraded by autophagy and has a protective effect on huntingtin-induced cell death. J. Cell Biol. *171*, 603–614.

Blommaart, E.F., Luiken, J.J., Blommaart, P.J., van Woerkom, G.M., and Meijer, A.J. (1995). Phosphorylation of ribosomal protein S6 is inhibitory for autophagy in isolated rat hepatocytes. J. Biol. Chem. *270*, 2320–2326.

Blommaart, E.F., Krause, U., Schellens, J.P., Vreeling-Sindelárová, H., and Meijer, A.J. (1997). The phosphatidylinositol 3-kinase inhibitors wortmannin and LY294002 inhibit autophagy in isolated rat hepatocytes. Eur. J. Biochem. *243*, 240–246.

Bonapace, L., Bornhauser, B.C., Schmitz, M., Cario, G., Ziegler, U., Niggli, F.K., Schäfer, B.W., Schrappe, M., Stanulla, M., and Bourquin, J.-P. (2010). Induction of autophagy-dependent necroptosis is required for childhood acute lymphoblastic leukemia cells to overcome glucocorticoid resistance. J. Clin. Invest. *120*, 1310–1323.

Boya, P., González-Polo, R.-A., Casares, N., Perfettini, J.-L., Dessen, P., Larochette, N., Métivier, D., Meley, D., Souquere, S., Yoshimori, T., et al. (2005). Inhibition of macroautophagy triggers apoptosis. Mol. Cell. Biol. *25*, 1025–1040.

Buytaert, E., Callewaert, G., Hendrickx, N., Scorrano, L., Hartmann, D., Missiaen, L., Vandenheede, J.R., Heirman, I., Grooten, J., and Agostinis, P. (2006). Role of endoplasmic reticulum depletion and multidomain proapoptotic BAX and BAK proteins in shaping cell death after hypericin-mediated photodynamic therapy. FASEB J. 20, 756–758.

Cárdenas-Aguayo, M. del C., Santa-Olalla, J., Baizabal, J.-M., Salgado, L.-M., and Covarrubias, L. (2003). Growth factor deprivation induces an alternative non-apoptotic death mechanism that is inhibited by Bcl2 in cells derived from neural precursor cells. J. Hematother. Stem Cell Res. 12, 735–748.

Clarki, S.L. (1957). Cellular differentiation in the kidneys of newborn mice studies with the electron microscope. J. Biophys. Biochem. Cytol. 3, 349–362.

Clarke, P.G. (1990). Developmental cell death: morphological diversity and multiple mechanisms. Anat. Embryol. (Berl). 181, 195–213.

Claude, A. (1946). Fractionation of mammalian liver cells by differential centrifugation: I. Problems, methods, and preparation of extract. J. Exp. Med. 84, 51–59.

Crighton, D., Wilkinson, S., O'Prey, J., Syed, N., Smith, P., Harrison, P.R., Gasco, M., Garrone, O., Crook, T., and Ryan, K.M. (2006). DRAM, a p53-Induced Modulator of Autophagy, Is Critical for Apoptosis. Cell 126, 121–134.

Degterev, A., Huang, Z., Boyce, M., Li, Y., Jagtap, P., Mizushima, N., Cuny, G.D., Mitchison, T.J., Moskowitz, M.A., and Yuan, J. (2005). Chemical inhibitor of nonapoptotic cell death with therapeutic potential for ischemic brain injury. Nat. Chem. Biol. 1, 112–119.

Deter, R.L., Baudhuin, P., and De Duve, C. (1967). Participation of lysosomes in cellular autophagy induced in rat liver by glucagon. J. Cell Biol. 35, C11–C16.

Doelling, J.H., Walker, J.M., Friedman, E.M., Thompson, A.R., and Vierstra, R.D. (2002). The APG8/12-activating enzyme APG7 is required for proper nutrient recycling and senescence in Arabidopsis thaliana. J. Biol. Chem. 277, 33105–33114.

De Duve, C. (1963). Ciba Foundation Symposium : Lysosomes. In The Lysosome Concept, A.V.S. de Reuck, and M.P. Cameron, eds. (Boston, Mass., U.S.A: Little, Brown & Co), pp. 362–383.

De Duve, C. (2005). [The lysosome turns fifty.]. Med. Sci. (Paris). 21, 12–15.

De Duve, C., and Wattiaux, R. (1966). Functions of lysosomes. Annu. Rev. Physiol. 28, 435–492.

De Duve, C., Pressman, B.C., Gianetto R., Wattoaix, R., and Appelmans, F. (1955). Tissue fractionation studies. 6. Intracellular distribution patterns of enzymes in rat-liver tissue. Biochem. J. 60, 604–617.

Eberhart, K., Oral, O., and Gozuacik, D. (2014). Autophagy: Cancer, Other Pathologies, Inflammation, Immunity, Infection, and Aging (Elsevier).

Effront J (1905). Sur l'autophagie de la levure. Monit. Sci.

Esclatine, A., Chaumorcel, M., and Codogno, P. (2009). Macroautophagy signaling and regulation. Curr. Top. Microbiol. Immunol. *335*, 33–70.

Galluzzi, L., Aaronson, S.A., Abrams, J., Alnemri, E.S., Andrews, D.W., Baehrecke, E.H., Bazan, N.G., Blagosklonny, M. V, Blomgren, K., Borner, C., et al. (2009). Guidelines for the use and interpretation of assays for monitoring cell death in higher eukaryotes. Cell Death Differ. *16*, 1093–1107.

Galluzzi, L., Vitale, I., Abrams, J.M., Alnemri, E.S., Baehrecke, E.H., Blagosklonny, M. V, Dawson, T.M., Dawson, V.L., El-Deiry, W.S., Fulda, S., et al. (2012). Molecular definitions of cell death subroutines: recommendations of the Nomenclature Committee on Cell Death 2012. Cell Death Differ. *19*, 107–120.

Gordon, P.B., and Seglen, P.O. (1988). Prelysosomal convergence of autophagic and endocytic pathways. Biochem. Biophys. Res. Commun. *151*, 40–47.

Gorski, S.M., Chittaranjan, S., Pleasance, E.D., Freeman, J.D., Anderson, C.L., Varhol, R.J., Coughlin, S.M., Zuyderduyn, S.D., Jones, S.J.M., and Marra, M.A. (2003). A SAGE approach to discovery of genes involved in autophagic cell death. Curr. Biol. *13*, 358–363.

Gozuacik, D., and Kimchi, A. (2004). Autophagy as a cell death and tumor suppressor mechanism. Oncogene *23*, 2891–2906.

Gozuacik, D., and Kimchi, A. (2006). DAPk protein family and cancer. Autophagy *2*, 74–79.

Gozuacik, D., and Kimchi, A. (2007). Autophagy and cell death. Curr. Top. Dev. Biol. *78*, 217–245.

Gozuacik, D., Bialik, S., Raveh, T., Mitou, G., Shohat, G., Sabanay, H., Mizushima, N., Yoshimori, T., and Kimchi, A. (2008). DAP-kinase is a mediator of endoplasmic reticulum stress-induced caspase activation and autophagic cell death. Cell Death Differ. *15*, 1875–1886.

Guilliermond AaT, F.W. (1920). The Yeasts (Norwood, Massachusets: The Plimpton Press).

Hanaoka, H., Noda, T., Shirano, Y., Kato, T., Hayashi, H., Shibata, D., Tabata, S., and Ohsumi, Y. (2002). Leaf senescence and starvation-induced chlorosis are accelerated by the disruption of an Arabidopsis autophagy gene. Plant Physiol. *129*, 1181–1193.

Hara, T., Nakamura, K., Matsui, M., Yamamoto, A., Nakahara, Y., Suzuki-Migishima, R., Yokoyama, M., Mishima, K., Saito, I., Okano, H., et al. (2006). Suppression of basal autophagy in neural cells causes neurodegenerative disease in mice. Nature *441*, 885–889.

Harding, T.M., Morano, K.A., Scott, S. V, and Klionsky, D.J. (1995). Isolation and characterization of yeast mutants in the cytoplasm to vacuole protein targeting pathway. J. Cell Biol. *131*, 591–602.

Hayashi-Nishino, M., Fujita, N., Noda, T., Yamaguchi, A., Yoshimori, T., and Yamamoto, A. (2009). A subdomain of the endoplasmic reticulum forms a cradle for autophagosome formation. Nat. Cell Biol. *11*, 1433–1437.

Hofius, D., Schultz-Larsen, T., Joensen, J., Tsitsigiannis, D.I., Petersen, N.H.T., Mattsson, O., Jørgensen, L.B., Jones, J.D.G., Mundy, J., and Petersen, M. (2009). Autophagic components contribute to hypersensitive cell death in Arabidopsis. Cell *137*, 773–783.

Holler, N., Zaru, R., Micheau, O., Thome, M., Attinger, A., Valitutti, S., Bodmer, J.L., Schneider, P., Seed, B., and Tschopp, J. (2000). Fas triggers an alternative, caspase-8-independent cell death pathway using the kinase RIP as effector molecule. Nat. Immunol. *1*, 489–495.

Hou, W., Han, J., Lu, C., Goldstein, L.A., and Rabinowich, H. (2010). Autophagic degradation of active caspase-8: a crosstalk mechanism between autophagy and apoptosis. Autophagy *6*, 891–900.

Hou, Y.C.C., Hannigan, A.M., and Gorski, S.M. (2009). An executioner caspase regulates autophagy. Autophagy *5*, 530–533.

Hou, Y.-C.C., Chittaranjan, S., Barbosa, S.G., McCall, K., and Gorski, S.M. (2008). Effector caspase Dcp-1 and IAP protein Bruce regulate starvation-induced autophagy during Drosophila melanogaster oogenesis. J. Cell Biol. *182*, 1127–1139.

Ichimura, Y., and Komatsu, M. (2011). Pathophysiological role of autophagy: lesson from autophagy-deficient mouse models. Exp. Anim. *60*, 329–345.

Inbal, B., Bialik, S., Sabanay, I., Shani, G., and Kimchi, A. (2002). DAP kinase and DRP-1 mediate membrane blebbing and the formation of autophagic vesicles during programmed cell death. J. Cell Biol. *157*, 455–468.

Jang, C.-W., Chen, C.-H., Chen, C.-C., Chen, J., Su, Y.-H., and Chen, R.-H. (2002). TGF-beta induces apoptosis through Smad-mediated expression of DAP-kinase. Nat. Cell Biol. *4*, 51–58.

Kabeya, Y. (2000). LC3, a mammalian homologue of yeast Apg8p, is localized in autophagosome membranes after processing. EMBO J. *19*, 5720–5728.

Kang, C., You, Y., and Avery, L. (2007). Dual roles of autophagy in the survival of Caenorhabditis elegans during starvation. Genes Dev. *21*, 2161–2171.

Kerr J. F. R., Wyllie A. H., C. a. R. (1972). Apoptosis : a Basic Biological Phenomenon With Wide-. Br. J. Cancer *26*, 239–257.

Khan, M.J., Rizwan Alam, M., Waldeck-Weiermair, M., Karsten, F., Groschner, L., Riederer, M., Hallstrom, S., Rockenfeller, P., Konya, V., Heinemann, A., et al. (2012). Inhibition of Autophagy Rescues Palmitic Acid-induced Necroptosis of Endothelial Cells. J. Biol. Chem. *287*, 21110–21120.

Klionsky, D.J., Cregg, J.M., Dunn, W.A., Emr, S.D., Sakai, Y., Sandoval, I. V, Sibirny, A., Subramani, S., Thumm, M., Veenhuis, M., et al. (2003). A unified nomenclature for yeast autophagy-related genes. Dev. Cell *5*, 539–545.

Klionsky, D.J., Cuervo, A.M., Dunn, W.A., Levine, B., van der Klei, I., and Seglen, P.O. (2007). How shall I eat thee? Autophagy *3*, 413–416.

Komatsu, M., Waguri, S., Chiba, T., Murata, S., Iwata, J., Tanida, I., Ueno, T., Koike, M., Uchiyama, Y., Kominami, E., et al. (2006). Loss of autophagy in the central nervous system causes neurodegeneration in mice. Nature *441*, 880–884.

Kovacs, J.R., Li, C., Yang, Q., Li, G., Garcia, I.G., Ju, S., Roodman, D.G., Windle, J.J., Zhang, X., and Lu, B. (2012). Autophagy promotes T-cell survival through degradation of proteins of the cell death machinery. Cell Death Differ. *19*, 144–152.

Kroemer, G., and Levine, B. (2008). Autophagic cell death: the story of a misnomer. Nat. Rev. Mol. Cell Biol. *9*, 1004–1010.

Kuma, A., Mizushima, N., Ishihara, N., and Ohsumi, Y. (2002). Formation of the approximately 350-kDa Apg12-Apg5.Apg16 multimeric complex, mediated by Apg16 oligomerization, is essential for autophagy in yeast. J. Biol. Chem. *277*, 18619–18625.

Kuma, A., Hatano, M., Matsui, M., Yamamoto, A., Nakaya, H., Yoshimori, T., Ohsumi, Y., Tokuhisa, T., and Mizushima, N. (2004). The role of autophagy during the early neonatal starvation period. Nature *432*, 1032–1036.

Kwon, S. Il, Cho, H.J., Jung, J.H., Yoshimoto, K., Shirasu, K., and Park, O.K. (2010). The Rab GTPase RabG3b functions in autophagy and contributes to tracheary element differentiation in Arabidopsis. Plant J. *64*, 151–164.

Lang, T., Schaeffeler, E., Bernreuther, D., Bredschneider, M., Wolf, D.H., and Thumm, M. (1998). Aut2p and Aut7p, two novel microtubule-associated proteins are essential for delivery of autophagic vesicles to the vacuole. EMBO J. *17*, 3597–3607.

Lee, C.Y., and Baehrecke, E.H. (2001). Steroid regulation of autophagic programmed cell death during development. Development *128*, 1443–1455.

Lee, C.-Y., Cooksey, B.A.K., and Baehrecke, E.H. (2002). Steroid regulation of midgut cell death during Drosophila development. Dev. Biol. *250*, 101–111.

Levine, B., and Yuan, J. (2005). Autophagy in cell death: an innocent convict? J. Clin. Invest. *115*, 2679–2688.

Li, F., and Vierstra, R.D. (2014). Arabidopsis ATG11, a scaffold that links the ATG1-ATG13 kinase complex to general autophagy and selective mitophagy. Autophagy *10*, 1466–1467.

Liang, X.H., Kleeman, L.K., Jiang, H.H., Gordon, G., Goldman, J.E., Berry, G., Herman, B., and Levine, B. (1998). Protection against fatal Sindbis virus encephalitis by beclin, a novel Bcl-2-interacting protein. J. Virol. *72*, 8586–8596.

Liang, X.H., Jackson, S., Seaman, M., Brown, K., Kempkes, B., Hibshoosh, H., and Levine, B. (1999). Induction of autophagy and inhibition of tumorigenesis by beclin 1. Nature *402*, 672–676.

Liu, Y., Shoji-Kawata, S., Sumpter, R.M., Wei, Y., Ginet, V., Zhang, L., Posner, B., Tran, K. a, Green, D.R., Xavier, R.J., et al. (2013). Autosis is a Na+,K+-ATPase-regulated form of cell death triggered by autophagy-inducing peptides, starvation, and hypoxia-ischemia. Proc. Natl. Acad. Sci. U. S. A. *110*, 20364–20371.

Lowe, S.W., Schmitt, E.M., Smith, S.W., Osborne, B.A., and Jacks, T. (1993). p53 is required for radiation-induced apoptosis in mouse thymocytes. Nature *362*, 847–849.

Lum, J.J., Bauer, D.E., Kong, M., Harris, M.H., Li, C., Lindsten, T., and Thompson, C.B. (2005). Growth factor regulation of autophagy and cell survival in the absence of apoptosis. Cell *120*, 237–248.

Luo, S., and Rubinsztein, D.C. (2010). Apoptosis blocks Beclin 1-dependent autophagosome synthesis: an effect rescued by Bcl-xL. Cell Death Differ. *17*, 268–277.

Maiuri, M.C., Zalckvar, E., Kimchi, A., and Kroemer, G. (2007). Self-eating and self-killing: crosstalk between autophagy and apoptosis. Nat. Rev. Mol. Cell Biol. *8*, 741–752.

Martin, D.N., and Baehrecke, E.H. (2004). Caspases function in autophagic programmed cell death in Drosophila. Development *131*, 275–284.

Maskey, D., Yousefi, S., Schmid, I., Zlobec, I., Perren, A., Friis, R., and Simon, H.-U. (2013). ATG5 is induced by DNA-damaging agents and promotes mitotic catastrophe independent of autophagy. Nat. Commun. *4*, 2130.

Meijer, A.J., and Codogno, P. (2004). Regulation and role of autophagy in mammalian cells. Int J Biochem Cell Biol *36*, 2445–2462.

Mizushima, N., Sugita, H., Yoshimori, T., and Ohsumi, Y. (1998). A new protein conjugation system in human. The counterpart of the yeast Apg12p conjugation system essential for autophagy. J. Biol. Chem. *273*, 33889–33892.

Mizushima, N., Yamamoto, A., Matsui, M., Yoshimori, T., and Ohsumi, Y. (2004). In vivo analysis of autophagy in response to nutrient starvation using transgenic mice expressing a fluorescent autophagosome marker. Mol. Biol. Cell *15*, 1101–1111.

Mortimore, G.E., and Ward, W.F. (1976). Behavior of the lysosomal system during organ perfusion. An inquiry into the mechanism of hepatic proteolysis. Front. Biol. *45*, 157–184.

Mortimore, G.E., Hutson, N.J., and Surmacz, C.A. (1983). Quantitative correlation between proteolysis and macro- and microautophagy in mouse hepatocytes during starvation and refeeding. Proc. Natl. Acad. Sci. U. S. A. *80*, 2179–2183.

Noda, T., and Ohsumi, Y. (1998). Tor, a phosphatidylinositol kinase homologue, controls autophagy in yeast. J. Biol. Chem. *273*, 3963–3966.

Novikoff, A.B., Beaufay, H., and de Duve, C. (1956). ELECTRON MICROSCOPY OF LYSOSOME-RICH FRACTIONS FROM RAT LIVER. J. Biophys. Biochem. Cytol. *2*, 179–184.

Ohsumi, Y. (2014). Historical landmarks of autophagy research. Cell Res. *24*, 9–23.

Oral, O., Oz-Arslan, D., Itah, Z., Naghavi, A., Deveci, R., Karacali, S., and Gozuacik, D. (2012). Cleavage of Atg3 protein by caspase-8 regulates autophagy during receptor-activated cell death. Apoptosis *17*, 810–820.

Otto, G.P., Wu, M.Y., Kazgan, N., Anderson, O.R., and Kessin, R.H. (2003). Macroautophagy is required for multicellular development of the social amoeba Dictyostelium discoideum. J. Biol. Chem. *278*, 17636–17645.

Pankiv, S., Clausen, T.H., Lamark, T., Brech, A., Bruun, J.-A., Outzen, H., Øvervatn, A., Bjørkøy, G., and Johansen, T. (2007). p62/SQSTM1 binds directly to Atg8/LC3 to facilitate degradation of ubiquitinated protein aggregates by autophagy. J. Biol. Chem. *282*, 24131–24145.

Pattingre, S., Tassa, A., Qu, X., Garuti, R., Liang, X.H., Mizushima, N., Packer, M., Schneider, M.D., and Levine, B. (2005). Bcl-2 antiapoptotic proteins inhibit Beclin 1-dependent autophagy. Cell *122*, 927–939.

Petiot, A., Ogier-Denis, E., Blommaart, E.F., Meijer, A.J., and Codogno, P. (2000). Distinct classes of phosphatidylinositol 3'-kinases are involved in signaling pathways that control macroautophagy in HT-29 cells [published erratum appears in J Biol Chem 2000 Apr 21;275(16):12360]. J. Biol. Chem. *275*, 992–998.

Pfeifer, U. (1977). Inhibition by insulin of the physiological autophagic breakdown of cell organelles. Acta Biol. Med. Ger. *36*, 1691–1694.

Pfeifer, U., and Strauss, P. (1981). Autophagic vacuoles in heart muscle and liver. A comparative morphometric study including circadian variations in meal-fed rats. J. Mol. Cell. Cardiol. *13*, 37–49.

Pfeifer, U., and Warmuth-Metz, M. (1983). Inhibition by insulin of cellular autophagy in proximal tubular cells of rat kidney. Am. J. Physiol. *244*, E109–E114.

Pyo, J.-O., Jang, M.-H., Kwon, Y.-K., Lee, H.-J., Jun, J.-I., Woo, H.-N., Cho, D.-H., Choi, B., Lee, H., Kim, J.-H., et al. (2005). Essential roles of Atg5 and FADD in autophagic cell death: dissection of autophagic cell death into vacuole formation and cell death. J. Biol. Chem. *280*, 20722–20729.

Qing, G., Yan, P., and Xiao, G. (2006). Hsp90 inhibition results in autophagy-mediated proteasome-independent degradation of IkappaB kinase (IKK). Cell Res. *16*, 895–901.

Qu, X., Yu, J., Bhagat, G., Furuya, N., Hibshoosh, H., Troxel, A., Rosen, J., Eskelinen, E.-L., Mizushima, N., Ohsumi, Y., et al. (2003). Promotion of tumorigenesis by heterozygous disruption of the beclin 1 autophagy gene. J. Clin. Invest. *112*, 1809–1820.

Raveh, T., Droguett, G., Horwitz, M.S., DePinho, R.A., and Kimchi, A. (2001). DAP kinase activates a p19ARF/p53-mediated apoptotic checkpoint to suppress oncogenic transformation. Nat. Cell Biol. *3*, 1–7.

Reef, S., Zalckvar, E., Shifman, O., Bialik, S., Sabanay, H., Oren, M., and Kimchi, A. (2006). A short mitochondrial form of p19ARF induces autophagy and caspase-independent cell death. Mol. Cell *22*, 463–475.

Roisin-Bouffay, C., Luciani, M.-F., Klein, G., Levraud, J.-P., Adam, M., and Golstein, P. (2004). Developmental cell death in dictyostelium does not require paracaspase. J. Biol. Chem. *279*, 11489–11494.

Rubinstein, A.D., Eisenstein, M., Ber, Y., Bialik, S., and Kimchi, A. (2011). The autophagy protein Atg12 associates with antiapoptotic Bcl-2 family members to promote mitochondrial apoptosis. Mol. Cell *44*, 698–709.

Saeki, K., Yuo, A., Okuma, E., Yazaki, Y., Susin, S.A., Kroemer, G., and Takaku, F. (2000). Bcl-2 down-regulation causes autophagy in a caspase-independent manner in human leukemic HL60 cells. Cell Death Differ. *7*, 1263–1269.

Sakai, Y., Koller, A., Rangell, L.K., Keller, G.A., and Subramani, S. (1998). Peroxisome degradation by microautophagy in Pichia pastoris: identification of specific steps and morphological intermediates. J. Cell Biol. *141*, 625–636.

Samara, C., Syntichaki, P., and Tavernarakis, N. (2008). Autophagy is required for necrotic cell death in Caenorhabditis elegans. Cell Death Differ. *15*, 105–112.

Samari, H.R., and Seglen, P.O. (1998). Inhibition of hepatocytic autophagy by adenosine, aminoimidazole-4-carboxamide riboside, and N6-mercaptopurine riboside. Evidence for involvement of amp-activated protein kinase. J. Biol. Chem. *273*, 23758–23763.

Schwarze, P.E., and Seglen, P.O. (1985). Reduced autophagic activity, improved protein balance and enhanced in vitro survival of hepatocytes isolated from carcinogen-treated rats. Exp. Cell Res. *157*, 15–28.

Schweichel, J.U., and Merker, H.J. (1973). The morphology of various types of cell death in prenatal tissues. Teratology *7*, 253–266.

Scott, R.C., Juhász, G., and Neufeld, T.P. (2007). Direct induction of autophagy by Atg1 inhibits cell growth and induces apoptotic cell death. Curr. Biol. *17*, 1–11.

Seglen, P.O., and Gordon, P.B. (1982). 3-Methyladenine: specific inhibitor of autophagic/lysosomal protein degradation in isolated rat hepatocytes. Proc. Natl. Acad. Sci. U. S. A. *79*, 1889–1892.

Seglen, P.O., Gordon, P.B., and Poli, A. (1980). Amino acid inhibition of the autophagic/lysosomal pathway of protein degradation in isolated rat hepatocytes. Biochim.Biophys.Acta *630*, 103–118.

Shani, G., Marash, L., Gozuacik, D., Bialik, S., Teitelbaum, L., Shohat, G., and Kimchi, A. (2004). Death-associated protein kinase phosphorylates ZIP kinase, forming a unique kinase hierarchy to activate its cell death functions. Mol. Cell. Biol. *24*, 8611–8626.

Shaw, R.J., Bardeesy, N., Manning, B.D., Lopez, L., Kosmatka, M., DePinho, R.A., and Cantley, L.C. (2004). The LKB1 tumor suppressor negatively regulates mTOR signaling. Cancer Cell *6*, 91–99.

Shen, H.-M., and Codogno, P. (2011). Autophagic cell death: Loch Ness monster or endangered species? Autophagy *7*, 457–465.

Shi, Y.-H., Jia Fan, C.-W.L., Ding, W.-X., and Yin, X.-M. (2011). Macroautophagy. In Molecular Pathology of Liver Diseases, S.P.S. Monga, ed. (Springer), pp. 389–396.

Shimizu, S., Kanaseki, T., Mizushima, N., Mizuta, T., Arakawa-Kobayashi, S., Thompson, C.B., and Tsujimoto, Y. (2004). Role of Bcl-2 family proteins in a non-apoptotic programmed cell death dependent on autophagy genes. Nat. Cell Biol. *6*, 1221–1228.

Sulston, J.E., and Horvitz, H.R. (1977). Post-embryonic cell lineages of the nematode, Caenorhabditis elegans. Dev. Biol. *56*, 110–156.

Suzuki, K., Kirisako, T., Kamada, Y., Mizushima, N., Noda, T., and Ohsumi, Y. (2001). The pre-autophagosomal structure organized by concerted functions of APG genes is essential for autophagosome formation. EMBO J. *20*, 5971–5981.

Takeshige, K., Baba, M., Tsuboi, S., Noda, T., and Ohsumi, Y. (1992). Autophagy in yeast demonstrated with proteinase-deficient mutants and conditions for its induction. J. Cell Biol. *119*, 301–311.

Tasdemir, E., Maiuri, M.C., Galluzzi, L., Vitale, I., Djavaheri-Mergny, M., D'Amelio, M., Criollo, A., Morselli, E., Zhu, C., Harper, F., et al. (2008). Regulation of autophagy by cytoplasmic p53. Nat. Cell Biol. *10*, 676–687.

Thorburn, J., Andrysik, Z., Staskiewicz, L., Gump, J., Maycotte, P., Oberst, A., Green, D.R., Espinosa, J.M., and Thorburn, A. (2014). Autophagy controls the kinetics and extent of mitochondrial apoptosis by regulating PUMA levels. Cell Rep. *7*, 45–52.

Thumm, M., Egner, R., Koch, B., Schlumpberger, M., Straub, M., Veenhuis, M., and Wolf, D.H. (1994). Isolation of autophagocytosis mutants of Saccharomyces cerevisiae. FEBS Lett. *349*, 275–280.

Trauth, B.C., Klas, C., Peters, A.M., Matzku, S., Möller, P., Falk, W., Debatin, K.M., and Krammer, P.H. (1989). Monoclonal antibody-mediated tumor regression by induction of apoptosis. Science *245*, 301–305.

Tsukada, M., and Ohsumi, Y. (1993). Isolation and characterization of autophagy-defective mutants of Saccharomyces cerevisiae. FEBS Lett. *333*, 169–174.

Vaux, D.L., Cory, S., and Adams, J.M. (1988). Bcl-2 gene promotes haemopoietic cell survival and cooperates with c-myc to immortalize pre-B cells. Nature *335*, 440–442.

Vande Velde, C., Cizeau, J., Dubik, D., Alimonti, J., Brown, T., Israels, S., Hakem, R., and Greenberg, A.H. (2000). BNIP3 and genetic control of necrosis-like cell death through the mitochondrial permeability transition pore. Mol. Cell. Biol. *20*, 5454–5468.

Wu, Y.-T., Tan, H.-L., Huang, Q., Kim, Y.-S., Pan, N., Ong, W.-Y., Liu, Z.-G., Ong, C.-N., and Shen, H.-M. (2008). Autophagy plays a protective role during zVAD-induced necrotic cell death. Autophagy *4*, 457–466.

Xie, J.-M., Li, B., Yu, H.-P., Gao, Q.G., Li, W., Wu, H.-R., and Qin, Z.-H. (2014). TIGAR has a dual role in cancer cell survival through regulating apoptosis and autophagy. Cancer Res.

Xue, L., Fletcher, G.C., and Tolkovsky, A.M. (2001). Mitochondria are selectively eliminated from eukaryotic cells after blockade of caspases during apoptosis. Curr. Biol. *11*, 361–365.

Yanagisawa, H., Miyashita, T., Nakano, Y., and Yamamoto, D. (2003). HSpin1, a transmembrane protein interacting with Bcl-2/Bcl-xL, induces a caspase-independent autophagic cell death. Cell Death Differ. *10*, 798–807.

Yang, H., Rudge, D.G., Koos, J.D., Vaidialingam, B., Yang, H.J., and Pavletich, N.P. (2013). mTOR kinase structure, mechanism and regulation. Nature *497*, 217–223.

Yin, X., Cao, L., Peng, Y., Tan, Y., Xie, M., Kang, R., Livesey, K.M., and Tang, D. (2011). A critical role for UVRAG in apoptosis. Autophagy *7*, 1242–1244.

Ylä-Anttila, P., Vihinen, H., Jokitalo, E., and Eskelinen, E.-L. (2009). 3D tomography reveals connections between the phagophore and endoplasmic reticulum. Autophagy *5*, 1180–1185.

Yonehara, S., Ishii, A., and Yonehara, M. (1989). A cell-killing monoclonal antibody (anti-Fas) to a cell surface antigen co-downregulated with the receptor of tumor necrosis factor. J. Exp. Med. *169*, 1747–1756.

Yonish-Rouach, E., Resnitzky, D., Lotem, J., Sachs, L., Kimchi, A., and Oren, M. (1991). Wild-type p53 induces apoptosis of myeloid leukaemic cells that is inhibited by interleukin-6. Nature *352*, 345–347.

Young, M.M., Takahashi, Y., Khan, O., Park, S., Hori, T., Yun, J., Sharma, A.K., Amin, S., Hu, C.-D., Zhang, J., et al. (2012). Autophagosomal membrane serves as platform for intracellular death-inducing signaling complex (iDISC)-mediated caspase-8 activation and apoptosis. J. Biol. Chem. *287*, 12455–12468.

Yousefi, S., Perozzo, R., Schmid, I., Ziemiecki, A., Schaffner, T., Scapozza, L., Brunner, T., and Simon, H.-U. (2006). Calpain-mediated cleavage of Atg5 switches autophagy to apoptosis. Nat. Cell Biol. *8*, 1124–1132.

Yu, L., Alva, A., Su, H., Dutt, P., Freundt, E., Welsh, S., Baehrecke, E.H., and Lenardo, M.J. (2004). Regulation of an ATG7-beclin 1 program of autophagic cell death by caspase-8. Science *304*, 1500–1502.

Yu, L., Wan, F., Dutta, S., Welsh, S., Liu, Z., Freundt, E., Baehrecke, E.H., and Lenardo, M. (2006). Autophagic programmed cell death by selective catalase degradation. Proc. Natl. Acad. Sci. U. S. A. *103*, 4952–4957.

Yue, Z., Jin, S., Yang, C., Levine, A.J., and Heintz, N. (2003). Beclin 1, an autophagy gene essential for early embryonic development, is a haploinsufficient tumor suppressor. Proc. Natl. Acad. Sci. U. S. A. *100*, 15077–15082.

Zalckvar, E., Berissi, H., Mizrachy, L., Idelchuk, Y., Koren, I., Eisenstein, M., Sabanay, H., Pinkas-Kramarski, R., and Kimchi, A. (2009). DAP-kinase-mediated phosphorylation on the BH3 domain of beclin 1 promotes dissociation of beclin 1 from Bcl-XL and induction of autophagy. EMBO Rep. *10*, 285–292.

Chapter 13: Autophagic flux and cell death (Loos)

Ben Loos

Department of Physiological Sciences, Stellenbosch University, Stellenbosch, South Africa

Abstract

Major developments have taken place in the past two decades that contributed to our understanding of cell death onset in a most fundamental manner. A key area of research that has grown tremendously is the field of macroautophagy (hereafter referred to as autophagy), and its impact on the regulation of apoptosis and also necrosis. Here, we highlight the role of autophagy in basal and diseased conditions, provide a brief outline of landmark papers in the context of autophagy and cell death with a particular focus on the cell's metabolic environment. By assessing how autophagy connects with the point of no return of cell death, we summarize our recent molecular understanding of the mitochondria in governing cellular fate, as most PONR modalities manifest there. The centrality of the rate of protein degradation through autophagy and the need to measure autophagic flux accurately is being highlighted and recent methodological approaches discussed. We conclude by asking fundamental questions that deserve major attention in the future. The advancements of techniques and tools to study autophagy and cell death have already allowed us to push the field significantly forward. Automated single cell acquisition systems and algorithms to analyse data as well as the recent introduction and commercialization of light sheet technology will allow the generation of invaluable data, derived through the imaging of living tissue in three dimensions over time. With these developments in mind, which may allow us to image flux and cell death dynamically *in vivo* and derive predictions on cell death onset, we can undoubtedly look forward to an illuminated next 20 years of cell death research.

Quantum leaps of development

Most impactful developments have taken place in the past 20 years that contributed to our understanding of cell death onset in a most fundamental manner. Although the process of macroautophagy (hereafter referred to as autophagy) had already been known since the early description by De Duve (Deter et al, 1967), its impact on the regulation of apoptosis and necrosis became only clear in the past 2 decades. It can be said that the field of autophagy in mammalian systems has been shaped in a most fundamental way in the past two decades, with annual major breakthroughs in the field, starting with the isolation of the target of rapamycin (TOR) gene in yeast and the mammalian system (Kunz et al, 1993). In 1995 the crucial observation was made that rapamycin inhibits mTOR, thereby inducing autophagy (Blommaart et al, 1995). In 1997 30 autophagy genes were identified in yeasts, and in 1998 Mizushima *et al* identify the first mammalian autophagy gene (Mizushima et al, 1998). In 1999 Beth Levine's laboratory identified beclin-1 (BECN1/ATG6) as a BCL-2 interacting protein (Liang et al, 1999) and in 2000 the development of the LC3-GFP assay made it possible to assess autophagy in the mammalian system (Kabeya et al, 2000). In addition, major application-based discoveries followed, such as the protective role of autophagy in Huntington's disease (Ravikumar et al, 2002) and its role in longevity in *Caenorhabditis* (Melendez et al, 2003). Later, a physiological role for autophagy in neonatal mice was provided by Mizushima's group (Kuma et al, 2002), and Komatsu *et al* demonstrated the involvement of basal autophagy in preventing neurodegeneration (Komatsu et al, 2006). Many of these articles became landmark papers and initiated major research focus areas.

Autophagy and cell death

A key landmark manuscript in this context was the generation of the LC3-GFP transgenic mouse model (Mizushima et al, 2004), which demonstrated for the first time that cells of all mammalian tissues undertake autophagy. Most striking was here the observation that the various tissue types are characterized by distinct autophagic activities, as well as distinct magnitudes of responses to induce autophagy upon starvation. This was fundamental. At this time, it was not generally accepted yet that the autophagic degradation rate can be assessed only when using lysosomal inhibitors concomitantly with assessments of numbers and types of organelles.

Autophagy was, historically, assessed morphologically, preferably through electron microscopy (EM), to identify the characteristic double membrane vacuoles (Figure 1). As some tissues, such as the brain, showed very few autophagic vacuoles under control conditions, it was assumed that autophagy played only a minor role in the brain. This was of course an assumption that was soon revisited. Autophagy was termed type II mode of cell death, taking position next to type I, apoptotic cell death, and type III, necrotic cell death (Edinger and Thompson, 2004, Festiens et al, 2006). This distinction was, for a time, recommended nomenclature, however, later it was advised to move away from such numeric distinction, primarily due to the large biochemical and morphological overlap between the cell death modalities. Autophagic cell death was initially defined morphologically, based on the appearance of massive cytoplasmic autophagic vacuolization in the absence of DNA condensation (Kroemer et al, 2005). The prominent role of autophagy as a pro-survival pathway was however already pointed out in 2009 (Kroemer et al, 2009), indicating the limited evidence for scenarios where the knockdown of genes required for autophagy would reduce cell death (Kroemer et al, 2009).

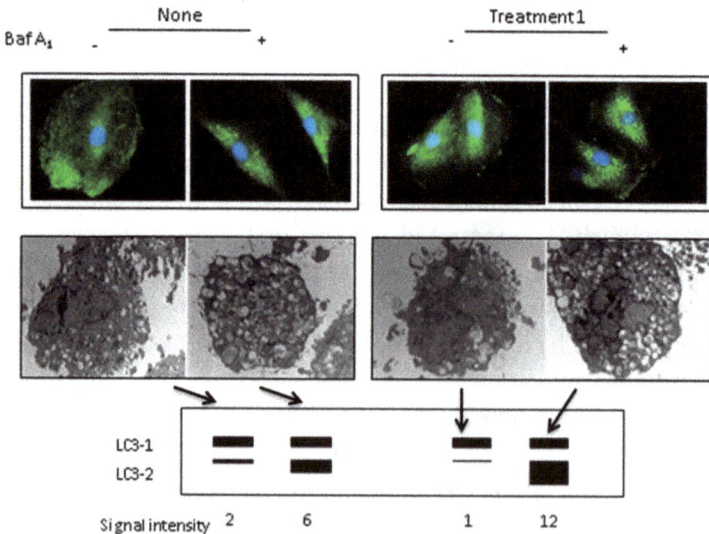

Figure 13.1: Fluorescence and electron microscopy as well as western blot analysis in the presence and absence of autophagosomal/lysosomal fusion inhibitors such as bafilomycin A1 (Baf) remain the key standard tools to assess autophagy. However, these techniques indicate only whether autophagic flux is increased or decreased. As they do not assess autophagosome pool size change in time, they do not measure flux *per se*.

At that time it was already stressed that care needs to be taken when assessing autophagy, and that the presence of autophagic vacuoles by no means indicates an induction of the autophagic pathway (Kroemer et al, 2009). Importantly, due to the development of genetic tools to disable the autophagic machinery, and the generation of, for example, Atg5$^{-/-}$ or Atg7$^{-/-}$ cells, it became clear that autophagy constitutes a stress response as an attempt of dying cells to withstand the cellular insult or metabolic perturbation. Since, considerable doubts have accumulated on the very existence of 'autophagic cell death' (Shensi et al, 2012). In 2006, another fundamental research performed by the Gottlieb laboratory had shown that ischemia induces autophagy, and that enhanced autophagic activity protects against ischemia/reperfusion injury (Hamacher-Brady et al, 2006). Hence, it became increasingly clear that autophagy may indeed operate as a survival pathway rather than a death mode per se. Subsequent experiments indicated that induction of autophagy through rapamycin treatment protects from zVAD induced necrosis while inhibition of autophagy through chloroquine treatment enhanced zVAD induced necrosis (Wu et al, 2008). Indeed, by employing live cell imaging and fluorescence resonance energy transfer (FRET) with a construct encoding a caspase 3 recognition site, it had later been demonstrated that upregulation of autophagy delays the point of no return (PONR) of both apoptotic and necrotic cell death (Loos et al, 2011). These findings connected the cell death modalities tighter to one another, and the known ATP dependent shift between apoptosis and necrosis (Leist et al, 1997) received a new meaning, since autophagy was able to catabolically influence this very ATP balance. It was hence proposed that the cell's autophagic activity may shift the time of apoptosis or necrosis induction backwards, depending on its effect on cellular energy homeostasis (Loos et al, 2013). mTOR, PKA (protein kinase, cAMP-dependent) and AMPK (protein kinase, AMPK-activated) signalling have here recently been proposed to form part of an energetic sensing system that integrates autophagy to modulate metabolic programming, metabolite generation and ATP production as well as to minimize ATP wasting by maintaining mitochondrial membrane potential (Loos et al, 2013). Taken together, such a cellular response would increase metabolic robustness and hence cell survival. It needs to, however, be stressed that metabolically, an increase in autophagic flux can only contribute through substrate provision if the substrate preference of the cell and its basal metabolism are favorable for oxidative phosphorylation. Hence, the cell's inherent metabolism dictates to what extent autophagy may contribute metabolically. It could therefore be said that ATP serves as 'currency' that connects the cell death modalities, with autophagy being able to generate ATP, apoptosis depending on ATP consumption for its execution and necrosis being characterized by the severe loss of ATP. It is therefore no surprise that current research questions aim to unravel the relationship between autophagic flux, ATP and the state of the mitochondria.

The rise, fall and re-birth of the point of no return in cell death

The existence and nature of a point of no return (PONR) in cell death has received major attention in the past 20 years. Although already described in the early 1960's (Majno et al, 1960), its role in cellular life-death decision making received increased consideration only from the 1990's onwards, due to the improved molecular biology techniques and microscopy systems. It has now, however, become an integral part of defining cell death modalities as the molecular and biochemical borders between the cell death modes fade. The PONR is defined as the point between induction and execution of cell death, between cell death as a process and cell death as an endpoint, and hence between a still living or already dead cell

(Loos and Engelbrecht, 2009). This contributed substantially to the molecular understanding of the mitochondria in governing cellular fate, as most PONR modalities manifest there. Caspase activation (Golbs et al, 2007; Youle and Strasser, 2008), dissipation of mitochondrial transmembrane potential (Galluzzi et al, 2007), cytochrome c release or mitochondrial membrane permeabilization (Leber et al, 2007; Galluzzi et al, 2008) have been considered as the PONR in programmed cell death. Due to our much better understanding of mitochondrial dynamics, such as fission and fusion activities (Karbowski et al, 2004), 'uniting' perinuclear as a more connected network (Blackstone and Chang, 2011), as well as network properties derived from modelling and biophysics (Chauhan et al, 2014; Loos et al, 2014), the question of the manifestation and molecular role players of the PONR can now be addressed with much greater care. It also remains to be elucidated to what extend the PONR position is affected by selective mitochondrial degradation through autophagy, a process termed mitophagy. Clinically, the PONR attracts major attention, as, when measured, it provides a tool to assess tissue damage and treatment modalities associated with cell death and cell survival. It is now recognized that the concept and definition of cell death has to incorporate a dynamic approach that describes the progression of molecular events that govern induction and execution of cell death. It becomes increasingly clear that the position of the actual point of no return in the signal transduction cascades are context dependent (Galluzzi et al, 2014). Hence, a major integration between intracellular stress response signalling systems and extracellular microenvironment parameters is currently being stressed (Loos et al, 2013), to define this cellular context better. It is however certain, that for the next few years, a crucial aspect in unravelling the control of the PONR has to include a dynamic approach that incorporates ATP and its homeostatic control systems. A systems approach that takes into account glycolysis and mitochondrial respiration rate as well as autophagic flux parameters may here pave the way towards a better evaluation of the causes that push cells beyond the PONR.

The current state of the field

One of the main questions in the field addresses the issue of the variability of autophagy in the context of cell death susceptibility (Loos et al, 2013). Autophagic flux affects cell death. Hence, understanding and measuring autophagy is now, more than ever, an intricate part of understanding cell death onset. It is becoming increasingly clear that autophagy is not only anchored within an energetic feedback loop (Loos et al, 2013) but that it is metabolically controlled by numerous metabolic circuits (Galluzzi et al, 2014). This stresses the need to assess the metabolic environment of the cell, such as ATP production systems and ATP consuming processes (Buttgereit and Brand, 1995) carefully and in context with autophagic flux.

Hence, a fine and reliable measure of autophagic flux that goes beyond the assessment of LC3 protein levels is becoming crucial. A major current question centers around a sensitive means to measure autophagic activity. Unlike in the yeast Saccharomyces cerevisae, where an elegant Pho8Δ80 assay has been designed to measure autophagic flux (Noda and Klionsky, 2008), in the mammalian system this remains a challenge. A major shift in the understanding and assessment of autophagic flux was achieved when the 'autophagic flux assay' was introduced in 2009 (Rubinsztein et al, 2009). Here it was stressed that the degradation rate of autophagosomes can be measured accurately only when concomitantly inhibiting the fusion process between autophagosomes and lysosomes. A combination of multiple techniques such as fluorescence microscopy, to probe for LC3-GFP positive puncta, western blot analysis and EM was recommended (Figure 1). Without such an intervention, the presence of autophagic vacuoles could be a result of either increased synthesis or decreased fusion with lysosomes, likewise, the absence of autophagic vacuoles may be a result of either decreased synthesis or significantly increased fusion with lysosomes. This was fundamental and led to a much better understanding of basal autophagic flux and the potency of autophagy modulators. However, since this system is based on the assessment of LC3-II protein levels, a dynamic approach seemed challenging.

Recent advances in the development of photoswitchable proteins have made it possible to report on the half life of the respective proteins in macroautophagy (Tsvetkov et al, 2013) as well as chaperone mediated autophagy (Koga et al, 2011). Such an approach provides invaluable information into the kinetics of the cellular system. In our laboratory we have taken this approach into the arena of whole cell poolsize assessment, and have developed a live cell imaging based approach with a component of metabolic control analysis, that allows us to assess the complete autophagosomal pool size (Figure 2) based on z-stack image acquisition in time, with and without the utilization of the H+-ATPase inhibitor bafilomycin A1 (Baf). Crucial in this approach is the utilization of saturating Baf concentrations, so as not to allow residual autophagic flux to take place. Next, the quantitative count of autophagosome structures is required, by using a modified watershed based algorhithm (Gniadek and Warren, 2007), which delivers the complete autophagosome pool size (nA). The cell needs to be selected through a region of interest (ROI) analysis, the puncta thresholded, and the puncta counted using a ImageJ software (Figure 2). Plotting of the pool size data in time under Baf treatment reveals the autophagic flux (J), i.e. the rate of autophagosomal degradation, and can be expressed as number of autophagosomes/cell/hour. Unlike with photoswitchable proteins, where the half life τ of the protein is reported, the ratio of pool size (nA) and flux (J) provides the transition time tau (τ), which is the measure of the time required by a cell to clear its complete autophagosomal pool. The advantage of this approach is the direct comparison of multiple cellular systems based on their basal autophagic flux J (Figure 3 A) as well as the response to treatment intervention to upregulate autophagic flux (Figure 3 B). Moreover, the time required to reach a defined autophagic flux can now be measured and compared between model systems. The ratio of pool size and flux provides here a robust numerical value.

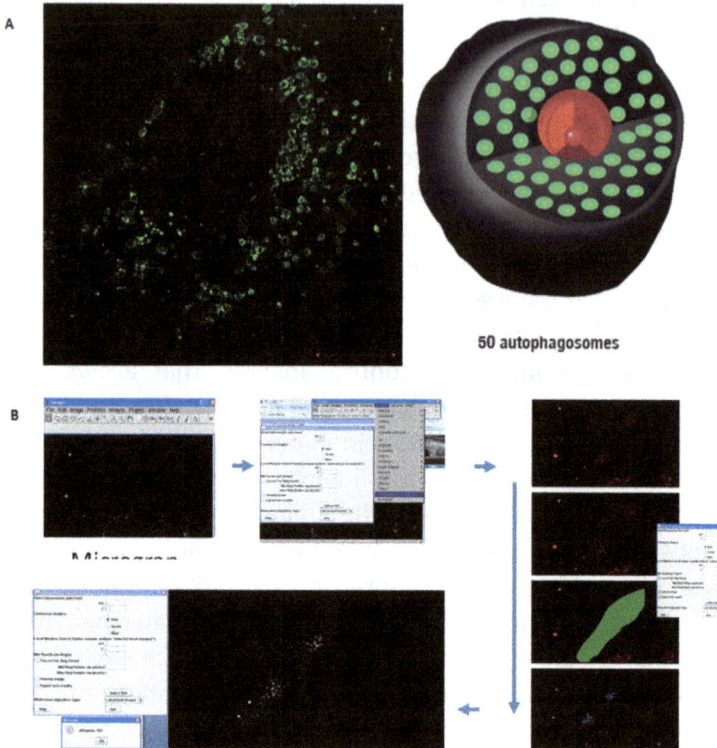

Figure 13.2: Counting the complete autophagosomal pool. A) The calculation of the autophagosomal pool per cell requires z-stack image data, 3 dimensional image acquisition and good signal resolution. The image and the mock-up illustration representation represent autophagosomes in a single cell. B) Automated counting of autophagosomes per cell based on z-stack image data using Image-J based Watershedcounting3D- plug-in is based on: thresholding (B-1), requires the selection of the cell to be analysed through region of interest (ROI) selection (B-2) and generates the count of intracellular LC3 positive structures as well as a new micrograph showing the counted structures (B-3 and B-4).

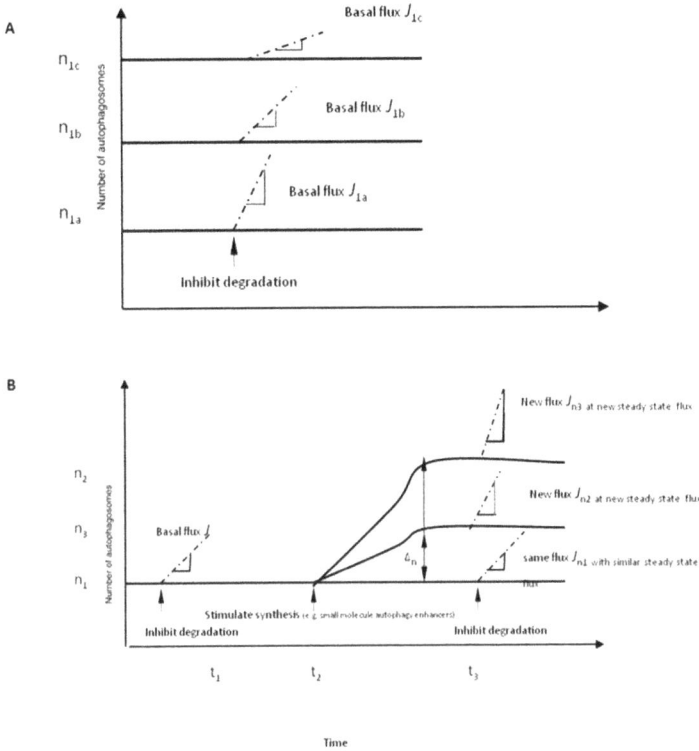

Figure 13.3: Plotting the change in autophgagosomal pool size: valuable information that can be gained when directly comparing cell lines in basal flux and steady state. A) High autophagosomal pool size (n) but low autophagic flux (J), intermediate n and intermediate flux or low n and high flux. The ratio of pool size and basal flux n/J1 per cell type provides the transition time tau (τ), which indicates the time required to clear the complete intracellular autophagosomal pool. B) Different responses to pharmacological regulators can now be assessed, each resulting in a defined pool size, steady state, and new flux.

At present, autophagic flux through time lapse microscopy in this manner is only being assessed in in vitro models. However, a main challenge and current question is how to provide autophagic flux data derived from tissue or biopsies? This is technically challenging, as it would require the immediate incubation of the tissue in a respective media in the presence and absence of Baf. It is likely that, at least on the protein level,

LC3 can be assessed in that manner, thereby quantifying whether autophagic flux increases or decreases significantly (note, not the flux J per se, as this will require live tissue imaging). However, the recent developments in fluorescence techniques such as light sheet microscopy (Pampaloni et al, 2013) may be one of the most suitable tools to achieve that. It is plausible to use, for example, live tissue from a LC3-GFP mouse model (Mizushima et al, 2004) and to capture 3 D data in real time.

The centrality of ATP and cell death

Major advancements have been made in the past two decades in assessing intracellular ATP consuming processes. Here, a hierarchy of ATP consumers was first described in 1995 (Buttgereit and Brand, 1995). The role of ATP in cell death became clear. Although it was already known that necrotic cell death is characterized by a fundamental loss in ATP, the relationship between apoptosis and necrosis in an ATP dependent manner was only more recently revealed (Leist et al, 1997). It had been elegantly demonstrated that as ATP is being depleted under conditions of an 'insult', the mode of cell death shifts from apoptosis to necrosis. Moreover, such necrotic cells could be rescued to complete apoptosis when their ATP generating systems were enabled. If ATP levels are preserved, above 50% of normal homeostatic levels of intracellular ATP, cytochrome c is released and caspases and endonucleases are activated. Importantly, at any time, ATP depletion can supervene to cause necrosis (Leist et al, 1997). Recently, through the development of a FRET-sensor based approach, ATP has been visualized and areas of differential ATP demands within the same cell assessed (Imamura et al, 2009). By using specific target sequences, even mitochondrial versus glycolytic derived ATP can be assessed in this manner (Kioka et al, 2013). This will allow us to understand much better the metabolic perturbations as well as metabolic 'Achilles heel' regions in single cells. Most interesting in this context is the recent discovery of autosis, a form of autophagic cell death (Liu et al,

2013). Here, autophagy inducing peptides have been utilized to induce autophagic cell death, which can be prevented by treating the cells with cardiac glycosides. When establishing the targets that are modified, it has been shown that autosis is a Na^+/K^+-ATPase–regulated process. As this ATPase utilizes highest quantities of ATP and is positioned high in the hierarchy of ATP consuming processes in the cell, the relationship between ATP consumption and autophagic activity in becoming increasingly apparent.

Expectations for further development of the field

There are a number of fundamental questions that deserve major attention in the future. The advancements of techniques and tools to study autophagy and cell death have already allowed us to push the field significantly forward. Automated single cell acquisition systems and algorithms to analyze data in a more powerful way have already started to shift many areas in the field. For example, the analysis of GFP-huntingtin expressing neurons over a duration of days and the observation of development of protein aggregates has revealed that, unlike previously assumed, the cells that manifest with huntingtin (Htt) positive inclusion bodies are the cells that undergo cell death last (Arrasate et al, 2004). Such an approach using automated microscopy, that allows the the experimenter to return to precisely the same neuron after a defined interval, allows also the description of cell death in a powerful manner, as cumulative survival and cumulative risk of death (Scrucca et al, 2010). Such data output is highly suitable to probe interventions against, and to assess how the risk of induction of cell death is being affected. It is moreover expected that a systems biology approach, as already initiated with some key cellular parameters such as ATP generating systems, oxidative phosphorylation, glycolysis and mitochondrial fission and fusion rate (Chauhan et al, 2014), will allow us to describe, model and subsequently predict cell death onset. With shared, cloud-based computing power, the complexity of such processes will be accommodated adequately. We can moreover expect an improved approach to measure autophagic flux *in vivo* (especially in the brain), and thereby to identify peripheral markers for diseases characterized by protein aggregation and proteotoxicity , such as Alzheimer's disease. The recent introduction and commercialization of the light sheet technology (Pampaloni et al, 2013) will allow the imaging of living tissue in three dimensions over time, with minimal phototoxicity. If combined with suitable transgenic model systems, *in vivo* imaging of autophagy and quantification of its flux as well as onset of cell death will become feasible. Correlation analysis of isolated cells, such as fibroblasts, with whole organ imaging may contribute to an advancement of such an approach. Another major area that will likely receive a major boost in the near future is that of searching for a suitable and applicable autophagic 'fluxometer'. Here, recent developments in genomics and proteomics of the human autophagy system and its network organization (Behrends et al, 2010) may assist in identifying flux indicator or flux response proteins that may be used for a microchip-based 'fluxometer' device. Incremental changes in flux with an accompanying modelling approach would be a crucial component to realize such data sets. This will allow us to utilize autophagy modulators for example in the setting of neurodegeneration or gliomas in a pharmacokinetically much more controlled and precise manner (Figure 4). Finally, the development of super-resolution techniques, such as structured illumination (SR-SIM), photoactivation localization microscopy (PALM) and stochastic optical reconstruction microscopy (STORM) will with no doubt unravel the molecular regulation and control of autophagy, apoptosis and necrosis in a most novel manner (Betzig et al, 2006). Taken together, we can expect a continuation of breakthroughs in our understanding of cell death, with improved targets for pharmaceutical intervention. The controlled manipulation of autophagy in this scenario will thereby play a cardinal role.

Conditions/Groups	n_A	J	τ
Control	n_A Control	J Control	τ Control
Treatment	n_A Treatment	J Treatment	τ Treatment
Δ Control/Treatment	Δn_A	ΔJ	$\Delta \tau$

Figure 13.4: The potential future assessment of autophagic flux in diseased cells and tissues. Basal flux and basal transition time (Control) have to first be established and compared to the diseased/to be treated (Treatment) condition. The differential Δ can then be used as guiding value required to maintain a beneficial and physiological autophagic flux to offset the pathological deviation.

References:

Arrasate, M., Mitra, S., Schweitzer, E.S., Segal, M.R., Finkbeiner, S. (2004). Inclusion body formation reduces levels of mutant huntingtin and the risk of neuronal death. Nature;431(7010):805-10.

Behrends, C., Sowa, M.E., Gygi, S.P., Harper, J.W. (2010). Network organization of the human autophagy system. Nature;466(7302):68-76.

Betzig, E., Patterson, G.H., Sougrat, R., Lindwasser, O.W., Olenych, S., Bonifacino, J.S., Davidson, M.W., Lippincott-Schwartz, J., Hess, H.F. (2006). Imaging intracellular fluorescent proteins at nanometer resolution. Science;313(5793):1642-5.

Blackstone, C. and Chang, C.R. (2011). Mitochondria unite to survive. Nat Cell Biol.;13:521-2.

Blommaart, E. F., Luiken, J. J., Blommaart, P. J.,van Woerkom, G. M. & Meijer, A. J. (1995). Phosphorylation of ribosomal protein S6 is inhibitory for autophagy in isolated rat hepatocytes. J. Biol. Chem. 270, 2320–2326.

Buttgereit, F. and Brand, M.D. (1995). A hierarchy of ATPconsuming processes in mammalian cells. Biochem J; 312:163-7.

Chauhan, A., Vera, J. and Wolkenhauer, O. (2014). The systems biology of mitochondrial fission and fusion and implications for disease and aging. Biogerontology; 15:1-12.

Deter, R.L., Baudhuin, P., De Duve, C. (1967). Participation of lysosomes in cellular autophagy induced in rat liver by glucagon. J Cell Biol, 35, C11-6.

Edinger, L.E., Thompson, C.B. (2004). Death by design: apoptosis, necrosis and autophagy. Curr Opin Cell Biol; 16:663-9.

Festiens, N., Vanden Berghe, T., Vandenabeele, P. (2006). Necrosis, a well-orchestrated form of cell demise: Signalling cascades, important mediators and concominant immuneresponse. Biochimica et Biophysica Acta; 1757:1371-87.

Galluzzi, L., Bravo-San Pedro, J.M., Vitale, I., Aaronson, S.A., Abrams, J.M., Adam, D. et al. (2015). Essential versus accessory aspects of cell death: recommendations of the NCCD. Cell Death Differ;22(1):58-73.

Galluzzi, L., Morselli, E., Kepp, O., Kroemer, G. (2008). Targeting post-mitochondrial effectors of apoptosis for neuroprotection. Biochim Biophys Acta; 1787:402-13.

Galluzzi, L., Pietrocola, F., Levine, B., Kroemer, G. (2014). Metabolic Control of Autophagy. Cell ;159(6):1263-1276.

Galluzzi, L., Zamzami, N., de La Motte Rouge, C.L., Brenner, C., Kroemer, G. (2007). Methods for the assessment of mitochondrial membrane permeabilization in apoptosis. Apoptosis; 12:803-13.

Gniadek, T.J. and Warren, G. (2007). WatershedCounting3D: A New Method for Segmenting and Counting Punctate Structures from Confocal Image Data. Traffic; 8: 339-346.

Golbs, A., Heck, N., Luhmann, H.J. (2007). A new technique for real-time analysis of caspase-3 dependent neuronal cell death. J Neurosci Methods; 161:234-43.

Hamacher-Brady, A., Brady, N.R., Gottlieb, R.A. (2006). Enhancing Macroautophagy protects

against ischemia/reperfusion injury in cardiac myocytes. J Biol Chem; 281:29776-87.

Imamura, H., Nhat, K.P., Togawa, H., Saito, K., Iino, R., Kato-Yamada, Y., Nagai, T., Noji, H. (2009). Visualization of ATP levels inside single living cells with fluorescence resonance energy transfer-based genetically encoded indicators. Proc Natl Acad Sci U S A.;106(37):15651-6.

Kabeya, Y., Mizushima, N., Ueno, T., Yamamoto, A., Kirisako, T., Noda, T., Kominami, E., Ohsumi, Y., Yoshimori, T. (2000). LC3, a mammalian homologue of yeast Apg8p, is localized in autophagosome membranes after processing. EMBO J. 19, 5720–5728.

Karbowski, M., Arnoult, D., Chen, H., Chan, D.C., Smith, C.L., Youle, R.J. (2004). Quantitation of mitochondrial dynamics by photolabeling of individual organelles shows that mitochondrial fusion is blocked during the Bax activation phase of apoptosis. J Cell Biol; 164:493-9.

Kioka, H., Kato, H., Fujikawa, M., Tsukamoto, O., Suzuki, T., Imamura, H., Nakano, A., Higo, S., Yamazaki, S., Matsuzaki, T., Takafuji, K., Asanuma, H., Asakura, M., Minamino, T., Shintani, Y., Yoshida, M., Noji, H., Kitakaze, M., Komuro, I., Asano, Y., Takashima, S. (2013). Evaluation of intramitochondrial

ATP levels identifies G0/G1 switch gene 2 as a positive regulator of oxidative phosphorylation.Proc Natl Acad Sci U S A.;111(1):273-8.

Koga, H., Martinez-Vicente, M., Macian, F., Verkusha, V.V. and Cuervo, A.M. (2011). A photoconvertible fluorescent reporter to track chaperone-mediated autophagy. Nat Commun; 2:386.

Komatsu, M., Waguri, S., Chiba, T., Murata, S., Iwata, J., Tanida, I., Ueno, T., Koike, M., Uchiyama, Y., Kominami, E., Tanaka, K. (2006). Loss of autophagy in the central nervous system causes neurodegeneration in mice. Nature 441, 880–884 .

Kroemer, G., El-Deiry, W.S., Golstein, P., Peter, M.E., Vaux, D., Vandenabeele, P. et al. (2005). Classification of cell death: recommendations of the Nomenclature Committee on Cell Death. Cell Death Differ; 12:1463-7.

Kroemer, G., Galluzzi, L., Vandenabeele, P., Abrams, J., Baehrecke, E.H., Blagosklonny, M.V.,et al. (2009). Classification of cell death: recommendations of the Nomenclature Committee on Cell Death. Cell Death Differ; 16:3-11.

Kuma, A., Hatano, M., Matsui, M., Yamamoto, A., Nakaya, H., Yoshimori, T., Ohsumi, Y., Tokuhisa, T., Mizushima, N. (2004). The role of autophagy during the early neonatal starvation period. Nature 432, 1032–1036.

Kunz , J. Henriquez, R., Schneider, U., Deuter-Reinhard, M., Movva, NR, Hall, M.N. (1993). Target of rapamycin in yeast, TOR2, is an essential phosphatidylinositol kinase homolog required for G1 progression. Cell 73, 585–596.

Leber, B., Lin, J., Andrews, D.W. (2007). Embedded together: The life and death consequences of interaction of the Bcl-2 family with membranes. Apoptosis; 12:897-911.

Leist, M., Single, B., Castoldi, A.F., Kuhnle, S., Nicotera, P. (1997). Intracellular adenosine triphosphase (ATP) concentration: A switch in the decision between apoptosis and necrosis. J Exp Med; 185:1481-6.

Liang, X. H., Jackson, S., Seaman, M., Brown, K., Kempkes, B., Hibshoosh, H., Levine, B. (1999). Induction of autophagy and inhibition of tumorigenesis by beclin 1. Nature 402, 672–676.

Liu, Y., Shoji-Kawata, S., Sumpter, R.M. Jr, Wei, Y., Ginet, V., Zhang, L., Posner, B., Tran, K.A., Green, D.R., Xavier, R.J., Shaw, S.Y., Clarke, P.G., Puyal, J., Levine, B. (2013). Autosis is a Na+,K+-ATPase-regulated form of cell death triggered by autophagy-inducing peptides, starvation, and hypoxia-ischemia. Proc Natl Acad Sci U S A.;110(51):20364-71.

Loos, B. and Engelbrecht, A.M. (2009). Cell death: a dynamic response concept. Autophagy; 5:590-603.

Loos, B., Engelbrecht, A.M., Lockshin, R.A., Klionsky, D.J. and Zakeri, Z. (2013). The variability of autophagy and cell death susceptibility: Unanswered questions. Autophagy; 9:1270-1285.

Loos, B., Genade, S., Ellis, B., Lochner, A., Engelbrecht, A.M. (2011). At the core of survival: autophagy delays the onset of both apoptotic and necrotic cell death in a model of ischemic cell injury. Exp Cell Res; 317:1437-53.

Loos, B., Hofmeyr, J.H., Müller-Nedebock, K., Boonzaaier, L. and Kinnear, C. (2014) Autophagic flux, fusion dynamics and cell death. Autophagy: Cancer, Other Pathologies, Inflammation, Immunity, Infection, and Aging, 1st Edition, Volume 3 - Mitophagy. Edited by E. Hayat. ISBN 9780124055292. Elsevier Inc.

Majno, G., La Gattuta, M., Thompson, T.E. (1960). Cellular death and necrosis: chemical, physical and morphologic changes in rat liver. Virchows Arch Pathol Anat; 333:421-465.

Melendez, A., Tallóczy, Z., Seaman, M., Eskelinen, E.L., Hall, D.H., Levine, B. (2003). Autophagy genes are essential for dauer development and life-span extension in C. elegans. Science 301, 1387–1391.

Mizushima, N., Sugita, H., Yoshimori, T. & Ohsumi, Y. (1998). A new protein conjugation system in human. The counterpart of the yeast Apg12p conjugation system essential for autophagy. J. Biol. Chem. 273, 33889–33892.

Mizushima, N., Yamamoto, A., Matsui, M., Yoshimori, T., Ohsumi, Y. (2004).In Vivo Analysis of Autophagy in Response to Nutrient Starvation Using Transgenic Mice Expressing a Fluorescent Autophagosome Marker. Molecular Biology of the Cell Vol. 15, 1101–1111.

Noda, T. and Klionsky, D.J. (2008). The quantitative pho8Δ60 assay of nonspecific autophagy. Meth Enzymol; 451:33-42.

Pampaloni, F., Ansari, N. and Stelzer, E.H.K. (2013). High-resolution deep imaging of live cellular spheroids with light-sheet-based fluorescence microscopy. Cell Tissue Res; 352: 161-177.

Ravikumar, B., Duden, R. & Rubinsztein, D. C. (2002). Aggregate-prone proteins with polyglutamine and polyalanine expansions are degraded by autophagy. Hum. Mol. Genet. 11, 1107–1117.

Rubinsztein, D.C., Cuervo, A.M., Ravikumar, B., Sarkar, S., Korolchuk, V., Kaushik, S., Klionsky, D.J. (2009). In search of an "autophagomometer". Autophagy; 5(5):585-9.

Scrucca, L., Santucci, A., Aversa, F. (2010). Regression modeling of competing risk using R: an in depth guide for clinicians. Bone Marrow Transplant;45(9):1388-95.

Shen, S. Kepp, O. and Kroemer, G. (2012). The end of autophagic cell death? Autophagy 8:1, 1-3.

Tsvetkov, A.S., Arrasate, M., Barmada, S., Ando, D.M., Sharma, P., Shaby, B.A. and Finkbeiner, S. (2013). Proteostasis of polyglutamine varies among neurons and predicts neurodegeneration. Nat Chem Biol; 9:586-592.

Wu Y.T., Tan, H.L., Huang, Q., Kim, Y.S., Pan, N., Ong, W.Y., Liu, Z.G., Ong, C.N., Shen, H.M. (2008). Autophagy plays a protective role during zVAD-induced necrotic cell death. Autophagy;4(4):457-66.

Youle, R.J., Strasser, A. (2008). The BCL-2 protein family: opposing activities that mediate celldeath. Nature Rev Mol Cel Biol; 9:47-59.

Chapter 14: Discovery of key mechanisms of cell death: from apoptosis to necroptosis (Yuan)

Junying Yuan. Department of Cell Biology, Harvard Medical School, 240 Longwood Ave. Boston, MA 02115. USA

Abstract:

This essay provides a brief account of my 30+ years of research in the field of cell death that has, thus far, led to the discovery of caspases in regulating apoptosis in mammals and a form of regulated necrosis known as necroptosis. My discoveries have demonstrated the molecular mechanisms of regulated cell death in mammalian cells, refuted the dogma that claimed necrosis to only be a passive form of cell death, and introduced the possibility of targeting necrosis in the treatment of human diseases.

Introduction

My interest in cell death was inspired by the specific loss of neuronal cells during normal embryonic development and during neurodegeneration that I had learned in the classes as a first-year graduate student at the Harvard Medical School in early 1980's. This interest compelled me to join the laboratory of Dr. H. R. Horvitz at MIT, where I studied mechanisms of programmed cell death in the nematode C. elegans as a part of my Ph.D. research. After obtaining my Ph.D. in Neurosciences from Harvard in 1989, I went on to establish my own laboratory at the Cardiovascular Research Center at Massachusetts General Hospital, intending to investigate the idea that mammalian cells might possess a genetically encoded, programmed cell death mechanism similar to what I had found in the nematode C. elegans. Little did I anticipate that my subsequent research would make pioneering discoveries t on key pathways involved not only in apoptosis, a mechanism of cell death that is evolutionarily conserved between nematodes and humans, but also in necroptosis, a mechanism that exists largely only in the vertebrate kingdom.

The first demonstration of caspases in regulating apoptosis of mammalian cells

The year 1994 witnessed the publication of our paper entitled, "Prevention of vertebrate neuronal death by the crmA gene" in Science (Gagliardini et al., 1994). This was the second paper that came out of my own laboratory, which was yet four years young. In this paper, we provided the first evidence for the role of caspases in regulating neuronal cell death. We demonstrated the ability of crmA, a viral inhibitor of caspase--1 known at the time as interleukin-1β converting enzyme, to block neuronal cell death induced by trophic factor deprivation. The year before, we had published a Cell paper entitled, "Induction of apoptosis in fibroblasts by IL-1 beta-converting enzyme, a mammalian homolog of the C. elegans cell death gene ced-3" in which we demonstrated the ability of caspase-1 and ced-3 to induce apoptosis in a manner dependent upon their cysteine protease activity in mammalian cells, as well as the ability of Bcl-2 to block caspase-mediated cell death (Miura et al., 1993). Caspase-1 mediated cell death, now known as pyroptosis, is a form of inflammatory caspase-dependent cell death similar to the basic mechanisms of apoptosis. Pyroptosis is morphologically distinguished by cell lysis that can be activated by several different molecules, including bacterial toxins as well as viral RNA (Henao-Mejia et al., 2012). These two papers – Miura et al. 1993 and Gagliardini et al. 1994 – provided the earliest evidence for the functional role of caspases in regulating apoptosis of mammalian cells. The model provided in Miura et al. 1993 (Figure 1) has been shown to hold true after more than two decades of research on caspases. These exciting first glimpses into cell death machinery regulated by caspases in mammalian cells generated tremendous interest in the world of research. Today, more than 70,000 papers have been published following these initial two.

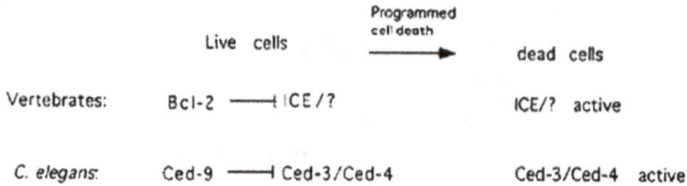

Figure 1. A model for the evolutionarily conserved molecular mechanisms of apoptosis (Taken from (Miura et al., 1993)).

A dual role of caspase-11 in regulating inflammation and apoptosis in mammalian cells

Following our discoveries regarding the role of caspases in regulating apoptosis in mammalian cells, we proceeded to identify and characterize additional members of the caspase family in higher organisms. We operated under the hypothesis that mammalian cells should have additional proteases that would comprise a family of caspases, and that caspase-1 was not unique. As this all took place prior to the completion of the Human Genome Project, our lab used low stringency hybridization, low stringency PCR, and other clever approaches in order to identify homologues of caspase-1. Our efforts led to the identification of caspase-2, known at the time as Ich-1 (Wang et al., 1994), and caspase-11, known at the time as Ich-3 (Wang et al., 1996; Wang et al., 1998), as well as caspase-12 (Nakagawa et al., 2000). We showed that caspase-2, an upstream caspase in mediating apoptosis, was required to mediate apoptosis of germ cells in mouse ovaries during aging and following exposure to chemotherapeutic drugs (Bergeron et al., 1998). The roles and mechanisms of caspase-2 as an upstream mediator of apoptosis would then become a subject of continuing investigation and debate over the next two decades. We also showed that caspase-12 was important in mediating ER stress-induced apoptosis (Nakagawa et al., 2000).

Caspase-11 we systematically studied, specifically regarding its roles in regulating inflammation and apoptosis. It was isolated as a homologue of human caspase-1 from a murine cDNA library by way of low stringency hybridization (Wang et al., 1996). We found that expression of caspase-11 was highly inducible upon stimulation by lipopolysaccharide (LPS), an endotoxin from Gram-negative bacteria. Since the expression of other caspases was largely unaffected by LPS, this finding suggested that caspase-11 might mediate the innate immune responses activated by LPS. Indeed, we found that caspase-11 knockout mice were highly resistant to LPS- induced septic shock and lethality. Further, the production of both IL-1α and IL-1β following LPS stimulation, a critical event during septic shock and an indication of caspase-1 activation, was blocked in caspase-11 mutant mice (Wang et al., 1998). We showed that caspase-11 interacted with caspase-1, which suggested that the former might regulate the activation of the latter. At the time, caspase-1 was known to directly cleave pro-IL1β into mature IL1β *in vitro* (Cerretti et al., 1992; Thornberry et al., 1992). We found, however, that both lines of caspase-1 knockout mice (Kuida et al., 1995; Li et al., 1995) also lacked caspase-11 expression while caspase-1 expression in our caspase-11 knockout mice was normal (Kang et al., 2000). The role of caspase-1 in the processing and maturation of pro- IL1β in vivo was not demonstrated conclusively until recently, with the generation of a line of caspase-1 knockout mice with preserved normal expression of caspase-11 by transgene expression (Kayagaki et al., 2011). Further, we found that the cleavage of cationic channel subunit transient receptor potential channel 1 (TRPC1) by caspase-11 promoted the secretion of IL-1β without modulating caspase-1 cleavage or cell death in macrophage. This suggested that caspase-11 could be involved in regulating inflammatory responses through the mediation of pathways that lack a conventional secretory signal peptide, such as the secretion of cytokines (Py et al., 2014). In addition to regulating caspase-1 activation and inflammatory responses, we found that caspase-11 could also mediate the activation of caspase-3 by direct cleavage (Kang et al., 2000). The activation of caspase-3 induced by ischemic brain injury and LPS stimulation *in vivo* was significantly reduced in caspase-11 knockout mice (Kang et al., 2000; Kang et al., 2002). We were also surprised to discover that caspase-11 was involved in regulating cell migration. We found that it interacted physically and functionally with actin interacting protein 1 (Aip1), an activator of cofilin-mediated actin depolymerization (Li et al., 2007). Caspase-11, therefore, served multiple roles in regulating inflammatory responses: by regulating the activation of caspase-1 to mediate the maturation of pro-ILβ, by activating caspase-3 to mediate apoptosis, by cleaving TRPC1 in the secretion of IL-1β, and by mediating cell migration in the recruitment of inflammatory cells.

The cleavage of BID as a mediator of mitochondrial damage in death receptor mediated apoptosis

Death receptor-mediated apoptosis is known as the "extrinsic apoptosis" pathway. This name highlights the initiation of apoptosis by extracellular death receptor ligands, although it is not especially precise – all types of apoptosis are mediated by cell--autonomous pathways, either in a mitochondria-dependent or –independent manner. Since death ligands initiate apoptotic signaling by interacting with death receptor located on the cytoplasmic membrane, we wondered how apoptotic signaling might have been transmitted from the cell periphery to the mitochondria. We discovered that the cleavage of BID, a BH-only pro- apoptotic member of the Bcl-2 family, by caspase-8, a member of the caspase family activated by the death receptor complex associated with the cell membrane, played a critical role in initiating mitochondrial damage and other downstream apoptotic mechanisms (Li et al., 1998). The cleaved product of BID, truncated BID (tBID), is a highly potent initiator of mitochondrial damage. We also showed that elevated expression of Bcl-2 could inhibit extrinsic apoptosis in a subset of cell types, as anti-apoptotic members of the Bcl-2 family can inhibit BID.

Discovery of the role of caspases in neural degeneration

Since my initial interest in cell death was in the context of neurodegenerative diseases, my next question pertained to how we might use our knowledge of caspases to reduce neuronal loss in patients. This led to a long and fruitful collaboration with Dr. Michael Moskowitz, a leading investigator in the field of stroke. Together we demonstrated the protective effect of blocking caspases in multiple animal models for ischemic brain injuries (Friedlander et al., 1997; Hara et al., 1997; Kang et al., 2000). Our work clearly illuminated the role of caspases in mediating ischemic brain injury and engaged the efforts of pharmaceutical companies towards the development of caspase inhibitors in the treatment of stroke. Unfortunately, the development of effective inhibitors of caspases was complicated by the fact that it was difficult, if not impossible, to develop specific chemical inhibitors for caspases with reasonable *in vivo* bioavailability.

Discovery of necroptosis as a form of regulated necrosis

An important proinflammatory cytokine was isolated as an endogenous factor in mice treated with an endotoxin that caused certain tumors to undergo hemorrhagic necrosis. This cytokine was given the name tumor necrosis factor (Carswell et al., 1975), or TNFα. TNFα purified from rabbit was shown to induce the death of murine fibrosarcoma-derived L929 cells (Ruff and Gifford, 1981), which was characterized as necrosis as it led to early permeabilization of the cytoplasmic membrane (Grooten et al., 1993). In late 1998, the field of cell death as a whole had largely reached the consensus that TNFα could induce extrinsic apoptosis through caspase activation. Vercammen et al., however, reported an interesting and paradoxical phenomenon. It was noted that inhibiting caspases made L929 cells much more sensitive to necrosis induced by TNFα (Vercammen et al., 1998). Vercammen et al. showed that inhibition of caspases by zVAD-fmk resulted in a rapid increase of TNFα-mediated production of oxygen radicals. Further, necrotic cell death could be inhibited by the addition of the oxygen radical scavenger butylated hydroxyanisole (BHA). These observations led the authors to conclude that caspases might be involved in protection against TNF-induced formation of oxygen radicals, and that inhibition of caspases would lead to increased production of reactive oxygen species which would then lead to necrosis.

In our own characterization of death receptor-mediated apoptosis, we found that though inhibiting caspases was sufficient to block apoptosis of HeLa cells activated by TNFα and CHX (Miura et al., 1995), it was not able to block cell death in other cell lines including Jurkat cells and U937 cells (Louise Bergeron and Junying Yuan, unpublished observation). Interestingly, we observed that cell death activated by ligands of the death receptor family in the presence of zVAD-fmk resembled necrosis, characterized by early plasma membrane permeabilization rather than the cell shrinkage characteristic of apoptosis. Though we were interested in exploring this mechanism further, in the era prior to an available siRNA library it was very difficult to investigate the molecular mechanism of a phenomenon observed in cultured cells alone. Fortunately, in 1998 Harvard Medical School established the Institute of Chemistry and Cell Biology (ICCB) under the co-directorship of Drs. Stuart Schreiber and Tim Mitchison and the leadership of Dr. Marc Kirschner. The ICCB unlocked large-scale small molecule libraries from the pharmaceutical industry and made them available to academic researchers. Dr. Alexei Degterev, an outstanding postdoctoral fellow with background in chemistry from his university studies in Russia, joined my lab in 1998. Together with Alexei, we decided to develop high-throughput cell-based screens for the identification of small molecule inhibitors of TNFα- mediated necrosis. However, we were advised that this type of cell-based screen was known as a "black-box" screen in the pharmaceutical industry as it was very difficult to identify the molecular targets isolated from cell-based assays. We were also concerned that, because TNF-induced necrosis of L929 cells was inhibited by oxygen radical scavengers (Vercammen et al., 1998), we might end up with many nonspecific hits if our cell-based models were similarly sensitive to redox status. It was well known at the time that chemical libraries were full of such molecules. Luckily, the oxygen radical scavenger BHA and a panel of general antioxidant compounds inhibited TNF-induced necrosis of neither U937 cells nor FADD-deficient Jurkat cells, our two cell-based models, with or without zVAD.fmk. This made it possible for us to explore key mechanisms involved in mediating necrosis using these cell-based models without having to worry about many nonspecific hits with general antioxidant properties (Degterev et al., 2005).

Our initial screen of ~15,000 compounds in a commercially available small molecule library led to the identification of necrostatin-1, Nec-1 for short, a small molecule inhibitor of necroptosis. By the time it was published in 2005, we had already found that Nec-1 could inhibit all models of necrosis published at the time, including those induced by ligands of death receptors such as TNF-induced necrosis of L929 cells, TNFα/zVAD-induced necrosis of 3T3 cells, embryonic fibroblast cells, HT29 cells, as well as FasL/zVAD/CHX-induced necrosis of Jurkat cells, and so on (Degterev et al., 2005). The ability of Nec-1 to inhibit necrosis in multiple cell types, regardless to their sensitivity to anti-oxidant, made it possible for us to propose that mammalian cells might have a common necrotic cell death pathway, termed necroptosis, in the absence of caspase activation. Nec-1 has become a popular tool for researchers in determining if their cell death models might involve necroptosis (Zhou and Yuan, 2014). We have since screened additional small molecule libraries, a total of about 500,000 compounds, in search of necroptosis inhibitors.

Unlike as in L929 cells, an increase in reactive oxygen species was observed only in a subpopulation of FADD-deficient Jurkat cells stimulated to die through necroptosis by TNFα. Further, necroptosis induced by dimerization of FADD, which is inhibited by Nec-1, was not accompanied by any sort of oxidative stress. Additionally, Nec-1 could not block the classic oxidative stress-induced cell death by menadione (Degterev et al., 2005). This was evidence that an increase in ROS was not an obligatory step in necroptosis, nor was Nec-1 likely to work by regulating the redox status of the cells. However, trying to pinpoint the exact molecular target of Nec-1 was extremely challenging, like "fishing for a needle in the ocean," a Chinese proverb describing an extremely challenging task.

The ability of Nec-1 to completely block necroptosis of FADD-deficient Jurkat cells stimulated by TNFα to allow normal proliferation (Degterev et al., 2005) suggests that the target of Nec-1 in the TNFα pathway controls a key upstream step in necroptosis. Given that the structure of Nec-1 contains an indole commonly found in kinase inhibitors, we suspected that the target of Nec-1 in necroptosis might be a kinase. Dr. Jurg Tschopp and his colleagues made an important discovery that the kinase activity of RIP1 was required for FasL and zVAD.fmk-induced necrosis of Jurkat cells (Holler et al., 2000). A general role of RIP1 kinase activity in regulating necrosis, however, was difficult to prove without generating individual RIP1 kinase knock-in mutant cell lines and a RIP1 kinase dead mutant mouse line. Importantly, we found that necrosis induced by the dimerization of RIP1 kinase was inhibited by Nec-1 (Degterev et al., 2005). Thus, we concluded that the target of Nec-1 might be RIP1 kinase itself or otherwise its key downstream event. The ease of applying a small molecule inhibitor Nec-1 to inhibit necrosis in a variety of cell-based models and animal models of human diseases turned out eventually to be critical for the rapid acceptance of necroptosis as an important form of regulated necrosis and for overturning the traditional dogma that necrosis can only be passive cell death (Christofferson and Yuan, 2010; Vandenabeele et al., 2010; Kaczmarek et al., 2013; Linkermann and Green, 2014; Pasparakis and Vandenabeele, 2015).

It took another three years of research to convince ourselves that the RIP1 kinase activity was indeed the target of Nec-1

(Degterev et al., 2008). We showed that Nec-1, but not its inactive analog, could inhibit the kinase activity of RIP1 kinase *in vitro*. Our structural modeling study showed that the activation T-loop segment of RIP1, which was very similar to that of B-RAF, played an important role in regulating RIP1 kinase activity. The mutations in Ser161 of RIP1, an autophosphorylation site corresponding to the Thr598 autophosphorylation site of B-RAF, made RIP1 insensitive to inhibition by Nec-1 while still maintaining the activity of RIP1 kinase.

Although the original necrostatin-1 isolated from the screen, methyl-thiohydantoin-Trp (MTH- Trp), can inhibit indoleamine 2,3-dioxygenase (IDO) activity at low level as well as RIP1 kinase activity, an improved analogue of necrostatin-1, R-7-Cl-O-Nec-1 [R- (5-(7-chloro-1H-indol-3-yl)methyl)-3-methylimidazolidine-2,4-dione)], developed by our chemistry collaborator Dr. Greg Cuny's lab, has no IDO inhibitory activity (Degterev et al., 2012; Takahashi et al., 2012). In an X-ray crystallography study, Dr. Yigong Shi's lab showed that R-7-Cl-O-Nec-1 binds to a hydrophobic pocket between the N- and C-lobes of the RIP1 kinase domain, and in close proximity to the activation loop but exterior to the ATP pocket. The ATP pocket is the common targeting site for most kinase inhibitors developed; these are known as type-I inhibitors. The binding of R-7-Cl-O-Nec-1 stabilizes RIP1 kinase in an inactive conformation, known as the Asp-Leu-Gly-out (or Asp-Phe-Gly out) conformation, through interactions with highly conserved amino acids in the activation loop and surrounding structural elements (Xie et al., 2013). Point mutations in RIP1 such as Ser161Ala and Ser161Glu that alter the conformation of the RIP1 kinase T-loop and block H-bond formation with R-7-Cl-O-Nec-1 prevent necrostatin-1-mediated inhibition of RIP1 activity and thereby necroptosis. Thus, R-7-Cl-O-Nec-1 is an excellent example of a type-II kinase inhibitor. 7-Cl-O-Nec-1 demonstrates exclusive selectivity towards RIP1, as it inhibits RIP1 kinase a thousand-fold more potently than 485 other human kinases, including several other members of the RIP family. It also binds to RIP1 with high affinity, with a dissociation constant (Kd) of ~3nM (Christofferson et al., 2012).

The ability of Nec-1 to inhibit RIP1 kinase specifically and effectively made it possible to conveniently test if necroptosis mediated by RIP1 kinase is involved in different cell and animal models of human disease. Degterev et al. demonstrated the ability of Nec-1 to reduce central nervous system injury induced by middle cerebral artery occlusion in an animal model of stroke. This provided the first clue to a possible pathophysiological role of necroptosis in mediating acute neurological damage *in vivo* (Degterev et al., 2005). Subsequently, inhibition of necroptosis by Nec-1 was shown to protect a variety of different animal models of ischemic injuries to the brain, heart, eye, and kidney (Lim et al., 2007; Rosenbaum et al., 2010; Trichonas et al., 2010; Northington et al., 2011; Linkermann et al., 2013). More recently, the list of human diseases that may involve necroptosis has been extended to include age-related macular degeneration of the eye (Murakami et al., 2014), amyotrophic lateral sclerosis (Re et al., 2014), Gaucher's disease (Vitner et al., 2014), Crohn's disease (Gunther et al., 2011; Welz et al., 2011), and Huntington's disease (Zhu et al., 2011). Since necroptosis is closely connected with the innate immunity signaling network, as was demonstrated by our genome-wide siRNA screen on genes regulating necroptosis (Figure 2) (Hitomi et al., 2008), we can expect that future research will demonstrate in the involvement of necroptosis in additional human diseases. The ability of Nec-1 to block death receptor- mediated necrosis *in vitro* and to reduce neurological and tissue injury *in vivo* provides strong evidence to disprove the notion that necrosis commonly observed in human pathology is merely a passive form of cell death caused by overwhelming stress and a proof-principle to block necrosis by pharmacological means.

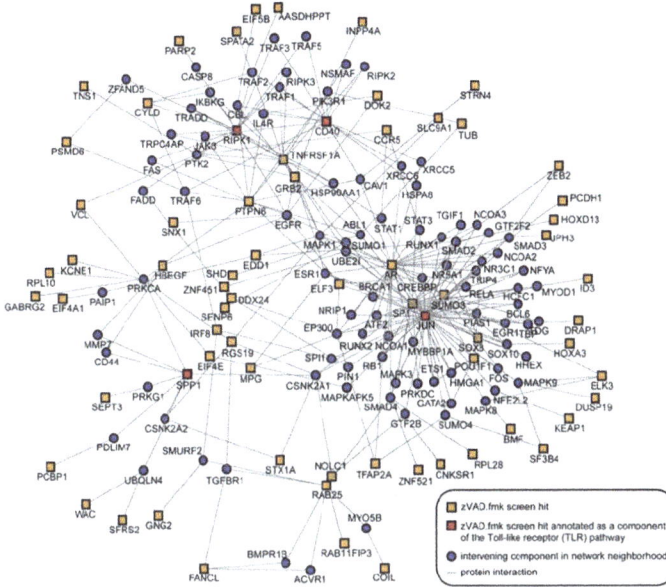

Figure 2. A closely connected network of necroptosis and innate immunity response (Taken from (Hitomi et al., 2008)).

Concluding remarks

Our research on cell death mechanisms in the past two decades has led us from apoptosis to necroptosis, two equally fascinating molecular mechanisms that regulate cellular destruction in mammals. Although both play important roles in regulating physiological and pathological cell death, the insights we have gained regarding their mechanisms show a number of interesting distinctions. First, the development of apoptosis inhibitors for the treatment of human disease has largely been hindered by the lack of known "druggable" targets in the pathway; in contrast, inhibitors of necroptosis have emerged as promising pharmaceutical candidates with the discovery of RIP1 kinase as the "druggable" target in the necroptosis pathway using small molecules like Nec-1 and the demonstrated involvement of necroptosis in the pathophysiology of human disease. Indeed, the development of RIP1 kinase inhibitors for the pharmaceutical treatment of human disease is already underway. Second, as a homologue of neither RIP1 nor RIP3 has been identified in lower organisms including nematode and fly, necroptosis may represent an evolutionary newcomer in the self-destructive arsenal of mammalian cells. Understanding molecular mechanisms unique to higher organisms could provide key insights into the biological mechanisms that maintain the homeostasis of complex multi-cellular organisms.

Acknowledgement:

I thank the past and present members of Yuan laboratory for their contributions to the works described here and for stimulating discussions. This work was supported in part by grants (to JY) from the National Institute on Neurological Diseases and Stroke (1R01NS082257), the National Institute on Aging (R37 AG012859 and 1R01AG047231-01) and a Senior Fellowship from the Ellison Foundation and the support from the Chinese Academy of Sciences for my collaborations with the Shanghai Institute of Organic Chemistry.

Reference:

Bergeron, L., Perez, G.I., Macdonald, G., Shi, L., Sun, Y., Jurisicova, A., Varmuza, S., Latham, K.E., Flaws, J.A., Salter, J.C., et al. (1998). Defects in regulation of apoptosis in caspase-2-deficient mice. Genes Dev 12, 1304-1314.

Carswell, E.A., Old, L.J., Kassel, R.L., Green, S., Fiore, N., and Williamson, B. (1975). An endotoxin-induced serum factor that causes necrosis of tumors. Proc Natl Acad Sci U S A 72, 3666-3670.

Cerretti, D.P., Kozlosky, C.J., Mosley, B., Nelson, N., Van Ness, K., Greenstreet, T.A., March, C.J., Kronheim, S.R., Druck, T., Cannizzaro, L.A., et al. (1992). Molecular cloning of the interleukin-1 beta converting enzyme. Science 256, 97-100.

Christofferson, D.E., Li, Y., Zhou, W., Hitomi, J., Upperman, C., Zhu, H., Gerber, S.A., Gygi, S., and Yuan, J. (2012). A Novel Role for RIP1 kinase in Mediating TNFα Production. Cell Death Diseases 3, e320.

Christofferson, D.E., and Yuan, J. (2010). Necroptosis as an alternative form of programmed cell death. Curr Opin Cell Biol 22, 263-268.

Degterev, A., Hitomi, J., Germscheid, M., Ch'en, I.L., Korkina, O., Teng, X., Abbott, D., Cuny, G.D., Yuan, C., Wagner, G., et al. (2008). Identification of RIP1 kinase as a specific cellular target of necrostatins. Nat Chem Biol 4, 313-321.

Degterev, A., Huang, Z., Boyce, M., Li, Y., Jagtap, P., Mizushima, N., Cuny, G.D., Mitchison, T.J., Moskowitz, M.A., and Yuan, J. (2005). Chemical inhibitor of nonapoptotic cell death with therapeutic potential for ischemic brain injury. Nat Chem Biol *1*, 112-119.

Degterev, A., Maki, J.L., and Yuan, J. (2012). Activity and specificity of necrostatin-1, small-molecule inhibitor of RIP1 kinase. Cell Death Differ *20*, 366.

Friedlander, R.M., Gagliardini, V., Hara, H., Fink, K.B., Li, W., MacDonald, G., Fishman, M.C., Greenberg, A.H., Moskowitz, M.A., and Yuan, J. (1997). Expression of a dominant negative mutant of interleukin-1 beta converting enzyme in transgenic mice prevents neuronal cell death induced by trophic factor withdrawal and ischemic brain injury. J Exp Med *185*, 933-940.

Gagliardini, V., Fernandez, P.A., Lee, R.K., Drexler, H.C., Rotello, R.J., Fishman, M.C., and Yuan, J. (1994). Prevention of vertebrate neuronal death by the crmA gene. Science *263*, 826-828.

Grooten, J., Goossens, V., Vanhaesebroeck, B., and Fiers, W. (1993). Cell membrane permeabilization and cellular collapse, followed by loss of dehydrogenase activity: early events in tumour necrosis factor-induced cytotoxicity. Cytokine *5*, 546-555.

Gunther, C., Martini, E., Wittkopf, N., Amann, K., Weigmann, B., Neumann, H., Waldner, M.J., Hedrick, S.M., Tenzer, S., Neurath, M.F., *et al.* (2011). Caspase-8 regulates TNF-alpha-induced epithelial necroptosis and terminal ileitis. Nature *477*, 335-339.

Hara, H., Friedlander, R.M., Gagliardini, V., Ayata, C., Fink, K., Huang, Z., Shimizu-Sasamata, M., Yuan, J., and Moskowitz, M.A. (1997). Inhibition of interleukin 1beta converting enzyme family proteases reduces ischemic and excitotoxic neuronal damage. Proc Natl Acad Sci U S A *94*, 2007-2012.

Henao-Mejia, J., Elinav, E., Strowig, T., and Flavell, R.A. (2012). Inflammasomes: far beyond inflammation. Nat Immunol *13*, 321-324.

Hitomi, J., Christofferson, D.E., Ng, A., Yao, J., Degterev, A., Xavier, R.J., and Yuan, J. (2008). Identification of a molecular signaling network that regulates a cellular necrotic cell death pathway. Cell *135,1311-1323*.

Holler, N., Zaru, R., Micheau, O., Thome, M., Attinger, A., Valitutti, S., Bodmer, J.L., Schneider, P., Seed, B., and Tschopp, J. (2000). Fas triggers an alternative, caspase-8-independent cell death pathway using the kinase RIP as effector molecule. Nat Immunol *1*, 489-495.

Kaczmarek, A., Vandenabeele, P., and Krysko, D.V. (2013). Necroptosis: the release of damage-associated molecular patterns and its physiological relevance. Immunity *38*, 209-223.

Kang, S.J., Wang, S., Hara, H., Peterson, E.P., Namura, S., Amin-Hanjani, S., Huang, Z., Srinivasan, A., Tomaselli, K.J., Thornberry, N.A., et al. (2000). Dual role of caspase-11 in mediating activation of caspase-1 and caspase-3 under pathological conditions. J Cell Biol 149, 613-622.

Kang, S.J., Wang, S., Kuida, K., and Yuan, J. (2002). Distinct downstream pathways of caspase-11 in regulating apoptosis and cytokine maturation during septic shock response. Cell Death Differ 9, 1115-1125.

Kayagaki, N., Warming, S., Lamkanfi, M., Vande Walle, L., Louie, S., Dong, J., Newton, K., Qu, Y., Liu, J., Heldens, S., et al. (2011). Non-canonical inflammasome activation targets caspase-11. Nature 479, 117-121.

Kuida, K., Lippke, J.A., Ku, G., Harding, M.W., Livingston, D.J., Su, M.S., and Flavell, R.A. (1995). Altered cytokine export and apoptosis in mice deficient in interleukin-1 beta converting enzyme. Science 267, 2000-2003.

Li, H., Zhu, H., Xu, C.J., and Yuan, J. (1998). Cleavage of BID by caspase 8 mediates the mitochondrial damage in the Fas pathway of apoptosis. Cell 94, 491-501.

Li, J., Brieher, W.M., Scimone, M.L., Kang, S.J., Zhu, H., Yin, H., von Andrian, U.H., Mitchison, T., and Yuan, J. (2007). Caspase-11 regulates cell migration by promoting Aip1-Cofilin-mediated actin depolymerization. Nat Cell Biol 9, 276-286.

Li, P., Allen, H., Banerjee, S., Franklin, S., Herzog, L., Johnston, C., McDowell, J., Paskind, M., Rodman, L., Salfeld, J., et al. (1995). Mice deficient in IL-1 beta-converting enzyme are defective in production of mature IL-1 beta and resistant to endotoxic shock. Cell 80, 401-411.

Lim, S.Y., Davidson, S.M., Mocanu, M.M., Yellon, D.M., and Smith, C.C. (2007). The cardioprotective effect of necrostatin requires the cyclophilin-D component of the mitochondrial permeability transition pore. Cardiovasc Drugs Ther 21, 467-469.

Linkermann, A., Brasen, J.H., Darding, M., Jin, M.K., Sanz, A.B., Heller, J.O., De Zen, F., Weinlich, R., Ortiz, A., Walczak, H., et al. (2013). Two independent pathways of regulated necrosis mediate ischemia-reperfusion injury. Proc Natl Acad Sci U S A 110, 12024-12029.

Linkermann, A., and Green, D.R. (2014). Necroptosis. N Engl J Med 370, 455-465.

Miura, M., Friedlander, R.M., and Yuan, J. (1995). Tumor necrosis factor-induced apoptosis is mediated by a CrmA-sensitive cell death pathway. Proc Natl Acad Sci U S A 92, 8318-8322.

Miura, M., Zhu, H., Rotello, R., Hartwieg, E.A., and Yuan, J. (1993). Induction of apoptosis in fibroblasts by IL-1 beta-converting enzyme, a mammalian homolog of the C. elegans cell death gene ced-3. Cell *75*, 653-660.

Murakami, Y., Matsumoto, H., Roh, M., Giani, A., Kataoka, K., Morizane, Y., Kayama, M., Thanos, A., Nakatake, S., Notomi, S., *et al.* (2014). Programmed necrosis, not apoptosis, is a key mediator of cell loss and DAMP-mediated inflammation in dsRNA-induced retinal degeneration. Cell Death Differ *21*, 270-277.

Nakagawa, T., Zhu, H., Morishima, N., Li, E., Xu, J., Yankner, B.A., and Yuan, J. (2000). Caspase-12 mediates endoplasmic-reticulum-specific apoptosis and cytotoxicity by amyloid-beta. Nature *403*, 98-103.

Northington, F.J., Chavez-Valdez, R., Graham, E.M., Razdan, S., Gauda, E.B., and Martin, L.J. (2011). Necrostatin decreases oxidative damage, inflammation, and injury after neonatal HI. J Cereb Blood Flow Metab *31*, 178-189.

Pasparakis, M., and Vandenabeele, P. (2015). Necroptosis and its role in inflammation. Nature *517*, 311-320.

Py, B.F., Jin, M., Desai, B.N., Penumaka, A., Zhu, H., Kober, M., Dietrich, A., Lipinski, M.M., Henry, T., Clapham, D.E., *et al.* (2014). Caspase-11 controls interleukin-1beta release through degradation of TRPC1. Cell reports *6*, 1122-1128.

Re, D.B., Le Verche, V., Yu, C., Amoroso, M.W., Politi, K.A., Phani, S., Ikiz, B., Hoffmann, L., Koolen, M., Nagata, T., *et al.* (2014). Necroptosis drives motor neuron death in models of both sporadic and familial ALS. Neuron *81*, 1001-1008.

Rosenbaum, D.M., Degterev, A., David, J., Rosenbaum, P.S., Roth, S., Grotta, J.C., Cuny, G.D., Yuan, J., and Savitz, S.I. (2010). Necroptosis, a novel form of caspase-independent cell death, contributes to neuronal damage in a retinal ischemia-reperfusion injury model. J Neurosci Res *88*, 1569-1576.

Ruff, M.R., and Gifford, G.E. (1981). Rabbit tumor necrosis factor: mechanism of action. Infection and immunity *31*, 380-385.

Smith, C.C., Davidson, S.M., Lim, S.Y., Simpkin, J.C., Hothersall, J.S., and Yellon, D.M. (2007). Necrostatin: a potentially novel cardioprotective agent? Cardiovasc Drugs Ther *21*, 227-233.

Takahashi, N., Duprez, L., Grootjans, S., Cauwels, A., Nerinckx, W., DuHadaway, J.B., Goossens, V., Roelandt, R., Van Hauwermeiren, F., Libert, C., *et al.* (2012). Necrostatin-1 analogues: critical issues on the specificity, activity and in vivo use in experimental disease models. Cell Death Dis *3*, e437.

Thornberry, N.A., Bull, H.G., Calaycay, J.R., Chapman, K.T., Howard, A.D., Kostura, M.J., Miller, D.K., Molineaux, S.M., Weidner, J.R., Aunins, J., et al. (1992). A novel heterodimeric cysteine protease is required for interleukin-1 beta processing in monocytes. Nature 356, 768-774.

Trichonas, G., Murakami, Y., Thanos, A., Morizane, Y., Kayama, M., Debouck, C.M., Hisatomi, T., Miller, J.W., and Vavvas, D.G. (2010). Receptor interacting protein kinases mediate retinal detachment-induced photoreceptor necrosis and compensate for inhibition of apoptosis. Proc Natl Acad Sci U S A.

Vandenabeele, P., Galluzzi, L., Vanden Berghe, T., and Kroemer, G. (2010). Molecular mechanisms of necroptosis: an ordered cellular explosion. Nat Rev Mol Cell Biol 11, 700-714.

Vercammen, D., Beyaert, R., Denecker, G., Goossens, V., Van Loo, G., Declercq, W., Grooten, J., Fiers, W., and Vandenabeele, P. (1998). Inhibition of caspases increases the sensitivity of L929 cells to necrosis mediated by tumor necrosis factor. J Exp Med 187, 1477-1485.

Vitner, E.B., Salomon, R., Farfel-Becker, T., Meshcheriakova, A., Ali, M., Klein, A.D., Platt, F.M., Cox, T.M., and Futerman, A.H. (2014). RIPK3 as a potential therapeutic target for Gaucher's disease. Nat Med 20, 204-208.

Wang, L., Miura, M., Bergeron, L., Zhu, H., and Yuan, J. (1994). Ich-1, an Ice/ced-3-related gene, encodes both positive and negative regulators of programmed cell death. Cell 78, 739-750.

Wang, S., Miura, M., Jung, Y., Zhu, H., Gagliardini, V., Shi, L., Greenberg, A.H., and Yuan, J. (1996). Identification and characterization of Ich-3, a member of the interleukin-1beta converting enzyme (ICE)/Ced-3 family and an upstream regulator of ICE. J Biol Chem 271, 20580-20587.

Wang, S., Miura, M., Jung, Y.K., Zhu, H., Li, E., and Yuan, J. (1998). Murine caspase-11, an ICE-interacting protease, is essential for the activation of ICE. Cell 92, 501-509.

Welz, P.S., Wullaert, A., Vlantis, K., Kondylis, V., Fernandez-Majada, V., Ermolaeva, M., Kirsch, P., Sterner-Kock, A., van Loo, G., and Pasparakis, M. (2011). FADD prevents RIP3-mediated epithelial cell necrosis and chronic intestinal inflammation. Nature 477, 330-334.

Xie, T., Peng, W., Liu, Y., Yan, C., Maki, J., Degterev, A., Yuan, J., and Shi, Y. (2013). Structural Basis of RIP1 Inhibition by Necrostatins. Structure 21, 493-499.

Zhou, W., and Yuan, J. (2014). SnapShot: Necroptosis. Cell 158, 464-464 e461.

Zhu, S., Zhang, Y., Bai, G., and Li, H. (2011). Necrostatin-1 ameliorates symptoms in R6/2 transgenic mouse model of Huntington's disease. Cell Death Dis 2, e115.

PART 4: VIRUSES AND CANCER

Chapter 15: Cell death and virus infection – a short review (Zakeri et al)

Zahra Zakeri[1, #], Sounak Ghosh-Roy[1,*], Emmanuel Datan[2,*], Jeffrey E. McLean[3], Keivan Zandi[4], and Gabrielle Germain[1]

[1] Department of Biology, Queens College and Graduate Center of the City University of New York, 65-30 Kissena Boulevard, Flushing, NY 11367, USA.

[2] Department of Pharmacology and Molecular Sciences, The Johns Hopkins School of Medicine, 725 N. Wolfe Street, Baltimore, MD 21205, USA.

[3] Rockland Community College, State University of New York, Suffern, NY 10901

[4] University of Malaya, Malaysia (http://malaya.academia.edu/), Department of Medical Microbiology (http://malaya.academia.edu/Departments/Department_of_M edical_Microbiology), Faculty Member

* Both contributed equally to this work.

Correspondence to Zahra Zakeri,

ABSTRACT

The ability of hosts to trigger cell death upon virus infection, especially apoptosis, is key in limiting the extent of viral propagation and damage to the organism. Many viruses through their own proteins have evolved around this hurdle by adapting their life cycles around the process of cell death where some viruses reproduce favorably when the infected cells are killed. It is generally accepted that most human viruses play with the cell death pathways, depending on the cells infected. Common targets of virus-induced cell death (apoptosis mostly) are cells of the immune system, and this can even determine the outcome and severity of viral infection. Viruses that reproduce less in cells that activate cell death pathways have viral proteins that turn on stress response signaling like autophagy to prolong the life of their host as viruses are produced. We also have our disposal knowledge about individual viral proteins (and in some cases, specific domains) inducing or inhibiting cell death pathways (apoptosis, autophagy) in different cells. Induction or repression of various cell survival pathways, therefore, plays an important role in viral pathogenicity apart from the canonical stress pathways.

A better understanding of the signaling pathways that viruses affect to kill or protect the infected cells will allow for the development of new antiviral therapies. This review focuses on key cell death and survival pathways manipulated during influenza, dengue and chikungunya infection, with special emphasis on the role of viral proteins, thus exploring the chance of using them for therapeutics.

Introduction

In the last two decades we have experienced a revolutionary breakthrough in the studies of viral infection and their immediate effect on cell death. The interactions are often exceedingly complex, and determine the outcome for both virus and host. Depending on virus and cell type, viruses may rapidly kill cells, prevent the host organism from killing infected cells, or anything in between (Hilleman 2004). Many viruses evade host immune responses and enhance their propagation by controlling the biochemistry and machinery of host cell death (Roulston A 1999)-(Casella CR 1999). They strongly inhibit cell death, allowing more time for viral replication before the host cell finally succumbs (Kumar A, Laurie M. Delmolino et al. 2002). The outcome for the host cell depends heavily upon the strain and initial dose of virus, as well as the type of cell infected (Hardwick 2001, Irusta, Chen et al. 2003, Hilleman 2004).

For example Human immunodeficiency virus 1 (HIV 1) infection results from virus-induced depletion of important cells of the immune system (Mbita, Hull et al. 2014). Originally this depletion was considered to derive directly from viral replication leading to the induction of apoptosis [2]. However, HIV-1 can induce apoptosis in bystander, relatively uninfected as well as uninfected CD4+ T cells, though not in CD8+ T cells (Martin, Matear et al. 1994). HIV-1 infection leads to programmed death of CD4+ lymphocytes expressing the precursor viral envelope glycoprotein gp160 (Lu, Koga et al. 1994; Jacotot, Callebaut et al. 1996). HIV-1 gp120 reduces expression of Bcl-2 in CD4+ cells, facilitating induction of apoptosis by Fas/Fas ligand interactions (Hashimoto, Oyaizu et al. 1997). However it can also kill independent of Fas pathway (Gandhi, Chen et al. 1998)-(Cooper, Garcia et al. 2013).

Another group of viruses that has been studied extensively with regard to induction of cell death is Herpes simplex 1 (HSV-1) and Herpes simplex 2 (HSV-2), which can infect the central nervous system (Widener and Whitley 2014) as well as cause cold sores and genital herpes. Their ability to induce or inhibit apoptosis depends on cell type (Perkins, Pereira et al. 2003). As determined by morphological change and caspase activity, HSV-1 inhibited apoptosis, but HSV-2 did not (Jerome, Fox et al. 2001). Both inhibited apoptosis in epithelial cells but led to apoptosis of lymphocytes. Expression of HSV-2 ICP10 is sufficient to induce apoptosis in Jurkat cells but the product of the same gene protects epithelial cells from apoptosis (Han, Miller et al. 2009). Similarly in hippocampal neurons, HSV-2, through ICP10, activated the ERK pathway and blocked apoptosis caused by withdrawal of nerve growth factor (NGF) (Perkins, Pereira et al. 2003). Thes alternatives exit because HSV can affect several processes. HSV-1 latency associated transcript LAT, which is abundant during the latency phase of infection, can inhibit caspase 8 and 9 induced apoptosis in neuro 2 A cells (Henderson, Peng et al. 2002). HSV-1 LAT also influences accumulation of transcripts such as Bcl-xL which encodes an antiapoptotic protein and Bcl-x(s) which encodes a proapoptotic protein. LAT may promote the accumulation of Bcl-xL in neurons, thus inhibiting apoptosis (Peng, Henderson et al. 2003). HSV-2 regulation of autophagy by ERK pathway leads to upregulation of a chaperone regulator, Bag-1 and an override of proapoptotic JNK/c-jun signal induced by other viral proteins. Thus this virus, depending on where it is and the type of cell with which it interacts, can kill by any of several pathways or inhibit cell death.

Hepatitis viruses have also been investigated for their roles in apoptosis. Chronic infection with hepatitis B virus leads to cirrhosis and hepatocellular carcinoma (Samal, Kandpal et al. 2012). Hepatitis B virus induces programmed cell death through its X gene. This virus provides another example of virus induction of cell death using different pathways and modes of cell death for its propagation. In humans, the protein HB is a multifunctional regulatory protein that affects both transcription and cell growth. Stimulation of X expression leads to p53 mediated induction of apoptosis (Chirillo, Pagano et al. 1997). Protein X sensitizes hepatocytes to p53-mediated apoptosis by activating the p38MAPK pathway (Wang, Hullinger et al. 2008). The hypothesis that the apoptosis of infected hepatocytes facilitates the propagation of HBV has been rejected, as HBV prevents apoptosis to ensure the release and propagation of infectious progeny (Arzberger, Hosel et al. 2010). HBx inhibits apoptosis by activating Akt (Rawat and Bouchard 2014). Akt regulates HBV replication by reducing the activity of the transcription factor Hepatocyte Nuclear Factor 4α [24]. In chronic infection with hepatitis C virus, hepatocyte apoptosis is enhanced and death inducing ligands CD965/Fas, and TNF-Related Apoptosis-Inducing Ligand (TRAIL) are upregulated (Fischer, Baumert et al. 2007). HCV core protein appears to be important in the pathogenesis of HCV infection (Ray, Meyer et al. 1996). The core protein has pro- and anti apoptotic effects in death ligand mediated apoptosis. The core protein together with other viral nonstructural proteins (NS4B and NS5B), enhances TNFα induced cell death by suppressing activation of NF-KB (Lim, El Khobar et al. 2014). HCV induces apoptosis, necroptosis and autophagy (Shrivastava, Bhanja Chowdhury et al. 2012, Richards and Jackson 2013). Hepatitis C virus induces autophagosome formation in hepatocytes favoring survival of the cells and thus further growth of virus (Richards and Jackson 2013). HCV induces autophagy by upregulating Beclin 1 and activating mTOR signaling pathways; the increased autophagy promotes hepatocyte growth. HCV NS5A and NS4B proteins are sufficient enough to induce the autophagic response (Marianneau, Cardona et al. 1997). However the specific role of autophagy during HCV remains controversial.

Thus the explanation of how viruses affect cell death is complex and somewhat contradictory description of how viruses affect cell death. To give a clearer sense of how the viruses work, it will help to follow in detail how specific viruses act. We have been working with three viruses that also control different paths of the cell death machinery: Influenza A (Orthomyxoviridae family), Dengue-2 (Flaviviridae family), and Chikungunya (Togaviridae family). Here we present in more detail how they regulate the cell death or cell survival machinery. The understanding of how viruses interact with host cell death pathways will ultimately lead to improved antiviral therapies.

Importance of Influenza pathogenicity

Influenza A virus (IAV) has caused at least four major pandemics in the last century (1889, 1918, 1957 and 1968). Influenza comes in three groups: A, B and C. Influenza A is most common in humans and principally affects the upper respiratory tract (nose, throat, bronchi and lungs). The effect of influenza infection on host cell survival has been studied since the early 1990s, when HeLa and MDCK cells were demonstrated to show signs of apoptotic death following infection (Takizawa, Matsukawa et al. 1993). Contemporary studies showed in vivo and in vitro depletion of lymphocytes and bronchial and alveolar epithelial cells, which exhibited membrane blebbing and nuclear condensation, typical hallmarks of apoptotic death (Hinshaw, Olsen et al. 1994)-(Mori, Komatsu et al. 1995). These studies laid the foundation of a new research arena that has led to the identification of several apoptotic pathways manipulated by the virus.

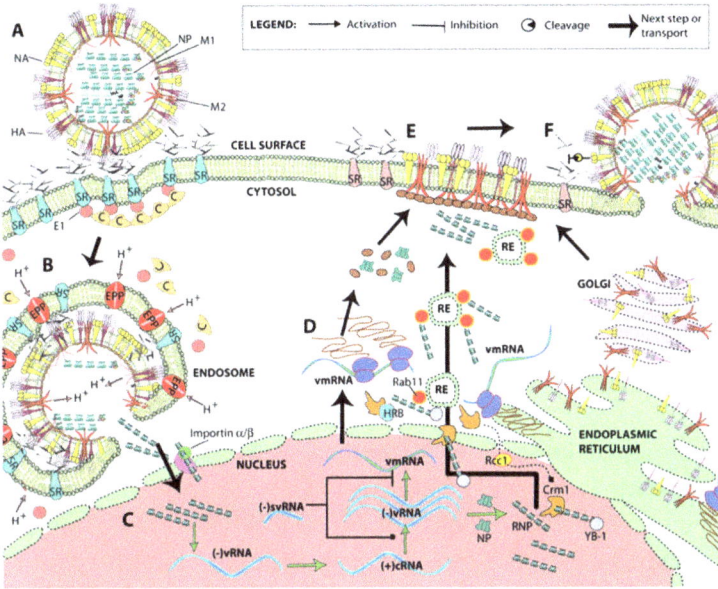

Figure 1. Life cycle of influenza virus. A mature virus is composed of hemaglutinin (HA), neuraminidase (NA), matrix protein 1 (M1), matrix protein 2 (M2) and nucleoprotein (NP), which complexes with viral RNA (vRNA). Hemaglutinin allows the virus to attach to cells through sialiated receptors (SR) on the host plasma membrane (A). Interaction between receptors and cathrin-mediated endocytosis proteins (clathrin (c) and epsin1 (E1)) allows virus entry. Influenza vRNA is then released in the cytosol by acidification of the virus carrying endosome through proton enrichment by endosomal protein pumps (EPP) and the viral M2 protein (B). The host proteins importin α and β shuttle the vRNA into the nucleus. To replicate the virus genome, negative single strand vRNA (-vRNA) is used as template for positive single strand vRNA that gives rise to more –vRNA (C). Production of small negative single strand vRNA reduces the production of viral messenger RNA from – vRNA while increasing –vRNA levels. Influenza proteins are produced from virus mRNA (vmRNA) in the cytosol or on the rough endoplasmic reticulum (D). Virus proteins and –vRNA-NP (RNP) complex are then brought to the plasma membrane where assembly of the virus occurs. RNP is shuttled out of the nucleus to the assembly sites by host YB-1, Crm1 and Rab11 proteins (E). HRB and Rcc1 allow Crm1 to be recycled back to the nucleus. Rabb11 tether RNP to recycled endosomes (RE) that are targeted to the assembly sites. The virus then buds off and detaches from the cell by the cleavage of SR by NA as it exits the cell for another round of infection (F).

The replication of the virus involves multiple organelles and modulates both intrinsic (caspase-8 mediated) and extrinsic (caspase-9 mediated) pathways of apoptosis (Lin, Holland et al. 2002, Wurzer, Planz et al. 2003). The effects exerted by these pathways on virus production (Fig.1) and host defense are unresolved (Lowy 2003, Ludwig, Pleschka et al. 2006). Caspase 3 has been reported both to improve (Wurzer, Planz et al. 2003, Wurzer, Ehrhardt et al. 2004, McLean, Datan et al. 2009) and restrict (Zhirnov, Konakova et al. 1999) viral replication. Moreover, the influenza A Matrix-1 (M1) protein specifically binds to caspase-8 and weakly to caspase-7, suggesting involvement of M1 in caspase-8 mediated apoptosis (Zhirnov, Konakova et al. 2002). Corroborating this argument, caspase inhibitors or knockdown of caspase 8 by siRNA impairs viral replication (Wurzer, Planz et al. 2003). However, non-specific inhibition of caspases does not improve survival of infected cells but changes how the cells die from apoptosis to non-apoptotic cell death (Matassov, Kagan et al. 2004, Datan, Shirazian et al. 2014). Still, ectopic expression of caspase-3 (the presumptive downstream effector of the activation of caspase-7 or -8) salvaged IAV replication in caspase-deficient cells (Wurzer, Ehrhardt et al. 2004). Caspase-deficient cells retain viral RNP complexes in their nuclei, an occurrence also witnessed in Bcl2-overexpressing cells (Olsen, Kehren et al. 1996, Wurzer, Planz et al. 2003, McLean, Datan et al. 2009). Overexpression of anti-apoptotic protein Bcl-2 interrupts proper glycosylation of the viral HA surface antigen, seriously limiting virus propagation (Olsen, Kehren et al. 1996).

Viruses can interact with the extrinsic (receptor-mediated) pathways of apoptosis. Pharmacological induction of apoptosis allows isolated NP or RNP complexes to translocate to the cytoplasm (Wurzer, Planz et al. 2003). Initial studies primarily focused on the extrinsic (receptor mediated) pathways of apoptosis. Influenza up-regulates TGF-β in MDCK cells and mice models, dependent on NA surface antigen and viral uncoating (Schultz-Cherry and Hinshaw 1996)-(Morris, Price et al. 1999). Tumor necrosis factor-α (TNF-α) is upregulated by influenza A inducing apoptosis of macrophages and B-cells (Nain, Hinder et al. 1990, Peschke, Bender et al. 1993, Sedger, Hou et al. 2002). Migration of T-cells to the lungs and concomitant TNF-α expression was delayed in neonatal mice, compared to adults, resulting in high mortality (Lines, Hoskins et al. 2010). Non-structural protein 1 (NS1) increases expression of the Tumor necrosis factor-Related Apoptosis-Inducing Ligand (TRAIL), and also inhibits interferon regulatory factor 3 (IRF-3), a key regulator of IFN (Talon, Horvath et al. 2000, Schultz-Cherry, Dybdahl-Sissoko et al. 2001, Donelan, Dauber et al. 2004, Wurzer, Ehrhardt et al. 2004, Brincks, Kucaba et al. 2008). Interferon type I-mediated apoptosis prevents the release of pro-inflammatory cytokines, lowering the presence of immune cells in infected tissues (Balachandran, Roberts et al. 2000, Brydon, Smith et al. 2003). The Fas/FasL apoptosis inducing receptor/ligand system is also activated during infection, which occurs in a PKR-dependent manner and results in cell killing via the extrinsic apoptosis pathway (Takizawa, Ohashi et al. 1996, Fujimoto, Takizawa et al. 1998, Balachandran, Roberts et al. 2000). The interplay between extrinsic pathways of apoptosis (Fig.2) and viral transport/release (Fig.1) is further indicated by in that Brefeldin A, an inhibitor of intracellular trafficking, is antiviral in MDCK cells (Saito, Tanaka et al. 1996) .

Influenza A also affects the intrinsic pathway of apoptosis. Viral activation of mitogen-activated kinases (MAPKs) has been linked to the onset of apoptosis, and MAPK kinase (MAPKK)-deficient MEF resist caspase 3-mediated apoptosis (Maruoka, Hashimoto et al. 2003, Hui, Lee et al. 2009). p53, another positive mediator of apoptosis, is also upregulated (Turpin, Luke et al. 2005). Neurovirulent Influenza A induces apoptosis though activation of JNK (but not p38) in the brains of infected mice (Mori, Goshima et al. 2003). Additionally, the otherwise antiviral NF-κB (Nuclear Factor Kappa-light-chain-enhancer of activated B cells, a transcription factor) promotes influenza A-induced apoptosis and replication, probably by activating proapoptotic factors such as TRAIL or FasL (Wurzer, Ehrhardt et al. 2004, Tsuchida, Kawai et al. 2009). NF-κB inhibitors are effective against viral infection (Ludwig, Planz et al. 2003). The influenza A genome segment coding for PB1 has an alternate reading frame that gives rise to a pro-apoptotic protein, PB1-F2, that can activate the intrinsic apoptosis pathway. PB1-F2 (containing a mitochondrial targeting sequence) interacts with inner mitochondrial membrane (ANT3) and outer mitochondrial membrane (VDAC-1) proteins resulting in cytochrome c release and subsequent activation of apoptosis (Yamada, Chounan et al. 2004) (Zamarin, Garcia-Sastre et al. 2005, Zell, Krumbholz et al. 2007, Chevalier, Al Bazzal et al. 2010). Infection with a recombinant virus lacking this protein lowers the rate of apoptosis in lymphocytes (Chen, Calvo et al. 2001). Further, PB1-F2-mediated apoptosis is most prevalent in different kinds of , leading to the hypothesis that the activity of this particular protein has evolved to cripple the host immune response (Zamarin, Ortigoza et al. 2006).

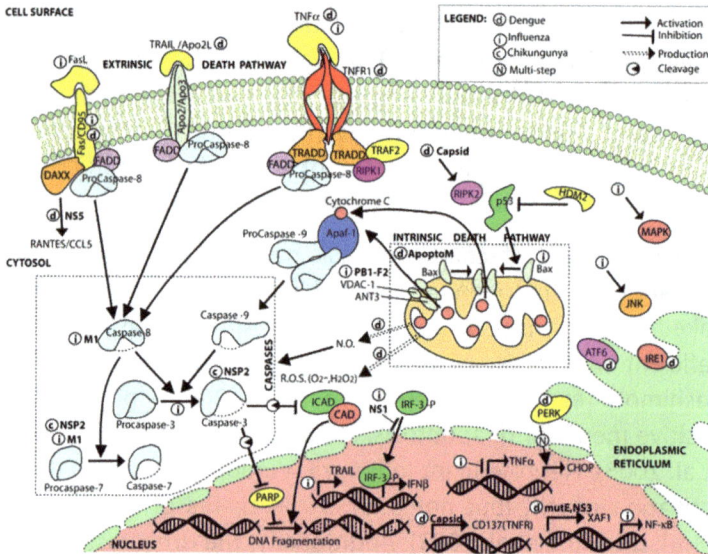

Figure 2. Manipulation of cell death pathways by dengue, influenza and chikungunya viruses. Apoptosis is induced after infection (d, i, and c dengue, influenza and chikungunya live viruses respectively) or expression of specific viral proteins (d, i and c with viral protein). Activation of extrinsic (Fas, TRAIL and TNFα) and intrinsic apoptotic pathways is observed in cells with dengue or

influenza. ER stress signaling is also triggered by dengue. Transcription of cell death genes is also increased upon infection by dengue or influenza. Chikungunya leads to cell death by activation of caspases 3 and 7.

Most recently evidence has accrued that the virus activates non-apoptotic cell death pathways. In Bax -/- MEF cells, influenza A infection leads to cell death by autophagy (PCD type II) (Matassov, Kagan et al. 2004, McLean, Datan et al. 2009, Lu, Masic et al. 2010). When pancaspase inhibitors are used in infected MDCK cells, autophagy emerges as a major alternate death pathway through mTORC1/2-p70S6K-dependent pathway. Although lethal autophagy is activated (Figure 3) when apoptosis is blocked, autophagy is also essential for viral replication. Maintenance-level mild autophagy that occurs with apoptosis does not rely on an active mTOR-p70S6K pathway (Datan, Shirazian et al. 2014).

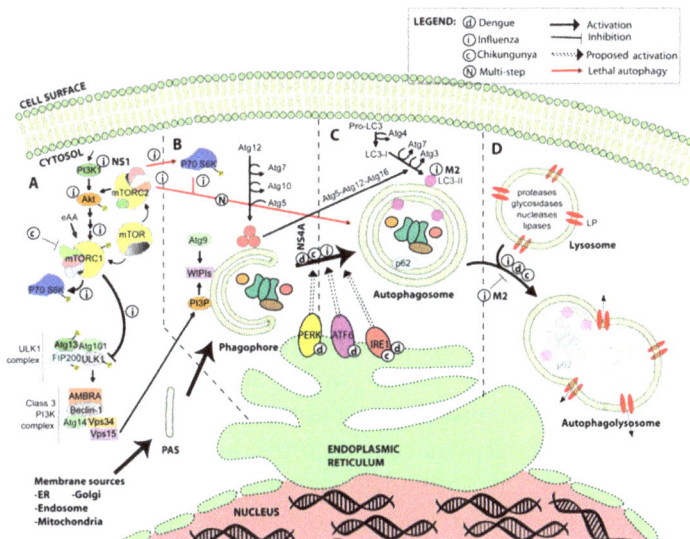

Figure 3. Manipulation of autophagy by dengue, influenza and chikungunya viruses. The process of autophagy comprises four steps namely induction (A), elongation (B), autophagosome formation (C) and lysosome turnover (D). Autophagy is increased during infection (d, i, and c dengue, influenza and chikungunya live viruses respectively) or expression of specific viral proteins (d, i and c with viral protein). Nutrient sensing protein complexes like mTORC1 regulate the induction stage. Influenza and chikungunya viruses affect mTOR-

dependent signaling during infection. It has been proposed that ER stress signaling (PERK, ATF6, IRE1) by dengue is also critical in dengue-induced autophagy. All three viruses increase autophagosome turnover to the lysosomes. Although autophagy helps the infected cell live longer, influenza infection during apoptosis leads to lethal autophagy that requires mTORC2-P70S6K dependent signaling.

Flaviviridae : Dengue virus

The most deadly arthropod-borne virus to come from Flaviviridae, Dengue, is responsible for the fastest spreading arboviral disease in the world. Infection leads to a flu-like illness which can often develop into a lethal hemorrhagic condition (hemorrhagic fever/severe shock). It is the primary cause of child mortality in south-east and south Asia, and in parts of Africa, Latin America and Mexico, though very recently a vaccine appears promising (Martina, Koraka et al. 2009). The intricate relationship between the virus and cell survival pathways has been studied for over a decade and we have learned how this virus can kill cells and facilitate infection.

Importance of Dengue virus structure

The virus comes in four serotypes (DENV 1-4), each comprising multiple phenotypes (Martina, Koraka et al. 2009). Compared to the 1970s, when only south Asia was home to all the serotypes, increased global trade and better communication have spread all serotypes to different regions of the world (Gubler 1998), further complicating the dynamics of dengue pathology, since the worst reactions occur when an individual previously infected with one serotype encounters a second serotype. Very little cross-immunity has been recorded among the four dengue serotypes, due to antibody-dependent enhancement. (ADE), ADE occurs when antibodies developed during a primary infection facilitate secondary infection by targeting the secondary virus (of a different serotype). This leads to infection and destruction of Fcγ receptor-bearing cells, such as macrophage/monocytes (Guzman, Kouri et al. 1990, Rothman 2011). Secondary infection triggers a massive, widespread and life-threatening inflammatory response. The ED3 domain of dengue E protein plays an important role in binding the neutralizing antibodies, thus contributing to ADE (Midgley, Bajwa-Joseph et al. 2011).

Dengue Structure

To understand how dengue affects cell death and apoptosis, it is useful to know the names and functions of its proteins. The enveloped icosahedral virion of dengue contains a 10.7 kb (+) sense single strand RNA, crowned with a type I cap at its 5' terminus (Cleaves and Dubin 1979). The genome uses host machinery to produce a 3391-amino acid polyprotein that is co- and post-translationally processed by viral (NS2B-3) and cellular proteases (Falgout, Pethel et al. 1991, Cahour, Falgout et al. 1992, Amberg, Nestorowicz et al. 1994). The first segment (from the 5' end) of the viral genome is translated into structural capsid (C), membrane (prM), and lipopolysaccharide envelope (E) proteins. The rest of the genome codes for eight non-structural proteins (NS1, 2A, 2B, 3, 4A, 2K peptide, 4B, 5) that are expressed only inside the host (Chambers, Hahn et al. 1990). The non-structural proteins NS1 and NS4A are involved in the formation of replication complex (RC), ensuring proper post-transcriptional modification (NS5) and mature virion assembly (NS2A) after replication (Fig.4). NS2B forms a complex with NS3 that acts as the viral protease, while NS4B serves as an agonist for the host interferon response (Guzman, Halstead et al. 2010).

Dengue entry and replication

The virions bind to the cell membrane, mediated by specific host receptors (DC-SIGN, Grp78/BiP, CD-14 associated molecules) and low-affinity co-receptors (heparin, glycosaminoglycan) (Mukhopadhyay, Kuhn et al. 2005). Upon internalization, the virus-associated vesicles fuse with the endosomal membrane and release viral genetic material (Figure 4B). The virus then takes over the cell machinery, engaging the ribosomes (on rough endoplasmic reticulum (RER)) to make a single polypeptide chain from the viral genome (Fig.4D). The polyprotein is then processed and spliced by viral and host proteases into ten individual proteins. The structural proteins localize at the ER surface along with progeny viral RNA that is replicated on intracellular membranes (Fig.4E). The immature viral particles then migrate through the trans-Golgi network (TGN), and are cleaved by a host protease (furin) to form the mature virion, and then subsequently released by exocytosis. (Allison, Schalich et al. 1995, Stadler, Allison et al. 1997, Corver, Ortiz et al. 2000, Allison, Schalich et al. 2001, Brinton 2002, Lindenbach and Rice 2003, Mukhopadhyay, Kuhn et al. 2005).

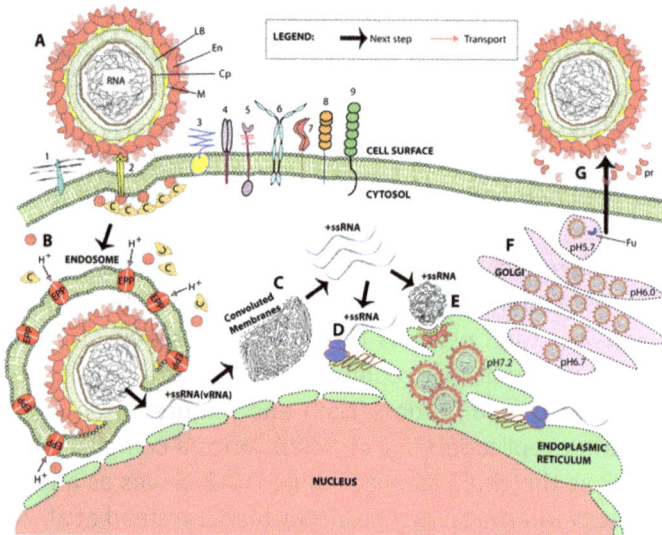

Figure 4. Life cycle of dengue virus. A mature virus is composed of a positive single strand RNA core (+ssRNA or vRNA) covered by capsid protein (Cp), then a lipid bilayer (LB) covered by membrane (M) and envelope (En) proteins. (A) Binding and entry into the cell involves surface proteins (1=heparin sulfate, 2=DC SIGN, 3= mannose receptor, 4=GRP78, 5=TIM, 6=TAM, 7=beta integrin, 8=HSP70, 9=HSP90) and clathrin (c). Increase of protons within the endosome by EPP then releases the dengue genome to the cytosol (B) for replication in convoluted membranes (C) and translation of the virus polyprotein on the rough endoplasmic reticulum (D). Host and viral proteases then cleave the polyprotein. Immature virus is then assembled on the endoplasmic reticulum where it is endocytosed to the ER lumen (E) and shuttled through the endoplasmic reticulum and the Golgi (F). Decreasing pH from the endoplasmic reticulum to the Golgi and cleavage of the virus pr protein by furin (Fu) are important in dengue maturation. Mature virus and pr are then release the cell (G).

Cell death and survival after infection with dengue:

How dengue affects cell death has been studied for over a decade. Although apoptosis has been the focus for most of these studies, autophagy is now been recognized as important. Dengue activates both extrinsic and intrinsic branches of apoptosis (Fig.2) in a wide range of human cells, viz., mast cells (KU812), macrophages, cerebral cells, umbilical cord vein endothelial cells, human microvascular endothelial cells, pulmonary microvascular endothelial cells (MECs) and kidney cells (HEK293). Different strains of the virus, from serotype 1 (human isolates of dengue type 1 virus FGA/89 and BR/90, neurovirulent variant FGA/NA d1d), 2 (strain NGC, 16681) and 3 (DENV3/5532) can kill human cells in an apoptosis-dependent fashion within 25 to 36 hours of infection (Ghosh Roy, Sadigh et al. 2014).

Normally one of the first cells to become infected are the keratinocytes (skin), followed by a concerted defense action orchestrated by interferon-stimulated activation of macrophages and monocytes, which unfortunately become infected as well. As these infected lymphocytes spread throughout the body, they cause viremia (Diamond 2003). In human dendritic cells, the virus is also successful in downregulating the interferon type I response (Rodriguez-Madoz, Bernal-Rubio et al. 2010). Apart from dendritic cells, the virus can also induce apoptotic cell death in peripheral blood mononuclear cells (PBMC) However, increased apoptosis does not seem to block production of virus; on the contrary, apoptosis is frequently associated with increased production (Myint, Endy et al. 2006, Jaiyen, Masrinoul et al. 2009).

The pathogenicity of dengue virus seems to vary among isolates. Clinical isolates of dengue outbreak from a fatal case (Paraguay 2007; DENV3/5532). These cultures had higher rate of replication in monocyte-derived human dendritic cells (mdDCs) compared to isolates from a non-fatal breakout (Brazil 2002; DENV3/290); the latter also induced more proinflammatory cytokines, commonly associated with apoptosis (Silveira, Meyer et al. 2011). he E and NS3 proteins are important in determining the toxicity of a viral strain (Duarte dos Santos, Frenkiel et al. 2000). pathogenicity attenuated in other cell types, in the presence of apoptosis-triggering molecules like staurosporine, cycloheximide, camptothecin, and even in the presence of influenza virus (McLean, Wudzinska et al. 2011).

Infection with both the whole virus or with individual viral proteins has triggered apoptosis pathways (intrinsic/extrinsic) in different cell types (Fig.2). Whole virus triggers activation of pro-apoptotic proinflammatory cytokines (TNF-α and interleukin–10), Apo2L/TRAIL, death receptors (FAS/CD 95), TNFR superfamily member 9/CD 137, TNFRI/TNF-α (caspase-independent) and IL-1β/ NFκB (caspase-dependent) pathways (Myint, Endy et al. 2006, Jaiyen, Masrinoul et al. 2009, Silveira, Meyer et al. 2011). Also, it is responsible for differential expression of interferon-inducible genes such as XAF1, which upregulate caspase-3 and mediate apoptosis within 36 hours of infection (Long, Li et al. 2013). The NS5 viral protein regulates the interaction of death protein 6 (Daxx) with death receptor FAS, which in turn activates the RANTES (CCL5) cytokine, which latter highly correlates with dengue hemorrhagic fever (DHF) (Nagila, Netsawang et al. 2011, Khunchai, Junking et al. 2012). As with the whole virus, transfection of capsid (C) protein increases expression of CD137 as well as the receptor-interacting serine/threonine protein kinase 2 (RIPK2), a master regulator of stress pathways (Morchang, Yasamut et al. 2011). Expression of M protein from all strains triggers intrinsic apoptosis in mouse neuroblastoma (Neuro2a) and human hepatoma (HepG2) cells, while a nine-residue ectodomain sequence ApoptoM (M-32 to -40) contributes to the cytotoxicity (Catteau, Kalinina et al. 2003). Infection also gives rise to toxic levels of mitochondrial reactive oxygen species (ROS) such as O2.- and H2O2, which activate calpains and cause apoptosis. Secondary messenger oxides like nitric oxide (NO) also mediate dengue-triggered apoptosis in a caspase-dependent manner (Lin, Lei et al. 2002).

As summarized above, several dengue strains are capable of inducing many branches and sub-branches of the apoptotic pathway (Fig.2). These events constitute an important stage in the virus life cycle, since the external flipping of phosphatidylserine (PS) on the cell membrane helps in the transmission of the virus by phagocytosis of these infected apoptotic cells (Alonzo, Lacuesta et al. 2012). Unlike lytic viruses, pro-apoptotic variants of dengue can also lose their pathogenicity in certain cells. The neurovirulent variant FGA/NA d1d, developed from the apoptosis inducing dengue 1 human isolate FGA/89, kills neuroblastoma but not hepatoma cells (Duarte dos Santos, Frenkiel et al. 2000). Dengue-2 virus strain 16681 triggers apoptotic cell death in human umbilical cord vein endothelial cells (ECV304) and Swiss Webster primary macrophages; however it protects MDCK, HeLa, HEK 293T, Vero or Swiss Webster primary MEF even after 6 days of infection (Avirutnan, Malasit et al. 1998, McLean, Wudzinska et al. 2011).

Autophagy and Dengue

Dengue pathogenicity also relies heavily on the cellular homeostasis (ATG5-dependent) macro-autophagy pathways, inducing pro-survival signaling in hepatic (Huh7) and fibroblast (MEF) cells. The consensus is that autophagy allows the propagation of mature virions by increasing the survival of host cells. However, it has no direct regulatory effect over infectivity, as downregulation of autophagy does not increase intracellular viral load (Lee, Lei et al. 2008, Panyasrivanit, Greenwood et al. 2011). However the downregulation of autophagy by spautin-1 in Huh7.a.1, BHK21 cell lines and AG129 mice rendered dengue virion that is heat-sensitive and non-infectious (Mateo, Nagamine et al. 2013).

We established a link between autophagy and pathogenicity. Dengue 2 16681 strain protects epithelial (MDCK) and fibroblast (MEF) cells against toxins such as camptothecin (CPT) by inducing PI3K-dependent autophagy. NS4A protein was the only component that contributed to the pro-survival effects of the whole virus. None of the other proteins were able to protect cells from the lethal toxins used in our experiment. Protection was diminished in an autophagy deficient background (Vps 34 inhibitor– Rapamycin/knockout MEFs–Beclin1-/-, ATG5-/-) providing a correlation between autophagy and cell protection (McLean, Wudzinska et al. 2011). Most recent studies, with ATM kinase inhibitor caffeine, point towards involvement of ATM kinase in eliciting autophagy during dengue infection. These conclusions were valid for infection with whole virus or transfection with NS4A.

Endoplasmic reticulum, the major site for protein folding and viral replication (Fig.4), plays an important role in virus-mediated autophagy (Fig.3). ER can release stress signals, like unfolded protein response (UPR), due to improper glycosylation, impaired ERAD (ER-associated degradation) machinery, oxidative stress, hypoxia, protein misfolding or viral infection. These pathways, namely, PERK, IRE-1α, ATF-6, can regulate different cell survival pathways (apoptosis, autophagy) (Price, Mannheim-Rodman et al. 1992, Carrell 2005, Schroder and Kaufman 2005, Obeng, Carlson et al. 2006, Turnbull, Rosser et al. 2007, Granell, Baldini et al. 2008). Dengue is an ER-tropic flavivirus and depends on the ER for its replication, assembly and maturation; it thus can trigger all of these UPR pathways at different stages of infection (Pena and Harris 2011). The closely related West Nile virus (WNV) induces ER stress while triggering apoptosis. Infection of human neuroblastoma (SK-N-MC) cells and primary rat hippocampal neurons with WNV induced ATF6 and PERK pathways, resulting in activation of CHOP and downstream apoptosis (Medigeshi, Lancaster et al. 2007). In wild type MEFs, the West Nile virus Kunjin strain (WNVKUN) blocked the PERK pathway and interferon-mediated STAT phosphorylation but not the ATF6 and IRE1 pathways. Knockout studies with ATF6-/-, IRE1-/- MEFs reveal the synergism of these pathways on WNVKUN pathogenesis, which they accomplish by restricting apoptotic cell death and increasing viral load (Ambrose and Mackenzie 2013).

In fibrosarcoma cells, dengue induces and represses the PERK pathway during early stages of infection, later activating IRE1-XBP1 and ATF6 pathways during the mid- and late- stages (Pena and Harris 2011). However, the time-dependent pattern varies with cell types. In epithelial cells, the flavivirus replication complex (RC) co-localizes with autophagosomes, which are derived from the ER membrane. The accumulation of viral proteins on ER membranes causes high ER stress that can activate one or more UPR pathways (Fig.4D). ER stress can also trigger autophagy by other means. Ca2+ release from the ER can trigger kinases (CaMKKthat can activate AMPK to induce autophagy (Hoyer-Hansen, Bastholm et al. 2007). Alternatively, ceramide-induced PP2A activates DAPK, which can dissociate Beclin1 from Bcl-2 and cause autophagy (Li, Wang et al. 2009). The IRE1 branch of UPR leads to JNK activation and increased Bcl-2 phosphorylation, thus releasing Beclin-1, which can then take part in formation of autophagosomes (Ogata, Hino et al. 2006, Ding, Ni et al. 2007, Wei, Sinha et al. 2008, Pattingre, Bauvy et al. 2009). The PERK pathway can trigger autophagy through ATF4-dependent expression of ATG12 (Talloczy, Jiang et al. 2002, Kouroku, Fujita et al. 2007, Carra, Brunsting et al. 2009, Salazar, Carracedo et al. 2009). As an alternate pathway, CHOP (mediated by ATF4) can activate the pseudokinase TRB3, which can cause autophagy through inhibition of Akt-mTOR (Carracedo, Gironella et al. 2006, Carracedo, Lorente et al. 2006, Salazar, Carracedo et al. 2009). In summary, cell death and survival pathways in the form of apoptosis and autophagy are activated as dengue replicates in various cell types. The toxic effect of dengue on cells depends on the type of cell it infects while autophagy consistently plays a role in keeping cells viable while the virus replicates. The activation of apoptosis and autophagy through ER stress or metabolic by products (ROS) appears beneficial for optimal dengue production and suppression of the antiviral response.

Chikungunya virus (CHIKV) importance and structure

Chikungunya, an arboviral disease first identified in Tanzania in 1952, is caused by chikungunya virus (CHIKV), genus Alphavirus (Togaviridae family) (Pialoux, Gauzere et al. 2007). The 11 kb CHIKV RNA genome has two open reading frames (5' and 3' ORFs); three structural proteins (capsid protein(C), precursor of membrane protein (Pr M), and envelope glycoprotein (E)). The 5' ORF encodes four virus non-structural proteins (nsP 1-4) (Solignat, Gay et al. 2009). It is carried by mosquitoes.

Chikungunya virus entry and replication

Alphavirus envelope glycoproteins are known as cell attachment proteins. CHIKV E1 and E2 glycoproteins are predicted to be viral entry elements and to form heterodimers that associate as trimeric spikes on the viral envelope. However, although the virus is able to replicate in the cells of both mosquitoes and higher vertebrates, the only known receptor, prohibitin (PHB), is found only on microglial cells (Wintachai, Wikan et al. 2012). Plasma membrane cholesterol is a key constituent for CHIKV entry. It was also suggested that the clathrin-dependent endocytic pathway mediates CHIKV entry into human cells, but this requires an intact early endosome compartment. The E1 protein mediates the fusion of Alphavirus with the endosomal membrane. Alphavirus capsid protein can bind the large ribosomal subunit, and this binding reaction could affect nucleocapsid uncoating (Solignat, Gay et al. 2009). Alphavirus replication then proceeds along a two-step pathway. It takes place in the host cell cytoplasm and is associated with membranous structures of the cytoplasm that co-purify with the mitochondrial fraction (Gomatos, Kaariainen et al. 1980). Early after infection, minus-strand RNAs (detected only at early stages of infection) and plus-strand (synthesized at a constant rate throughout the remainder of infection cycle) are both transcribed under control of non-structural proteins (nsP). The subgenomic positive-strand mRNA of alphaviruses, referred to as 26S RNA, serves as the mRNA for the synthesis of the viral structural proteins. Alphavirus virion assembly starts with nucleocapsid assembly in the cytoplasm. The nucleocapsid of CHIKV has icosahedral symmetry through forming of RNA-bound dimers with the capsid protein. The lateral interactions with other capsid proteins determine the structure of the icosahedral shell. CHIKV nucleocapsid complexes assembled in the cytoplasm are thought to diffuse freely to the plasma membrane (Perera, Owen et al. 2001). The members of Alphavirus then bud from the cell membrane. Virions acquire a lipid bilayer envelope containing the virus-encoded E1 and E2 glycoproteins. The mature envelope glycoprotein spikes are composed of trimers of E1–E2 (Schuffenecker, Iteman et al. 2006).

Cell death, autophagy and CHIKV

CHIKV replication, described in detail above, leads to rapid syncytium formation and apoptosis of infected cells (Sourisseau, Schilte et al. 2007, Dhanwani, Khan et al. 2012). Nevertheless, CHIKV can infect neighboring cells by phagocytosis of apoptotic vesicles (Tang 2012), or it may block, delay, or even induce apoptosis. NsP2 for CHIKV, an essential and multifunctional component of viral replicase complex, can block or induce apoptosis (Fig.2). The location of nsP2 in nucleus is related to its role in apoptosis, for which its carboxy-terminal domain is required (Tamm, Merits et al. 2008, Bourai, Lucas-Hourani et al. 2012).

Autophagy, a catabolic process that sequesters cytosolic components within double-membrane vescicles and targets them for degradation in lysosomes (Yang and Klionsky 2010), can be used to destroy pathogens (Joubert, Werneke et al. 2012). Nevertheless, CHIKV, like dengue and influenza, (Jackson, Giddings et al. 2005) can subvert autophagy to trigger an autophagic response that enhances viral replication in human cells (Krejbich-Trotot, Gay et al. 2011). This is also observed in mice and HeLa cells as shown by the decline of p62 protein and increased conversion of LC3-II to LC3-II (Judith, Mostowy et al. 2013). Autophagy is then followed by induction of ER and oxidative stress (Joubert, Werneke et al. 2012). Among the ER stress pathways that are induced, only IRE1α/XBP1 is involved in regulation of autophagy (Fig.3). Increased ROS during CHIKV replication blocks mTORC1, inducing autophagy. Autophagy can promote cell survival by limiting apoptotic death but may also limit release of virus, as apoptotic cell death is important for CHIKV replication and propagation.

The autophagy inhibitors wortmannin or 3-methyladenine (3-MA) can block the release of CHIKV particles. In contrast rapamycin, an inducer of autophagy, enhances viral replication. Depletion of p62 significantly increased CHIKV replication in contrast to NDP52 (autophagy receptor for ubiquitin-decorated cytosolic bacteria). However, NDP52 depletion significantly increased CHIKV-induced cell death. Thus p62 and NDP52 exert contrasting effects on viral infection in human cells. Interestingly, the cytoprotective effect of p62 derives from its ability to bind ubiquitinated CHIKV-capsid and the proviral role of NDP52 on CHIKV replication links to its interaction with CHIKV-nsP2. The mechanism by which NDP52 promotes viral infection is species-specific, perhaps explaining the previous controversy about the role of autophagy on CHIKV replication (Krejbich-Trotot, Gay et al. 2011, Joubert, Werneke et al. 2012). The absence of NDP52–nsP2 interaction in mouse cells might account for the lower permissiveness of mice for CHIKV replication compared to humans (Judith, Mostowy et al. 2013).

Conclusions
In this chapter, we summarize some of the most important findings in the field of clinical virology during the last twenty years. We briefly present the research done with some of the most relevant human viruses and how they affect mortality in different cell types. It is now evident that the effects are quite diverse, depending upon both the virus and the cell typevirus always regulates the death pathways according to its need to replicate and release mature virions, be it the induction of apoptosis or the upregulation of autophagy. The information

presented here should contribute to a comprehensive understanding of the way some of the viruses act inside the host cell. This information can help in designing antiviral therapies. More studies should be devoted to identify the target proteins that help in mediating the effects of virus entry/ replication to cell death.

References

Allison, S. L., J. Schalich, K. Stiasny, C. W. Mandl and F. X. Heinz (2001). "Mutational evidence for an internal fusion peptide in flavivirus envelope protein E." J Virol **75**(9): 4268-4275.

Allison, S. L., J. Schalich, K. Stiasny, C. W. Mandl, C. Kunz and F. X. Heinz (1995). "Oligomeric rearrangement of tick-borne encephalitis virus envelope proteins induced by an acidic pH." J Virol **69**(2): 695-700.

Alonzo, M. T., T. L. Lacuesta, E. M. Dimaano, T. Kurosu, L. A. Suarez, C. A. Mapua, Y. Akeda, R. R. Matias, D. J. Kuter, S. Nagata, F. F. Natividad and K. Oishi (2012). "Platelet apoptosis and apoptotic platelet clearance by macrophages in secondary dengue virus infections." J Infect Dis **205**(8): 1321-1329.

Amberg, S. M., A. Nestorowicz, D. W. McCourt and C. M. Rice (1994). "NS2B-3 proteinase-mediated processing in the yellow fever virus structural region: in vitro and in vivo studies." J Virol **68**(6): 3794-3802.

Ambrose, R. L. and J. M. Mackenzie (2013). "ATF6 signaling is required for efficient West Nile virus replication by promoting cell survival and inhibition of innate immune responses." J Virol **87**(4): 2206-2214.

Arzberger, S., M. Hosel and U. Protzer (2010). "Apoptosis of hepatitis B virus-infected hepatocytes prevents release of infectious virus." J Virol **84**(22): 11994-12001.

Avirutnan, P., P. Malasit, B. Seliger, S. Bhakdi and M. Husmann (1998). "Dengue virus infection of human endothelial cells leads to chemokine production, complement activation, and apoptosis." J Immunol **161**(11): 6338-6346.

Balachandran, S., P. C. Roberts, T. Kipperman, K. N. Bhalla, R. W. Compans, D. R. Archer and G. N. Barber (2000). "Alpha/beta interferons potentiate virus-induced apoptosis through activation of the FADD/Caspase-8 death signaling pathway." J Virol **74**(3): 1513-1523.

Bourai, M., M. Lucas-Hourani, H. H. Gad, C. Drosten, Y. Jacob, L. Tafforeau, P. Cassonnet, L. M. Jones, D. Judith, T. Couderc, M. Lecuit, P. Andre, B. M. Kummerer, V. Lotteau, P. Despres, F. Tangy and P. O. Vidalain (2012).

"Mapping of Chikungunya virus interactions with host proteins identified nsP2 as a highly connected viral component." J Virol **86**(6): 3121-3134.

Brincks, E. L., T. A. Kucaba, K. L. Legge and T. S. Griffith (2008). "Influenza-induced expression of functional tumor necrosis factor-related apoptosis-inducing ligand on human peripheral blood mononuclear cells." Hum Immunol **69**(10): 634-646.

Brinton, M. A. (2002). "The molecular biology of West Nile Virus: a new invader of the western hemisphere." Annu Rev Microbiol **56**: 371-402.

Brydon, E. W., H. Smith and C. Sweet (2003). "Influenza A virus-induced apoptosis in bronchiolar epithelial (NCI-H292) cells limits pro-inflammatory cytokine release." J Gen Virol **84**(Pt 9): 2389-2400.

Cahour, A., B. Falgout and C. J. Lai (1992). "Cleavage of the dengue virus polyprotein at the NS3/NS4A and NS4B/NS5 junctions is mediated by viral protease NS2B-NS3, whereas NS4A/NS4B may be processed by a cellular protease." J Virol **66**(3): 1535-1542.

Carra, S., J. F. Brunsting, H. Lambert, J. Landry and H. H. Kampinga (2009). "HspB8 participates in protein quality control by a non-chaperone-like mechanism that requires eIF2{alpha} phosphorylation." J Biol Chem **284**(9): 5523-5532.

Carracedo, A., M. Gironella, M. Lorente, S. Garcia, M. Guzman, G. Velasco and J. L. Iovanna (2006). "Cannabinoids induce apoptosis of pancreatic tumor cells via endoplasmic reticulum stress-related genes." Cancer Res **66**(13): 6748-6755.

Carracedo, A., M. Lorente, A. Egia, C. Blazquez, S. Garcia, V. Giroux, C. Malicet, R. Villuendas, M. Gironella, L. Gonzalez-Feria, M. A. Piris, J. L. Iovanna, M. Guzman and G. Velasco (2006). "The stress-regulated protein p8 mediates cannabinoid-induced apoptosis of tumor cells." Cancer Cell **9**(4): 301-312.

Carrell, R. W. (2005). "Cell toxicity and conformational disease." Trends Cell Biol **15**(11): 574-580.

Casella CR, R. E., Finkel TH (1999). "Vpu Increases Susceptibility of Human Immunodeficiency Virus Type 1-Infected Cells to Fas Killing." JOURNAL OF VIROLOGY **73**(1): 92–100.

Catteau, A., O. Kalinina, M. C. Wagner, V. Deubel, M. P. Courageot and P. Despres (2003). "Dengue virus M protein contains a proapoptotic sequence referred to as ApoptoM." J Gen Virol **84**(Pt 10): 2781-2793.

Chambers, T. J., C. S. Hahn, R. Galler and C. M. Rice (1990). "Flavivirus genome organization, expression, and replication." Annu Rev Microbiol **44**: 649-688.

Chen, W., P. A. Calvo, D. Malide, J. Gibbs, U. Schubert, I. Bacik, S. Basta, R. O'Neill, J. Schickli, P. Palese, P. Henklein, J. R. Bennink and J. W. Yewdell

(2001). "A novel influenza A virus mitochondrial protein that induces cell death." Nat Med **7**(12): 1306-1312.

Chevalier, C., A. Al Bazzal, J. Vidic, V. Fevrier, C. Bourdieu, E. Bouguyon, R. Le Goffic, J. F. Vautherot, J. Bernard, M. Moudjou, S. Noinville, J. F. Chich, B. Da Costa, H. Rezaei and B. Delmas (2010). "PB1-F2 influenza A virus protein adopts a beta-sheet conformation and forms amyloid fibers in membrane environments." J Biol Chem **285**(17): 13233-13243.

Chirillo, P., S. Pagano, G. Natoli, P. L. Puri, V. L. Burgio, C. Balsano and M. Levrero (1997). "The hepatitis B virus X gene induces p53-mediated programmed cell death." Proc Natl Acad Sci U S A **94**(15): 8162-8167.

Cleaves, G. R. and D. T. Dubin (1979). "Methylation status of intracellular dengue type 2 40 S RNA." Virology **96**(1): 159-165.

Cooper, A., M. Garcia, C. Petrovas, T. Yamamoto, R. A. Koup and G. J. Nabel (2013). "HIV-1 causes CD4 cell death through DNA-dependent protein kinase during viral integration." Nature **498**(7454): 376-379.

Corver, J., A. Ortiz, S. L. Allison, J. Schalich, F. X. Heinz and J. Wilschut (2000). "Membrane fusion activity of tick-borne encephalitis virus and recombinant subviral particles in a liposomal model system." Virology **269**(1): 37-46.

Datan, E., A. Shirazian, S. Benjamin, D. Matassov, A. Tinari, W. Malorni, R. A. Lockshin, A. Garcia-Sastre and Z. Zakeri (2014). "mTOR/p70S6K signaling distinguishes routine, maintenance-level autophagy from autophagic cell death during influenza A infection." Virology **452-453**: 175-190.

Dhanwani, R., M. Khan, A. S. Bhaskar, R. Singh, I. K. Patro, P. V. Rao and M. M. Parida (2012). "Characterization of Chikungunya virus infection in human neuroblastoma SH-SY5Y cells: role of apoptosis in neuronal cell death." Virus Res **163**(2): 563-572.

Diamond, M. S. (2003). "Evasion of innate and adaptive immunity by flaviviruses." Immunol Cell Biol **81**(3): 196-206.

Ding, W. X., H. M. Ni, W. Gao, T. Yoshimori, D. B. Stolz, D. Ron and X. M. Yin (2007). "Linking of autophagy to ubiquitin-proteasome system is important for the regulation of endoplasmic reticulum stress and cell viability." Am J Pathol **171**(2): 513-524.

Donelan, N. R., B. Dauber, X. Wang, C. F. Basler, T. Wolff and A. Garcia-Sastre (2004). "The N- and C-terminal domains of the NS1 protein of influenza B virus can independently inhibit IRF-3 and beta interferon promoter activation." J Virol **78**(21): 11574-11582.

Duarte dos Santos, C. N., M. P. Frenkiel, M. P. Courageot, C. F. Rocha, M. C. Vazeille-Falcoz, M. W. Wien, F. A. Rey, V. Deubel and P. Despres (2000). "Determinants in the envelope E protein and viral RNA helicase NS3 that influence the induction of apoptosis in response to infection with dengue type 1 virus." Virology **274**(2): 292-308.

Falgout, B., M. Pethel, Y. M. Zhang and C. J. Lai (1991). "Both nonstructural proteins NS2B and NS3 are required for the proteolytic processing of dengue virus nonstructural proteins." J Virol **65**(5): 2467-2475.

Fischer, R., T. Baumert and H. E. Blum (2007). "Hepatitis C virus infection and apoptosis." World J Gastroenterol **13**(36): 4865-4872.

Fujimoto, I., T. Takizawa, Y. Ohba and Y. Nakanishi (1998). "Co-expression of Fas and Fas-ligand on the surface of influenza virus-infected cells." Cell Death Differ **5**(5): 426-431.

Gandhi, R. T., B. K. Chen, S. E. Straus, J. K. Dale, M. J. Lenardo and D. Baltimore (1998). "HIV-1 directly kills CD4+ T cells by a Fas-independent mechanism." J Exp Med **187**(7): 1113-1122.

Ghosh Roy, S., B. Sadigh, E. Datan, R. A. Lockshin and Z. Zakeri (2014). "Regulation of cell survival and death during Flavivirus infections." World J Biol Chem **5**(2): 93-105.

Gomatos, P. J., L. Kaariainen, S. Keranen, M. Ranki and D. L. Sawicki (1980). "Semliki Forest virus replication complex capable of synthesizing 42S and 26S nascent RNA chains." J Gen Virol **49**(1): 61-69.

Granell, S., G. Baldini, S. Mohammad, V. Nicolin, P. Narducci, B. Storrie and G. Baldini (2008). "Sequestration of mutated alpha1-antitrypsin into inclusion bodies is a cell-protective mechanism to maintain endoplasmic reticulum function." Mol Biol Cell **19**(2): 572-586.

Gubler, D. J. (1998). "Dengue and dengue hemorrhagic fever." Clin Microbiol Rev **11**(3): 480-496.

Guzman, M. G., S. B. Halstead, H. Artsob, P. Buchy, J. Farrar, D. J. Gubler, E. Hunsperger, A. Kroeger, H. S. Margolis, E. Martinez, M. B. Nathan, J. L. Pelegrino, C. Simmons, S. Yoksan and R. W. Peeling (2010). "Dengue: a continuing global threat." Nat Rev Microbiol **8**(12 Suppl): S7-16.

Guzman, M. G., G. P. Kouri, J. Bravo, M. Soler, S. Vazquez and L. Morier (1990). "Dengue hemorrhagic fever in Cuba, 1981: a retrospective seroepidemiologic study." Am J Trop Med Hyg **42**(2): 179-184.

Han, J. Y., S. A. Miller, T. M. Wolfe, H. Pourhassan and K. R. Jerome (2009). "Cell type-specific induction and inhibition of apoptosis by Herpes Simplex virus type 2 ICP10." J Virol **83**(6): 2765-2769.

Hardwick, J. M. (2001). "Apoptosis in viral pathogenesis." Cell Death Differ **8**(2): 109-110.

Hashimoto, F., N. Oyaizu, V. S. Kalyanaraman and S. Pahwa (1997). "Modulation of Bcl-2 protein by CD4 cross-linking: a possible mechanism for lymphocyte apoptosis in human immunodeficiency virus infection and for rescue of apoptosis by interleukin-2." Blood 90(2): 745-753.

Henderson, G., W. Peng, L. Jin, G. C. Perng, A. B. Nesburn, S. L. Wechsler and C. Jones (2002). "Regulation of caspase 8- and caspase 9-induced apoptosis by the herpes simplex virus type 1 latency-associated transcript." J Neurovirol 8 Suppl 2: 103-111.

Hilleman, M. R. (2004). "Strategies and mechanisms for host and pathogen survival in acute and persistent viral infections." PNAS 101(suppl. 2): 14560–14566.

Hilleman, M. R. (2004). "Strategies and mechanisms for host and pathogen survival in acute and persistent viral infections." Proc Natl Acad Sci U S A 101 Suppl 2: 14560-14566.

Hinshaw, V. S., C. W. Olsen, N. Dybdahl-Sissoko and D. Evans (1994). "Apoptosis: a mechanism of cell killing by influenza A and B viruses." J Virol 68(6): 3667-3673.

Hoyer-Hansen, M., L. Bastholm, P. Szyniarowski, M. Campanella, G. Szabadkai, T. Farkas, K. Bianchi, N. Fehrenbacher, F. Elling, R. Rizzuto, I. S. Mathiasen and M. Jaattela (2007). "Control of macroautophagy by calcium, calmodulin-dependent kinase kinase-beta, and Bcl-2." Mol Cell 25(2): 193-205.

Hui, K. P., S. M. Lee, C. Y. Cheung, I. H. Ng, L. L. Poon, Y. Guan, N. Y. Ip, A. S. Lau and J. S. Peiris (2009). "Induction of proinflammatory cytokines in primary human macrophages by influenza A virus (H5N1) is selectively regulated by IFN regulatory factor 3 and p38 MAPK." J Immunol 182(2): 1088-1098.

Irusta, P. M., Y. B. Chen and J. M. Hardwick (2003). "Viral modulators of cell death provide new links to old pathways." Curr Opin Cell Biol 15(6): 700-705.

Jackson, W. T., T. H. Giddings, Jr., M. P. Taylor, S. Mulinyawe, M. Rabinovitch, R. R. Kopito and K. Kirkegaard (2005). "Subversion of cellular autophagosomal machinery by RNA viruses." PLoS Biol 3(5): e156.

Jacotot, E., C. Callebaut, J. Blanco, Y. Riviere, B. Krust and A. G. Hovanessian (1996). "HIV envelope glycoprotein-induced cell killing by apoptosis is enhanced with increased expression of CD26 in CD4+ T cells." Virology 223(2): 318-330.

Jaiyen, Y., P. Masrinoul, S. Kalayanarooj, R. Pulmanausahakul and S. Ubol (2009). "Characteristics of dengue virus-infected peripheral blood mononuclear cell death that correlates with the severity of illness." Microbiol Immunol 53(8): 442-450.

Jerome, K. R., R. Fox, Z. Chen, P. Sarkar and L. Corey (2001). "Inhibition of apoptosis by primary isolates of herpes simplex virus." Arch Virol **146**(11): 2219-2225.

Joubert, P. E., S. W. Werneke, C. de la Calle, F. Guivel-Benhassine, A. Giodini, L. Peduto, B. Levine, O. Schwartz, D. J. Lenschow and M. L. Albert (2012). "Chikungunya virus-induced autophagy delays caspase-dependent cell death." J Exp Med **209**(5): 1029-1047.

Judith, D., S. Mostowy, M. Bourai, N. Gangneux, M. Lelek, M. Lucas-Hourani, N. Cayet, Y. Jacob, M. C. Prevost, P. Pierre, F. Tangy, C. Zimmer, P. O. Vidalain, T. Couderc and M. Lecuit (2013). "Species-specific impact of the autophagy machinery on Chikungunya virus infection." EMBO Rep **14**(6): 534-544.

Khunchai, S., M. Junking, A. Suttitheptumrong, U. Yasamut, N. Sawasdee, J. Netsawang, A. Morchang, P. Chaowalit, S. Noisakran, P. T. Yenchitsomanus and T. Limjindaporn (2012). "Interaction of dengue virus nonstructural protein 5 with Daxx modulates RANTES production." Biochem Biophys Res Commun **423**(2): 398-403.

Kouroku, Y., E. Fujita, I. Tanida, T. Ueno, A. Isoai, H. Kumagai, S. Ogawa, R. J. Kaufman, E. Kominami and T. Momoi (2007). "ER stress (PERK/eIF2alpha phosphorylation) mediates the polyglutamine-induced LC3 conversion, an essential step for autophagy formation." Cell Death Differ **14**(2): 230-239.

Krejbich-Trotot, P., B. Gay, G. Li-Pat-Yuen, J. J. Hoarau, M. C. Jaffar-Bandjee, L. Briant, P. Gasque and M. Denizot (2011). "Chikungunya triggers an autophagic process which promotes viral replication." Virol J **8**: 432.

Kumar A, Y. Z., Gaoyuan Meng, Musheng Zeng, Seetha Srinivasan,, Q. G. Laurie M. Delmolino, Goberdhan Dimri, Georg F. Weber, and H. B. a. V. B. David E. Wazer (2002). "Human Papillomavirus Oncoprotein E6 Inactivates the Transcriptional Coactivator Human ADA3." Molec. Cell. Biol. **22**(16): 5801-5812.

Lee, Y. R., H. Y. Lei, M. T. Liu, J. R. Wang, S. H. Chen, Y. F. Jiang-Shieh, Y. S. Lin, T. M. Yeh, C. C. Liu and H. S. Liu (2008). "Autophagic machinery activated by dengue virus enhances virus replication." Virology **374**(2): 240-248.

Li, D. D., L. L. Wang, R. Deng, J. Tang, Y. Shen, J. F. Guo, Y. Wang, L. P. Xia, G. K. Feng, Q. Q. Liu, W. L. Huang, Y. X. Zeng and X. F. Zhu (2009). "The pivotal role of c-Jun NH2-terminal kinase-mediated Beclin 1 expression during anticancer agents-induced autophagy in cancer cells." Oncogene **28**(6): 886-898.

Lim, E. J., K. El Khobar, R. Chin, L. Earnest-Silveira, P. W. Angus, C. T. Bock, U. Nachbur, J. Silke and J. Torresi (2014). "Hepatitis C virus-induced hepatocyte cell death and protection by inhibition of apoptosis." J Gen Virol **95**(Pt 10): 2204-2215.

Lin, C., R. E. Holland, Jr., J. C. Donofrio, M. H. McCoy, L. R. Tudor and T. M. Chambers (2002). "Caspase activation in equine influenza virus induced apoptotic cell death." Vet Microbiol **84**(4): 357-365.

Lin, C. F., H. Y. Lei, A. L. Shiau, H. S. Liu, T. M. Yeh, S. H. Chen, C. C. Liu, S. C. Chiu and Y. S. Lin (2002). "Endothelial cell apoptosis induced by antibodies against dengue virus nonstructural protein 1 via production of nitric oxide." J Immunol **169**(2): 657-664.

Lindenbach, B. D. and C. M. Rice (2003). "Molecular biology of flaviviruses." Adv Virus Res **59**: 23-61.

Lines, J. L., S. Hoskins, M. Hollifield, L. S. Cauley and B. A. Garvy (2010). "The migration of T cells in response to influenza virus is altered in neonatal mice." J Immunol **185**(5): 2980-2988.

Long, X., Y. Li, Y. Qi, J. Xu, Z. Wang, X. Zhang, D. Zhang, L. Zhang and J. Huang (2013). "XAF1 contributes to dengue virus-induced apoptosis in vascular endothelial cells." FASEB J **27**(3): 1062-1073.

Lowy, R. J. (2003). "Influenza virus induction of apoptosis by intrinsic and extrinsic mechanisms." Int Rev Immunol **22**(5-6): 425-449.

Lu, X., A. Masic, Y. Li, Y. Shin, Q. Liu and Y. Zhou (2010). "The PI3K/Akt pathway inhibits influenza A virus-induced Bax-mediated apoptosis by negatively regulating the JNK pathway via ASK1." J Gen Virol **91**(Pt 6): 1439-1449.

Lu, Y. Y., Y. Koga, K. Tanaka, M. Sasaki, G. Kimura and K. Nomoto (1994). "Apoptosis induced in CD4+ cells expressing gp160 of human immunodeficiency virus type 1." J Virol **68**(1): 390-399.

Ludwig, S., O. Planz, S. Pleschka and T. Wolff (2003). "Influenza-virus-induced signaling cascades: targets for antiviral therapy?" Trends Mol Med **9**(2): 46-52.

Ludwig, S., S. Pleschka, O. Planz and T. Wolff (2006). "Ringing the alarm bells: signalling and apoptosis in influenza virus infected cells." Cell Microbiol **8**(3): 375-386.

Marianneau, P., A. Cardona, L. Edelman, V. Deubel and P. Despres (1997). "Dengue virus replication in human hepatoma cells activates NF-kappaB which in turn induces apoptotic cell death." J Virol **71**(4): 3244-3249.

Martin, S. J., P. M. Matear and A. Vyakarnam (1994). "HIV-1 infection of human CD4+ T cells in vitro. Differential induction of apoptosis in these cells." J Immunol **152**(1): 330-342.

Martina, B. E., P. Koraka and A. D. Osterhaus (2009). "Dengue virus pathogenesis: an integrated view." Clin Microbiol Rev **22**(4): 564-581.

Maruoka, S., S. Hashimoto, Y. Gon, H. Nishitoh, I. Takeshita, Y. Asai, K. Mizumura, K. Shimizu, H. Ichijo and T. Horie (2003). "ASK1 regulates influenza virus infection-induced apoptotic cell death." Biochem Biophys Res Commun **307**(4): 870-876.

Matassov, D., T. Kagan, J. Leblanc, M. Sikorska and Z. Zakeri (2004). "Measurement of apoptosis by DNA fragmentation." Methods Mol Biol **282**: 1-17.

Mateo, R., C. M. Nagamine, J. Spagnolo, E. Mendez, M. Rahe, M. Gale, Jr., J. Yuan and K. Kirkegaard (2013). "Inhibition of cellular autophagy deranges dengue virion maturation." J Virol **87**(3): 1312-1321.

Mbita, Z., R. Hull and Z. Dlamini (2014). "Human immunodeficiency virus-1 (HIV-1)-mediated apoptosis: new therapeutic targets." Viruses **6**(8): 3181-3227.

McLean, J. E., E. Datan, D. Matassov and Z. F. Zakeri (2009). "Lack of Bax prevents influenza A virus-induced apoptosis and causes diminished viral replication." J Virol **83**(16): 8233-8246.

McLean, J. E., A. Wudzinska, E. Datan, D. Quaglino and Z. Zakeri (2011). "Flavivirus NS4A-induced autophagy protects cells against death and enhances virus replication." J Biol Chem **286**(25): 22147-22159.

Medigeshi, G. R., A. M. Lancaster, A. J. Hirsch, T. Briese, W. I. Lipkin, V. Defilippis, K. Fruh, P. W. Mason, J. Nikolich-Zugich and J. A. Nelson (2007). "West Nile virus infection activates the unfolded protein response, leading to CHOP induction and apoptosis." J Virol **81**(20): 10849-10860.

Midgley, C. M., M. Bajwa-Joseph, S. Vasanawathana, W. Limpitikul, B. Wills, A. Flanagan, E. Waiyaiya, H. B. Tran, A. E. Cowper, P. Chotiyarnwong, J. M. Grimes, S. Yoksan, P. Malasit, C. P. Simmons, J. Mongkolsapaya and G. R. Screaton (2011). "An in-depth analysis of original antigenic sin in dengue virus infection." J Virol **85**(1): 410-421.

Morchang, A., U. Yasamut, J. Netsawang, S. Noisakran, W. Wongwiwat, P. Songprakhon, C. Srisawat, C. Puttikhunt, W. Kasinrerk, P. Malasit, P. T. Yenchitsomanus and T. Limjindaporn (2011). "Cell death gene expression profile: role of RIPK2 in dengue virus-mediated apoptosis." Virus Res **156**(1-2): 25-34.

Mori, I., F. Goshima, T. Koshizuka, N. Koide, T. Sugiyama, T. Yoshida, T. Yokochi, Y. Nishiyama and Y. Kimura (2003). "Differential activation of the c-Jun N-terminal kinase/stress-activated protein kinase and p38 mitogen-activated protein kinase signal transduction pathways in the mouse brain upon infection with neurovirulent influenza A virus." J Gen Virol **84**(Pt 9): 2401-2408.

Mori, I., T. Komatsu, K. Takeuchi, K. Nakakuki, M. Sudo and Y. Kimura (1995). "In vivo induction of apoptosis by influenza virus." Journal of general virology **76**(11): 2869.

Morris, S. J., G. E. Price, J. M. Barnett, S. A. Hiscox, H. Smith and C. Sweet (1999). "Role of neuraminidase in influenza virus-induced apoptosis." J Gen Virol 80 (Pt 1): 137-146.

Mukhopadhyay, S., R. J. Kuhn and M. G. Rossmann (2005). "A structural perspective of the flavivirus life cycle." Nat Rev Microbiol 3(1): 13-22.

Myint, K. S., T. P. Endy, D. Mongkolsirichaikul, C. Manomuth, S. Kalayanarooj, D. W. Vaughn, A. Nisalak, S. Green, A. L. Rothman, F. A. Ennis and D. H. Libraty (2006). "Cellular immune activation in children with acute dengue virus infections is modulated by apoptosis." J Infect Dis 194(5): 600-607.

Nagila, A., J. Netsawang, C. Srisawat, S. Noisakran, A. Morchang, U. Yasamut, C. Puttikhunt, W. Kasinrerk, P. Malasit, P. T. Yenchitsomanus and T. Limjindaporn (2011). "Role of CD137 signaling in dengue virus-mediated apoptosis." Biochem Biophys Res Commun 410(3): 428-433.

Nain, M., F. Hinder, J. H. Gong, A. Schmidt, A. Bender, H. Sprenger and D. Gemsa (1990). "Tumor necrosis factor-alpha production of influenza A virus-infected macrophages and potentiating effect of lipopolysaccharides." J Immunol 145(6): 1921-1928.

Obeng, E. A., L. M. Carlson, D. M. Gutman, W. J. Harrington, Jr., K. P. Lee and L. H. Boise (2006). "Proteasome inhibitors induce a terminal unfolded protein response in multiple myeloma cells." Blood 107(12): 4907-4916.

Ogata, M., S. Hino, A. Saito, K. Morikawa, S. Kondo, S. Kanemoto, T. Murakami, M. Taniguchi, I. Tanii, K. Yoshinaga, S. Shiosaka, J. A. Hammarback, F. Urano and K. Imaizumi (2006). "Autophagy is activated for cell survival after endoplasmic reticulum stress." Mol Cell Biol 26(24): 9220-9231.

Olsen, C. W., J. C. Kehren, N. R. Dybdahl-Sissoko and V. S. Hinshaw (1996). "bcl-2 alters influenza virus yield, spread, and hemagglutinin glycosylation." J Virol 70(1): 663-666.

Panyasrivanit, M., M. P. Greenwood, D. Murphy, C. Isidoro, P. Auewarakul and D. R. Smith (2011). "Induced autophagy reduces virus output in dengue infected monocytic cells." Virology 418(1): 74-84.

Pattingre, S., C. Bauvy, S. Carpentier, T. Levade, B. Levine and P. Codogno (2009). "Role of JNK1-dependent Bcl-2 phosphorylation in ceramide-induced macroautophagy." J Biol Chem 284(5): 2719-2728.

Pena, J. and E. Harris (2011). "Dengue virus modulates the unfolded protein response in a time-dependent manner." J Biol Chem 286(16): 14226-14236.

Peng, W., G. Henderson, G. C. Perng, A. B. Nesburn, S. L. Wechsler and C. Jones (2003). "The gene that encodes the herpes simplex virus type 1 latency-associated transcript influences the accumulation of transcripts (Bcl-x(L) and

Bcl-x(S)) that encode apoptotic regulatory proteins." J Virol **77**(19): 10714-10718.

Perera, R., K. E. Owen, T. L. Tellinghuisen, A. E. Gorbalenya and R. J. Kuhn (2001). "Alphavirus nucleocapsid protein contains a putative coiled coil alpha-helix important for core assembly." J Virol **75**(1): 1-10.

Perkins, D., E. F. Pereira and L. Aurelian (2003). "The herpes simplex virus type 2 R1 protein kinase (ICP10 PK) functions as a dominant regulator of apoptosis in hippocampal neurons involving activation of the ERK survival pathway and upregulation of the antiapoptotic protein Bag-1." J Virol **77**(2): 1292-1305.

Peschke, T., A. Bender, M. Nain and D. Gemsa (1993). "Role of macrophage cytokines in influenza A virus infections." Immunobiology **189**(3-4): 340-355.

Pialoux, G., B. A. Gauzere, S. Jaureguiberry and M. Strobel (2007). "Chikungunya, an epidemic arbovirosis." Lancet Infect Dis **7**(5): 319-327.

Price, B. D., L. A. Mannheim-Rodman and S. K. Calderwood (1992). "Brefeldin A, thapsigargin, and AlF4- stimulate the accumulation of GRP78 mRNA in a cycloheximide dependent manner, whilst induction by hypoxia is independent of protein synthesis." J Cell Physiol **152**(3): 545-552.

Rawat, S. and M. Bouchard (2014). "The Hepatitis B Virus HBx protein activates AKT to simultaneously regulate HBV replication and hepatocyte survival." J Virol.

Ray, R. B., K. Meyer and R. Ray (1996). "Suppression of apoptotic cell death by hepatitis C virus core protein." Virology **226**(2): 176-182.

Richards, A. L. and W. T. Jackson (2013). "How positive-strand RNA viruses benefit from autophagosome maturation." J Virol **87**(18): 9966-9972.

Rodriguez-Madoz, J. R., D. Bernal-Rubio, D. Kaminski, K. Boyd and A. Fernandez-Sesma (2010). "Dengue virus inhibits the production of type I interferon in primary human dendritic cells." J Virol **84**(9): 4845-4850.

Rothman, A. L. (2011). "Immunity to dengue virus: a tale of original antigenic sin and tropical cytokine storms." Nat Rev Immunol **11**(8): 532-543.

Roulston A, M. R., Branton PE (1999). "Viruses and Apoptosis." Annu. Rev. Microbiol. **53**: 577-628.

Saito, T., M. Tanaka and I. Yamaguchi (1996). "Effect of brefeldin A on influenza A virus-induced apoptosis in vitro." J Vet Med Sci **58**(11): 1137-1139.

Salazar, M., A. Carracedo, I. J. Salanueva, S. Hernandez-Tiedra, M. Lorente, A. Egia, P. Vazquez, C. Blazquez, S. Torres, S. Garcia, J. Nowak, G. M. Fimia, M. Piacentini, F. Cecconi, P. P. Pandolfi, L. Gonzalez-Feria, J. L. Iovanna, M. Guzman, P. Boya and G. Velasco (2009). "Cannabinoid action induces autophagy-mediated cell death through stimulation of ER stress in human glioma cells." J Clin Invest **119**(5): 1359-1372.

Samal, J., M. Kandpal and P. Vivekanandan (2012). "Molecular mechanisms underlying occult hepatitis B virus infection." Clin Microbiol Rev **25**(1): 142-163.

Schroder, M. and R. J. Kaufman (2005). "The mammalian unfolded protein response." Annu Rev Biochem **74**: 739-789.

Schuffenecker, I., I. Iteman, A. Michault, S. Murri, L. Frangeul, M. C. Vaney, R. Lavenir, N. Pardigon, J. M. Reynes, F. Pettinelli, L. Biscornet, L. Diancourt, S. Michel, S. Duquerroy, G. Guigon, M. P. Frenkiel, A. C. Brehin, N. Cubito, P. Despres, F. Kunst, F. A. Rey, H. Zeller and S. Brisse (2006). "Genome microevolution of chikungunya viruses causing the Indian Ocean outbreak." PLoS Med **3**(7): e263.

Schultz-Cherry, S., N. Dybdahl-Sissoko, G. Neumann, Y. Kawaoka and V. S. Hinshaw (2001). "Influenza virus ns1 protein induces apoptosis in cultured cells." J Virol **75**(17): 7875-7881.

Schultz-Cherry, S. and V. S. Hinshaw (1996). "Influenza virus neuraminidase activates latent transforming growth factor beta." J Virol **70**(12): 8624-8629.

Sedger, L. M., S. Hou, S. R. Osvath, M. B. Glaccum, J. J. Peschon, N. van Rooijen and L. Hyland (2002). "Bone marrow B cell apoptosis during in vivo influenza virus infection requires TNF-alpha and lymphotoxin-alpha." J Immunol **169**(11): 6193-6201.

Shrivastava, S., J. Bhanja Chowdhury, R. Steele, R. Ray and R. B. Ray (2012). "Hepatitis C virus upregulates Beclin1 for induction of autophagy and activates mTOR signaling." J Virol **86**(16): 8705-8712.

Silveira, G. F., F. Meyer, A. Delfraro, A. L. Mosimann, N. Coluchi, C. Vasquez, C. M. Probst, A. Bafica, J. Bordignon and C. N. Dos Santos (2011). "Dengue virus type 3 isolated from a fatal case with visceral complications induces enhanced proinflammatory responses and apoptosis of human dendritic cells." J Virol **85**(11): 5374-5383.

Solignat, M., B. Gay, S. Higgs, L. Briant and C. Devaux (2009). "Replication cycle of chikungunya: a re-emerging arbovirus." Virology **393**(2): 183-197.

Sourisseau, M., C. Schilte, N. Casartelli, C. Trouillet, F. Guivel-Benhassine, D. Rudnicka, N. Sol-Foulon, K. Le Roux, M. C. Prevost, H. Fsihi, M. P. Frenkiel, F. Blanchet, P. V. Afonso, P. E. Ceccaldi, S. Ozden, A. Gessain, I. Schuffenecker, B. Verhasselt, A. Zamborlini, A. Saib, F. A. Rey, F. Arenzana-Seisdedos, P. Despres, A. Michault, M. L. Albert and O. Schwartz (2007). "Characterization of reemerging chikungunya virus." PLoS Pathog **3**(6): e89.

Stadler, K., S. L. Allison, J. Schalich and F. X. Heinz (1997). "Proteolytic activation of tick-borne encephalitis virus by furin." J Virol **71**(11): 8475-8481.

Takizawa, T., S. Matsukawa, Y. Higuchi, S. Nakamura, Y. Nakanishi and R. Fukuda (1993). "Induction of programmed cell death (apoptosis) by influenza virus infection in tissue culture cells." J Gen Virol **74 (Pt 11)**: 2347-2355.

Takizawa, T., K. Ohashi and Y. Nakanishi (1996). "Possible involvement of double-stranded RNA-activated protein kinase in cell death by influenza virus infection." J Virol **70**(11): 8128-8132.

Talloczy, Z., W. Jiang, H. W. t. Virgin, D. A. Leib, D. Scheuner, R. J. Kaufman, E. L. Eskelinen and B. Levine (2002). "Regulation of starvation- and virus-induced autophagy by the eIF2alpha kinase signaling pathway." Proc Natl Acad Sci U S A **99**(1): 190-195.

Talon, J., C. M. Horvath, R. Polley, C. F. Basler, T. Muster, P. Palese and A. Garcia-Sastre (2000). "Activation of interferon regulatory factor 3 is inhibited by the influenza A virus NS1 protein." J Virol **74**(17): 7989-7996.

Tamm, K., A. Merits and I. Sarand (2008). "Mutations in the nuclear localization signal of nsP2 influencing RNA synthesis, protein expression and cytotoxicity of Semliki Forest virus." J Gen Virol **89**(Pt 3): 676-686.

Tang, B. L. (2012). "The cell biology of Chikungunya virus infection." Cell Microbiol **14**(9): 1354-1363.

Tsuchida, T., T. Kawai and S. Akira (2009). "Inhibition of IRF3-dependent antiviral responses by cellular and viral proteins." Cell Res **19**(1): 3-4.

Turnbull, E. L., M. F. Rosser and D. M. Cyr (2007). "The role of the UPS in cystic fibrosis." BMC Biochem **8 Suppl 1**: S11.

Turpin, E., K. Luke, J. Jones, T. Tumpey, K. Konan and S. Schultz-Cherry (2005). "Influenza virus infection increases p53 activity: role of p53 in cell death and viral replication." J Virol **79**(14): 8802-8811.

Wang, W. H., R. L. Hullinger and O. M. Andrisani (2008). "Hepatitis B virus X protein via the p38MAPK pathway induces E2F1 release and ATR kinase activation mediating p53 apoptosis." J Biol Chem **283**(37): 25455-25467.

Wei, Y., S. Sinha and B. Levine (2008). "Dual role of JNK1-mediated phosphorylation of Bcl-2 in autophagy and apoptosis regulation." Autophagy **4**(7): 949-951.

Widener, R. W. and R. J. Whitley (2014). "Herpes simplex virus." Handb Clin Neurol **123**: 251-263.

Wintachai, P., N. Wikan, A. Kuadkitkan, T. Jaimipuk, S. Ubol, R. Pulmanausahakul, P. Auewarakul, W. Kasinrerk, W. Y. Weng, M. Panyasrivanit, A. Paemanee, S. Kittisenachai, S. Roytrakul and D. R. Smith (2012).

"Identification of prohibitin as a Chikungunya virus receptor protein." J Med Virol **84**(11): 1757-1770.

Wurzer, W. J., C. Ehrhardt, S. Pleschka, F. Berberich-Siebelt, T. Wolff, H. Walczak, O. Planz and S. Ludwig (2004). "NF-kappaB-dependent induction of tumor necrosis factor-related apoptosis-inducing ligand (TRAIL) and Fas/FasL is crucial for efficient influenza virus propagation." J Biol Chem **279**(30): 30931-30937.

Wurzer, W. J., O. Planz, C. Ehrhardt, M. Giner, T. Silberzahn, S. Pleschka and S. Ludwig (2003). "Caspase 3 activation is essential for efficient influenza virus propagation." EMBO J **22**(11): 2717-2728.

Yamada, H., R. Chounan, Y. Higashi, N. Kurihara and H. Kido (2004). "Mitochondrial targeting sequence of the influenza A virus PB1-F2 protein and its function in mitochondria." FEBS Lett **578**(3): 331-336.

Yang, Z. and D. J. Klionsky (2010). "Eaten alive: a history of macroautophagy." Nat Cell Biol **12**(9): 814-822.

Zamarin, D., A. Garcia-Sastre, X. Xiao, R. Wang and P. Palese (2005). "Influenza virus PB1-F2 protein induces cell death through mitochondrial ANT3 and VDAC1." PLoS Pathog **1**(1): e4.

Zamarin, D., M. B. Ortigoza and P. Palese (2006). "Influenza A virus PB1-F2 protein contributes to viral pathogenesis in mice." J Virol **80**(16): 7976-7983.

Zell, R., A. Krumbholz, A. Eitner, R. Krieg, K. J. Halbhuber and P. Wutzler (2007). "Prevalence of PB1-F2 of influenza A viruses." J Gen Virol **88**(Pt 2): 536-546.

Zhirnov, O. P., T. E. Konakova, W. Garten and H. Klenk (1999). "Caspase-dependent N-terminal cleavage of influenza virus nucleocapsid protein in infected cells." J Virol **73**(12): 10158-10163.

Zhirnov, O. P., T. E. Konakova, T. Wolff and H. D. Klenk (2002). "NS1 protein of influenza A virus down-regulates apoptosis." J Virol **76**(4): 1617-1625.

Chapter 16: Harnessing apoptosis pathways for childhood cancer (Fulda)

Simone Fulda

Institute for Experimental Cancer Research in Pediatrics, Goethe-University Frankfurt, Komturstr. 3a, 60528 Frankfurt, Germany

Abstract

Since apoptosis is frequently disturbed in cancers including childhood malignancies, therapeutic targeting of this form of programmed cell death provides a promising strategy in oncology. Based on the concept that signaling to cell death can be disturbed by either the dominance of inhibitory factors and/or loss of expression or function of apoptosis-promoting factors, a number of strategies have been developed over the last decades in order to trigger cell death in cancer cells. In the years to come the challenge resides in successfully transferring this knowledge into the development of more effective approaches for the treatment of children with cancer.

Introduction

Apoptosis represents one of the best characterized forms of programmed cell death that is typically disturbed in human cancers (Fulda, 2009b). Evasion of apoptosis can not only contribute to tumor formation but also to progression and metastasis of neoplastic diseases. In addition, the anticancer activity of many currently used treatment options in oncology critically depends on the engagement of apoptosis pathways in tumor cells (Fulda and Debatin, 2006b). The death receptor (extrinsic) pathway and the mitochondrial (intrinsic) pathway constitute the core apoptotic machinery (Fulda and Debatin, 2006b). Both signaling cascades eventually result in the activation of caspases as cell death effector molecules (Fulda and Debatin, 2006b). In principle, apoptosis can be disabled by loss or inactivation of proapoptotic components of the network or, alternatively, by the relative dominance of antiapoptotic constituents (Fulda, 2009b). Primary or acquired treatment resistance of human cancers can be caused by defective apoptosis programs. Since resistance to currently available therapies represents one of the major unsolved problems in oncology, further insights into the signaling network that governs apoptosis sensitivity will open new opportunities for the identification of novel therapeutic targets and the rational development of new therapeutics. This concept is of general relevance for human malignancies including pediatric cancers. The current review will provide prototypic examples of therapeutic targeting of apoptosis programs in childhood cancers.

How to harness apoptosis signaling pathways for the treatment of cancer?

There are several possibilities to target apoptosis signaling pathways for therapeutic purposes in order to elicit cell death in cancer cells. In principle, these strategies can be divided into those that aim at enhancing proapoptotic signals and those that neutralize antiapoptotic breaks. Cancer cells evade apoptosis by either inactivation of proapoptotic molecules and/or increased expression or activity of antiapoptotic factors.

Strategies to potentiate proapoptotic signals

One of the best examples of engaging apoptotic programs via stimulation of proapoptotic signals is the death receptor ligand system. Death receptors are a family of transmembrane proteins that are expressed at the cell surface and that provide a link between extracellular signals and the intracellular signaling machinery that eventually leads to the induction of apoptosis (Ashkenazi, 2008). Within the death receptor ligand family, the Tumor-Necrosis-Factor-related apoptosis-inducing ligand (TRAIL)/TRAIL receptor system is of particular relevance for cancer therapy, since it has been reported to preferentially engage apoptosis in malignant compared to non-malignant cells (Ashkenazi and Herbst, 2008). This opens the perspective of tumor-selective targeting of apoptosis-inducing therapies, while sparing normal cells and tissues. In order to trigger the TRAIL system via receptors on the cell surface, agonistic antibodies that are selectively directed against one of the two proapoptotic TRAIL receptors, i.e. TRAIL receptor 1 or 2, have been designed (Ashkenazi and Herbst, 2008). In addition to agonistic humanized TRAIL receptor antibodies, soluble TRAIL has been developed as well (Ashkenazi and Herbst, 2008). Engagement of proapoptotic TRAIL receptors by either agonistic antibodies or by soluble recombinant TRAIL has been described to trigger apoptosis in various childhood malignancies, including leukemia, sarcoma and neuroblastoma (Petak et al., 2000; Van Valen et al., 2000; Kontny et al., 2001; Fulda et al., 2002a; Ehrhardt et al., 2003; Yang and Thiele, 2003; Fulda, 2008; Kang et al., 2011; Yang03 26Abhari et al., 2013). Evaluation of the fully humanized monoclonal antibody Mapatumumab ®, which targets TRAIL receptor 1, by the Pediatric Preclinical Testing Program, an initiative of the National Cancer Institute, to identify drugs for prioritization for further clinical testing, revealed limited antitumor activity of single agent treatment with Mapatumumab against childhood cancer. This finding is in line with many preclinical studies using TRAIL receptor agonists as monotherapy and highlights the importance of designing rational combinations of TRAIL receptor agonists together with other cytotoxic principles. Accordingly, several concepts have been proposed to use TRAIL receptor agonists in combination treatments to enhance the antitumor activity of TRAIL. One approach is the combined use of TRAIL receptor agonists together with classical anticancer drugs that engage the DNA damage response (Van Valen et al., 2003; Komdeur et al., 2004; Muhlethaler-Mottet et al., 2004; Wang et al., 2007; Wang et al., 2010). In addition, TRAIL receptor agonists have also been used in combination with a variety of signal transduction modulators including histone deacetylase inhibitors, proteasome inhibitors and demethylating agents (Fulda et al., 2001; Hacker et al., 2009; Naumann et al., 2011). Also, simultaneous administration of TRAIL receptor agonists to engage the death receptor pathway of apoptosis together with agents that trigger mitochondrial signaling, such as BH3 mimetics, was found to synergize in the induction of apoptosis (Cristofanon and Fulda, 2012). Moreover, releasing the breaks on caspase activation, e.g. by using small-molecule inhibitors of Inhibitors of Apoptosis (IAP) proteins such as Second mitochondria-derived activator of caspase (Smac) mimetics, was found to synergize together with TRAIL receptor agonists to elicit apoptotic cell death in pediatric cancers (Fulda et al., 2002b; Fakler et al., 2009; Basit et al., 2012; Abhari et al., 2013).

Another strategy to enhance proapoptotic signals in pediatric cancers resides in the restoration of caspase-8 expression. Caspase-8 represents a key molecule in the apoptosis signaling network that has been reported to be critical for the engagement of death receptor-induced apoptosis in a variety of cancers (Fulda, 2009a). Of note, caspase-8 expression is frequently silenced in particular in pediatric malignancies including neuroblastoma, medulloblastoma and Ewing sarcoma. Since epigenetic mechanisms including aberrant DNA methylation or disturbance of the acetylation status of histones contribute to epigenetic silencing of caspase-8, epigenetic modifiers can provide a mean to restore caspase-8 expression in pediatric malignancies. Indeed, the use of demethylating agents, histone deacetylase inhibitors as well as interferons, has been reported to successfully upregulate caspase-8 in pediatric cancer cells, thereby sensitizing cells to death receptor- as well as chemotherapy-mediated apoptosis (Fulda et al., 2001; Fulda and Debatin, 2002; Fulda and Debatin, 2006a; Hacker et al., 2009). Since loss of caspase-8 expression correlates with poor prognosis for example in medulloblastoma (Pingoud-Meier et al., 2003), restoration of caspase-8 by small-molecule modulators including epigenetic drugs represents a promising approach that warrants further investigation.

Strategies to antagonize antiapoptotic factors

Another concept to engage apoptosis pathways in pediatric cancers resides in the neutralization of antiapoptotic proteins that prevent initiation or execution of apoptotic cell death. Overexpression of antiapoptotic proteins belongs to the hallmarks of human cancers including pediatric malignancies (Fulda, 2009b). This implies that targeting antiapoptotic proteins may also provide a means to preferentially engage apoptosis in malignant versus normal cells. IAP proteins represent a family of antiapoptotic factors that are frequently expressed at high levels in human neoplasms (Fulda and Vucic, 2012). This family comprises eight human analogs including, for example, X-linked IAP protein (XIAP), cellular inhibitor of apoptosis (cIAP) proteins and survivin (Fulda and Vucic, 2012). All IAP proteins harbor a baculoviral IAP repeat (BIR) domain that is also required for binding to and inhibition of caspases (Fulda and Vucic, 2012). In addition, some IAP proteins express a Really Interesting New Gene (RING) domain that possesses E3 ubiquitin ligase activity (Fulda and Vucic, 2012). Thus, cIAP proteins can modulate ubiquitination events that are involved in the control of cell survival pathways. Since IAP proteins block apoptosis pathways at a central node by preventing activation of caspases, they are considered as promising targets for therapeutic intervention. Consequently, a number of intervention strategies have been developed in the last decades including small-molecule inhibitors that neutralize IAP proteins as well as antisense strategies to downregulate XIAP. Most small molecules that aim at neutralizing IAP proteins mimic the endogenous Smac proteins, as Smac represents an endogenous antagonist of IAP proteins that is released from the mitochondrial interspace into the cytosol upon the induction of apoptosis. Since the interaction domain of Smac that is required for its binding to IAP proteins has been identified, this motif has been used to design a variety of small-molecule inhibitors that target IAP proteins (Fulda and Vucic, 2012). Of note, Smac mimetics have been shown to induce apoptosis either as single agent or in combination with additional cytotoxic principles, including death receptor ligands, anticancer drugs or γ-irradiation in a number of pediatric cancers, e.g. leukemia, rhabdomyosarcoma, neuroblastoma and glioblastoma (Fulda et al., 2002b; Fakler et al., 2009; Basit et al., 2012; Loeder et al., 2012; Abhari et al., 2013; Belz et al., 2014).

Conclusions

Since apoptosis represents a key program that is critical for the regulation of tissue homeostasis and typically disturbed in human cancers, therapeutic targeting of apoptosis represents a promising approach in oncology. In recent years, the discovery of apoptosis signaling pathways and molecules as well as the identification of molecular mechanisms that permit evasion of apoptosis in pediatric cancers has led to the identification of critical targets for therapeutic intervention. In principle, the concept of targeting apoptosis pathways for the treatment of human cancers has been demonstrated in preclinical in vitro and in vivo models. In addition, clinical studies exploiting apoptosis-targeting drugs have recently entered the clinical stage. It is expected that harnessing apoptosis pathways for the treatment of pediatric malignancies will pave the avenue for more effective treatment options in the future.

Conflict of interest

The research was conducted in the absence of any commercial or financial relationships that could be construed as a potential conflict of interest.

Acknowledgements

The expert secretarial assistance of C. Hugenberg is greatly appreciated. Work in the author's laboratory is supported by grants from the Deutsche Forschungsgemeinschaft, the Deutsche Krebshilfe, the Bundesministerium für Forschung und Technologie (01GM1104C), Wilhelm-Sander Stiftung, IUAP and the European Community.

References:

Abhari BA, Cristofanon S, Kappler R, von Schweinitz D, Humphreys R, Fulda S (2013). RIP1 is required for IAP inhibitor-mediated sensitization for TRAIL-induced apoptosis via a RIP1/FADD/caspase-8 cell death complex. *Oncogene* **32:** 3263-73.

Ashkenazi A (2008). Targeting the extrinsic apoptosis pathway in cancer. *Cytokine Growth Factor Rev* **19:** 325-31.

Ashkenazi A, Herbst RS (2008). To kill a tumor cell: the potential of proapoptotic receptor agonists. *J Clin Invest* **118**: 1979-90.

Basit F, Humphreys R, Fulda S (2012). RIP1 Protein-dependent Assembly of a Cytosolic Cell Death Complex Is Required for Inhibitor of Apoptosis (IAP) Inhibitor-mediated Sensitization to Lexatumumab-induced Apoptosis. *J Biol Chem* **287**: 38767-77.

Belz K, Schoeneberger H, Wehner S, Weigert A, Bonig H, Klingebiel T *et al* (2014). Smac mimetic and glucocorticoids synergize to induce apoptosis in childhood ALL by promoting ripoptosome assembly. *Blood* **124**: 240-50.

Cristofanon S, Fulda S (2012). ABT-737 promotes tBid mitochondrial accumulation to enhance TRAIL-induced apoptosis in glioblastoma cells. *Cell Death Dis* **3**: e432.

Ehrhardt H, Fulda S, Schmid I, Hiscott J, Debatin KM, Jeremias I (2003). TRAIL induced survival and proliferation in cancer cells resistant towards TRAIL-induced apoptosis mediated by NF-kappaB. *Oncogene* **22**: 3842-52.

Fakler M, Loeder S, Vogler M, Schneider K, Jeremias I, Debatin KM *et al* (2009). Small molecule XIAP inhibitors cooperate with TRAIL to induce apoptosis in childhood acute leukemia cells and overcome Bcl-2-mediated resistance. *Blood* **113**: 1710-22.

Fulda S (2008). Targeting inhibitor of apoptosis proteins (IAPs) for cancer therapy. *Anticancer Agents Med Chem* **8**: 533-9.

Fulda S (2009a). Caspase-8 in cancer biology and therapy. *Cancer Lett* **281**: 128-33.

Fulda S (2009b). Tumor resistance to apoptosis. *Int J Cancer* **124**: 511-5.

Fulda S, Debatin KM (2002). IFNgamma sensitizes for apoptosis by upregulating caspase-8 expression through the Stat1 pathway. *Oncogene* **21**: 2295-308.

Fulda S, Debatin KM (2006a). 5-Aza-2'-deoxycytidine and IFN-gamma cooperate to sensitize for TRAIL-induced apoptosis by upregulating caspase-8. *Oncogene* **25**: 5125-33.

Fulda S, Debatin KM (2006b). Extrinsic versus intrinsic apoptosis pathways in anticancer chemotherapy. *Oncogene* **25**: 4798-811.

Fulda S, Kufer MU, Meyer E, van Valen F, Dockhorn-Dworniczak B, Debatin KM (2001). Sensitization for death receptor- or drug-induced apoptosis by re-expression of caspase-8 through demethylation or gene transfer. *Oncogene* **20**: 5865-77.

Fulda S, Meyer E, Debatin KM (2002a). Inhibition of TRAIL-induced apoptosis by Bcl-2 overexpression. *Oncogene* **21**: 2283-94.

Fulda S, Vucic D (2012). Targeting IAP proteins for therapeutic intervention in cancer. *Nat Rev Drug Discov* **11:** 109-24.

Fulda S, Wick W, Weller M, Debatin KM (2002b). Smac agonists sensitize for Apo2L/TRAIL- or anticancer drug-induced apoptosis and induce regression of malignant glioma in vivo. *Nat Med* **8:** 808-15.

Hacker S, Dittrich A, Mohr A, Schweitzer T, Rutkowski S, Krauss J *et al* (2009). Histone deacetylase inhibitors cooperate with IFN-gamma to restore caspase-8 expression and overcome TRAIL resistance in cancers with silencing of caspase-8. *Oncogene* **28:** 3097-110.

Kang Z, Chen JJ, Yu Y, Li B, Sun SY, Zhang B *et al* (2011). Drozitumab, a human antibody to death receptor 5, has potent antitumor activity against rhabdomyosarcoma with the expression of caspase-8 predictive of response. *Clin Cancer Res* **17:** 3181-92.

Komdeur R, Meijer C, Van Zweeden M, De Jong S, Wesseling J, Hoekstra HJ *et al* (2004). Doxorubicin potentiates TRAIL cytotoxicity and apoptosis and can overcome TRAIL-resistance in rhabdomyosarcoma cells. *Int J Oncol* **25:** 677-84.

Kontny HU, Hammerle K, Klein R, Shayan P, Mackall CL, Niemeyer CM (2001). Sensitivity of Ewing's sarcoma to TRAIL-induced apoptosis. *Cell Death Differ* **8:** 506-14.

Loeder S, Fakler M, Schoeneberger H, Cristofanon S, Leibacher J, Vanlangenakker N *et al* (2012). RIP1 is required for IAP inhibitor-mediated sensitization of childhood acute leukemia cells to chemotherapy-induced apoptosis. *Leukemia* **26:** 1020-9.

Muhlethaler-Mottet A, Bourloud KB, Auderset K, Joseph JM, Gross N (2004). Drug-mediated sensitization to TRAIL-induced apoptosis in caspase-8-complemented neuroblastoma cells proceeds via activation of intrinsic and extrinsic pathways and caspase-dependent cleavage of XIAP, Bcl-xL and RIP. *Oncogene* **23:** 5415-25.

Naumann I, Kappler R, von Schweinitz D, Debatin KM, Fulda S (2011). Bortezomib primes neuroblastoma cells for TRAIL-induced apoptosis by linking the death receptor to the mitochondrial pathway. *Clin Cancer Res* **17:** 3204-18.

Petak I, Douglas L, Tillman DM, Vernes R, Houghton JA (2000). Pediatric rhabdomyosarcoma cell lines are resistant to Fas-induced apoptosis and highly sensitive to TRAIL-induced apoptosis. *Clin Cancer Res* **6:** 4119-27.

Pingoud-Meier C, Lang D, Janss AJ, Rorke LB, Phillips PC, Shalaby T *et al* (2003). Loss of caspase-8 protein expression correlates with unfavorable survival outcome in childhood medulloblastoma. *Clin Cancer Res* **9:** 6401-9.

Van Valen F, Fulda S, Schafer KL, Truckenbrod B, Hotfilder M, Poremba C *et al* (2003). Selective and nonselective toxicity of TRAIL/Apo2L combined with chemotherapy in human bone tumour cells vs. normal human cells. *Int J Cancer* **107**: 929-40.

Van Valen F, Fulda S, Truckenbrod B, Eckervogt V, Sonnemann J, Hillmann A *et al* (2000). Apoptotic responsiveness of the Ewing's sarcoma family of tumours to tumour necrosis factor-related apoptosis-inducing ligand (TRAIL). *Int J Cancer* **88**: 252-9.

Wang MJ, Liu S, Liu Y, Zheng D (2007). Actinomycin D enhances TRAIL-induced caspase-dependent and -independent apoptosis in SH-SY5Y neuroblastoma cells. *Neurosci Res* **59**: 40-6.

Wang S, Ren W, Liu J, Lahat G, Torres K, Lopez G *et al* (2010). TRAIL and doxorubicin combination induces proapoptotic and antiangiogenic effects in soft tissue sarcoma in vivo. *Clin Cancer Res* **16**: 2591-604.

Yang X, Thiele CJ (2003). Targeting the tumor necrosis factor-related apoptosis-inducing ligand path in neuroblastoma. *Cancer Lett* **197**: 137-43.

Chapter 17: South African medicinal plants inducing apoptosis in cancer cells: a treasure trove of anti-cancer agents? (van der Walt and Cronjé)

Nicola B. van der Walt and Marianne J. Cronjé*

Department of Biochemistry, University of Johannesburg, P.O. Box 524, Auckland Park, 2006, South Africa

Abstract

Apoptosis is important in the regulation of cell populations and impairment of this process contributes to carcinogenesis. Cancer remains one of the leading causes of death worldwide. Many cancer chemotherapies have unpleasant, toxic side effects. There remains a need to identify new anti-cancer drugs with fewer or no side effects. Medicinal plants have been used for the prevention and treatment of cancer for centuries by a wide variety of cultures. Many of the anti-cancer drugs currently in clinical use have been derived from plant sources. South Africa is home to a diverse range of plants with many being used for their medicinal properties, including those that are culturally used as 'anti-cancer' agents. Reviewed here is a selection of medicinal plants indigenous to southern Africa having anti-cancer activity due to the presence of compounds able to induce apoptosis. Clearly this research is in its infancy with much work required to not only identify active ingredients but also to elucidate mechanisms of action. Nevertheless, it is believed that these indigenous medicinal plants represent a wealth of unexplored novel anti-cancer drugs.

Introduction

Cancer is the leading cause of death in economically developed countries and the second leading cause of death in developing countries, including South Africa. In 2008, there were approximately 12.7 million reported cancer cases and about 7.6 million cancer-related deaths worldwide (Jemal et al., 2011). Cancer treatment typically includes the use of chemotherapeutic drugs. The mechanism of action of many of these drugs involves the induction of apoptosis (Kim et al., 2002).

For centuries, medicinal plants have been used for the treatment of cancer. They contain mixtures of different phytochemicals, which can affect the cell through a variety of different mechanisms. These phytochemicals may act individually, additively, or in synergy to produce a combined effect. A number of anti-cancer drugs in clinical use today have originated from medicinal plants. These include *inter alia* vinblastine, vincristine, etoposide, taxol and camptothecin (Cragg and Newman, 2005; Treasure, 2005; Mahomoodally, 2013; Millimouno *et al.*, 2014).

In South Africa, approximately 3000 plant species are regularly consumed for medicinal purposes (van Wyk, 2008). A number of these are reported to have anti-cancer activity. In the last 20 years there has been an increase in the number of publications relating to the South African medicinal plants with purported activity against cancer. This review aims to highlight a select few indigenous South African medicinal plants with anti-cancer activity that may be attributed to the induction of apoptosis. *Sutherlandia frutescens* is commonly known as the "cancer bush" (Fig. 1 a). It is a shrub which is indigenous to South Africa, Lesotho, southern Namibia as well as south-eastern Botswana (van Wyk and Albrecht, 2008). *S. frutescens* was first used by the Khoi-San people and early Cape settlers as a general medicine and to treat stomach complaints, 'internal cancers', wounds and infections. Today the plant is used by many cultural groups for a number of ailments including fevers, coughs and colds, diabetes, kidney 'conditions', stress and anxiety (van Wyk and Albrecht 2008). However, the plant is best known for its anti-cancer activity, hence its popular name 'cancer bush'. It is commercially available as tablets or as a tincture (Fig. 1 c and d).

Sutherlandia frutescens

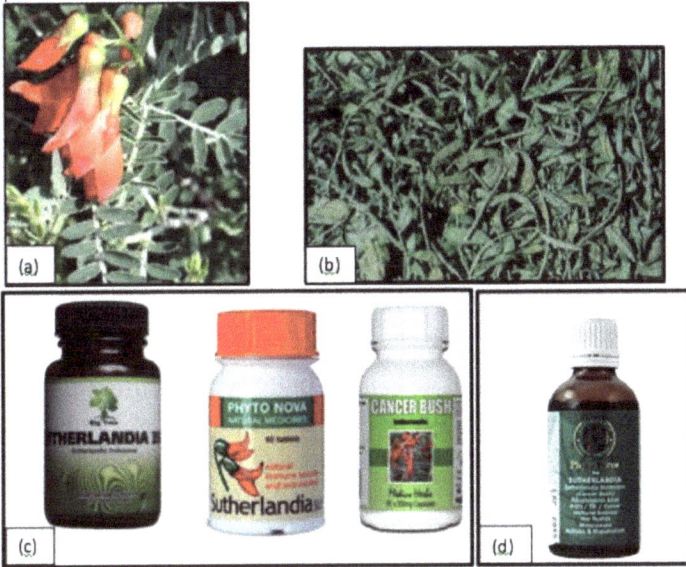

Fig. 1 (a) Flowering *S. frutescens* plant (from van Wyk and Albrecht, 2008); (b) dried *S. frutescens* leaves (from van Wyk and Albrecht, 2008); (c) *S. frutescens* tablets commercially available from Big Tree, Phyto Nova and Medico Herbs and (d) *S. frutescens* tincture from Phyto Force.

Previous studies have found that aqueous and ethanolic extracts of *S. frutescens* do, in fact, have anti-proliferative effects on cancerous cell lines including breast cancer (MFC-7, MDA-MB-468, MDA-MB-231) (Tai *et al.*, 2004; Steenkamp and Gouws, 2006; Stander *et al.*, 2007; Stander *et al.*, 2009;), cervical cancer (Caski) (Chinkwo, 2005), oesophageal cancer (SNO) (Skerman *et al.*, 2011) and leukaemia (Jurkat and HL 60) (Tai *et al.*, 2004) cell lines. The extracts have also demonstrated anti-proliferative effects on normal cell lines such as breast (MCF12A) (Stander *et al.*, 2009; Vorster *et al.,* 2012), blood (PBMC) (Korb *et al.*, 2010) and kidney (MDBK and LLC-PKI) (Phulukdaree *et al.*, 2010) cells. However, it has been shown that *S. frutescens* extracts were more toxic to cancerous cells than to non-cancerous cells (Stander *et al.*, 2009; Skerman *et al.*, 2011; Vorster *et al.,* 2012).

S. frutescens extracts appear to exert their anti-proliferative effects by inducing apoptosis. This was evidenced by typical apoptotic morphology (Chinkwo, 2005; Stander *et al.*, 2007; Stander *et al.*, 2009; Skerman *et al.*, 2011; Vorster *et al.*, 2012), translocation of phosphatidylserine residues (Chinkwo, 2005; Stander *et al.*, 2009; Korb *et al.*, 2010; Skerman *et al.*, 2011; Vorster *et al.*, 2012) , caspase-3 and/or -7 activity (Korb *et al.*, 2010; Phulukdaree *et al.*, 2010; Skerman *et al.*, 2011) mitochondrial membrane depolarization (Korb *et al.*, 2010; Phulukdaree *et al.*, 2010) and the expression of apoptosis-related genes (Stander *et al.*, 2007).

One of the phytochemicals found in *S. frutescens* is L-canavanine (Fig. 2). The levels of L-canavanine in the leaves of the different species and populations of *Sutherlandia* have been found to vary between 0.42 and 14.5 mg/g (Moshe, 1998).

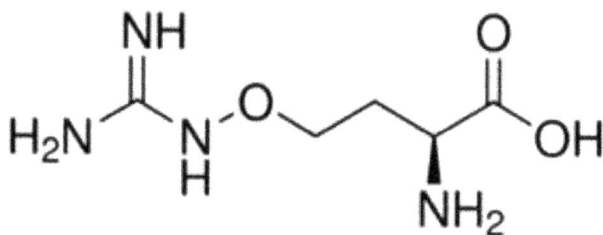

Fig. 2 Chemical structure of L-canavanine

L-canavanine was shown to be cytotoxic to a variety of cancerous cell lines including lung (A549) (Ding *et al.* 1999), uterine (MES-SA, Dx-5) (Worthen *et al.*, 1998), leukaemia (K562, K562-R7 and Jurkat) (Worthen *et al.*, 1998; Jang *et al.*, 2002), pancreatic (MIA PaCa-2) (Swaffar *et al.*, 1994) and breast (Walker 256) (Kruse *et al.*, 1958) cancer cell lines. L-canavanine was reported to induce apoptosis in Jurkat cells via a cytochrome *c*-independent caspase-3 activation pathway (Jang *et al.*, 2002).

Hypoxis species

The Hypoxidaceae family of plants is one of the most popularly used in southern Africa. Hypoxis hemerocallidea is sometimes referred to as the "African potato" (Fig. 3 a). This species is widely distributed in the eastern part of southern Africa, from the Eastern Cape of South Africa to Botswana and Mozambique. Hypoxis stellipilis and Hypoxis sobolifera var sobolifera are endemic to South Africa (Ncube et al., 2013).

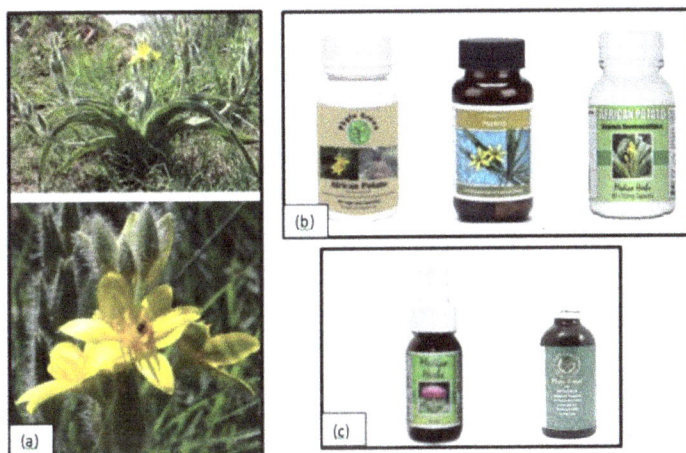

Fig. 3 (a) *H. hemerocallidea* plant (Hölscher, 2009); (b) *H. hemerocallidea* tablets and capsules available from Phyto Green, Afrigenics and Medico Herbs and (c) *H. hemerocallidea* tincture from Phyto Force.

H. hemerocallidea has been used to treat a variety of ailments including treatment for tuberculosis, cancer, diabetes, high blood pressure, urinary tract infections, headaches, stomach complaints and dermatitis. It is also consumed as a general tonic for good health (van Wyk, 2008; Ncube *et al.*, 2013). *H. hemerocallidea* is commercially available in tablet form and as a tincture (Fig 3. b and c).

H. hemerocallidea and *H. sobolifera* var *sobolifera* extracts were found to exhibit cytotoxic effects against cervical (HeLa) (Boukes and van de Venter, 2011), colorectal (HT-29) (Boukes and van de Venter, 2011) and breast (MCF-7) (Steenkamp and Gouws,

2006; Boukes and van de Venter, 2011) cancer cells. *H. stellipilis*, however, did not exhibit any cytotoxicity. The *H. sobolifera* var *sobolifera* extract caused cell cycle arrest at the late stages of G1 and/or early S phase. An increase in caspase-7 activity and as well as DNA fragmentation were also observed, indicating the induction of apoptosis by *H. sobolifera* var *sobolifera* extract (Boukes and van de Venter, 2011).

Hypoxis species contain varying levels of hypoxoside and β-sitosterol . *H. sobolifera* var. *sobolifera* was reported to contain the highest levels of β-sitosterol (74.69 µg/ 5 mg chloroform extract), while *H. hemerocallidea* was richest in hypoxoside (12.27 µg/5 mg chloroform extract) (Boukes *et al.*, 2008).

Hypoxoside is a non-toxic diglucoside. In the presence of β-glucosidase, hypoxoside is converted to cytotoxic rooperol (see Fig. 4). Rooperol has patented anti-cancer activity (Drewes and Liebenberg, 1983). In melanoma cells (BL6 and UCT-Mel 1), rooperol was found to cause disturbances in the chromosome structural integrity and segregation during mitosis. It also led to an increase in vacuole formation in the cytoplasm and the appearance of blebs on the outer membrane. Some cells were also observed to have condensed cytoplasm and chromatin (Albrecht *et al.*, 1995). Further studies revealed that rooperol acts as an inducer of apoptosis in cancer cells (Boukes *et al.*, 2009).

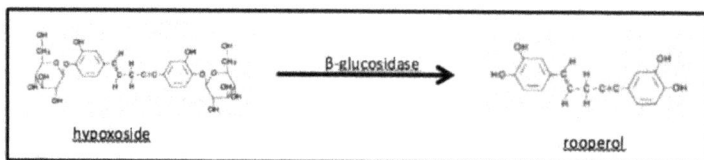

Fig. 4 The conversion of hypoxoside to rooperol in the presence of β-glucosidase (extracted from Albrecht *et al.*, 1995).

Interestingly, β-sitosterol was also found to induce apoptosis in cancer cells, activating both the intrinsic and extrinsic pathways (Awad *et al.*, 2003; Awad *et al.*, 2007). The chemical structure is indicated in Fig. 5.

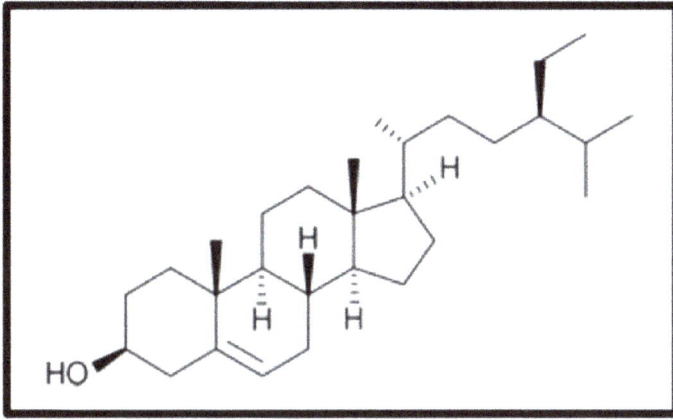

Fig. 5 Chemical structure of β-sitosterol

Centella asiatica

C. asiatica, or commonly known as Pennywort (Fig. 6), is indigenous to South Africa and can be found in all nine provinces. It is also distributed in Southeast Asia, Sri Lanka, parts of China, in the western South Sea Islands, Madagascar, in the southeast of the U.S.A., Mexico, Venezuela and Columbia, as well as in the eastern regions of South America (van Wyk *et al.*, 1997; Brinkhaus *et al.*, 2000).

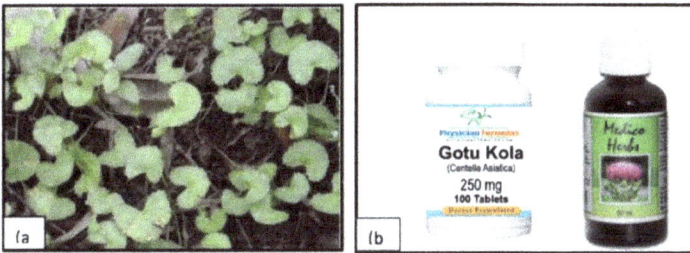

Fig. 6 (a) *C. asiatica* plant (from African Traditional Medicine pharmacopoeia and resources for research, no date); (b) *C. asiatica* tablets from Advanced Physician Formulas and (c) *C. asiatica* tincture from Medico Herbs.

The main use of *C. asiatica* is for wound healing and the treatment of skin diseases including leprosy. It has also used for the treatment of tuberculosis, cancer, inflammation, asthma, hypertension, rheumatism, syphilis and epilepsy (van Wyk *et al.*, 1997; Brinkhaus *et al.*, 2000). *C. asiatica* is commercially available as tablets or as a tincture (Fig. 6 b).

Extracts of *C. asiatica* were found to be cytotoxic to against human breast cancer (MDA-MB-231 and MCF-7) (Babykutty *et al.*, 2009; Pittella *et al.*, 2009), human hepatocellular (Hep G2) (Hussin *et al.*, 2014), mouse melanoma (B16F1) (Pittella *et al.*, 2009) and rat glioma (C6) (Pittella *et al.*, 2009) cell lines but not to human lung carcinoma (A549) (Pittella *et al.*, 2009), normal human liver (Chang) (Hussin *et al.*, 2014) or normal hamster kidney (BHK-21) (Pittella *et al.*, 2009) cells. A methanolic extract of *C. asiatica* was found to induce apoptosis in MCF-7 cells as evidenced by morphology characteristics, nuclear condensation, phosphatidylserine translocation and the loss of mitochondrial membrane integrity (Babykutty *et al.*, 2009).

A number of the phytochemicals found in *C. asiatica* have documented anti-cancer activity. These include the saponin-containing triterpene acids, asiatic acid and madecassic acid, as well as their sugar esters, asiaticoside and madecassoside (Fig. 7) (Brinkhaus *et al.*, 2000; Hussin *et al.*, 2014).

Asiatic acid (Fig. 7a) reduced cellular viability of a number of cancerous cell lines including hepatocarcinoma (HepG2) (Lee *et al.*, 2002), colon cancer (SW480) (Tang *et al.*, 2009) breast cancer (MCF-7 and MDA-MB-231) (Hsu *et al.*, 2005) glioblastoma (U-87) (Cho *et al.*, 2006) and melanoma (SK-MEL-2)(Park *et al.*, 2005). It also resulted in the translocation of phosphatidylserine (Park *et al.*, 2005; Cho *et al.*, 2006) and nuclear condensation (Tang *et al.*, 2009). Furthermore, treatment with asiatic acid caused the protein levels of Bax (Hsu *et al.*, 2005; Park *et al.*, 2005) and p53 to increase (Lee *et al.*, 2002) and those of Bcl-2 and Bcl-XL to decrease (Hsu *et al.*, 2005). Increased mitochondrial membrane permeability was observed (Cho *et al.*, 2006; Tang *et al.*, 2009) and cytochrome *c* was released from the mitochondria (Hsu *et al.*, 2005; Tang *et al.*, 2009). Caspase-9 and caspase-3 were activated (Hsu *et al.*,

2005; Park *et al.*, 2005; Cho *et al.*, 2006; Tang *et al.*, 2009) and cleaved poly(ADP-ribose) polymerase was detected (Tang *et al.*, 2009). Together, these results indicate that asiatic acid triggers the intrinsic mitochondrial apoptotic pathway in cancer cells.

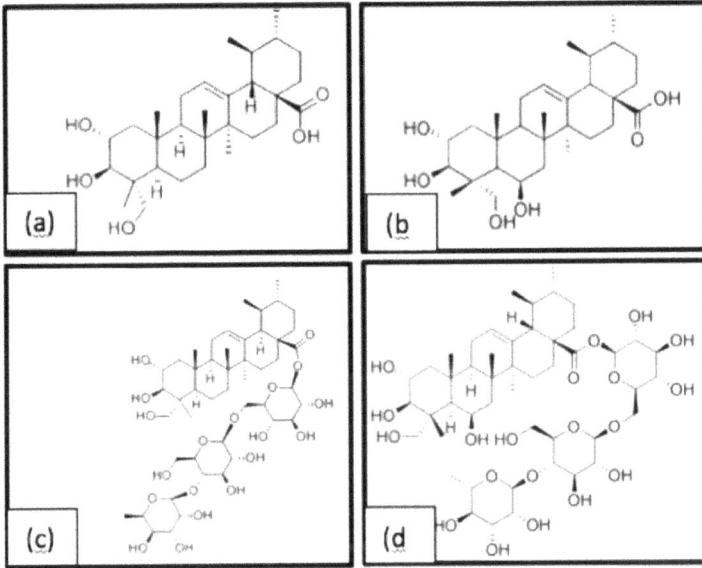

Fig. 7 Chemical structures of (a) asiatic acid; (b) madecassic acid; (c) asiaticoside and (d) madecassoside

Madecassic acid (Fig. 7b) has been reported to inhibit the growth of colon cancer (C26) cells *in vitro* as well as *in vivo*. This inhibition was ascribed to the induction of apoptosis (Zhang *et al.*, 2014).

Madecassoside (Fig. 7d) was found to induce instrinsic, mitochondrion-dependent apoptosis in keloid fibroblasts. This was evidenced by the translocation of phosphatidylserine, depolarization of the mitochondrial membrane potential, increased Bax and decreased Bcl-2 mRNA expression as well as protein levels and increased caspase-9 and caspase-3 protein levels (Song *et al.*, 2011).

Tulbaghia species

T. alliaceae and *T. violacea* are two species of wild garlic that are indigenous to South Africa (Fig. 8 a and b). They occur mainly in the Eastern Cape and Southern KwaZulu-Natal regions of South Africa (van Wyk *et al.*, 1997).

Fig. 8 (a) *T. alliaceae* flowers (from South African National Biodiversity Institute, no date); (b) flowering *T. violacea* plant (from Harris, 2004); (c) *T. violacea* tincture from Medico Herbs

Tulbaghia species have been reportedly used in the treatment fever and colds, asthma, tuberculosis, stomach problems and esophageal cancer (van Wyk *et al.*, 1997). *T. violacea* is commercially available as a tincture (Fig. 8 c).

Aqueous and chloroform extracts of *T. alliacea* were found to induce apoptosis in breast cancer (MCF-7), leukaemia (Jurkat) and bone osteosarcoma (MG63) cell lines. Treatment of Jurkat cells with *T. alliacea* extracts resulted in mitochondrial depolarization, caspase-3 cleavage and DNA fragmentation. Increased gene expression levels of *Bax*, *caspase-3* and *caspase-9* were also observed, suggesting that *T. alliacea* extracts induced apoptosis via the mitochondrial pathway (Thamburan, 2009).

Methanol extracts of *T. violacea* inhibited the growth of breast (MCF-7), esophageal (WHCO3), colon (HT29) and cervical (HeLa) cancers. These growth inhibitory effects were found to be a result of the induction of apoptosis (Bungu *et al.*, 2006).

A crude aqueous extract of *T. violacea* demonstrated cytotoxicity against breast cancer (MCF-7), cervical cancer (HeLa) and normal hamster (CHO) cells but not against oral squamous cell carcinoma (H157) or osteosarcoma (MG63) cells. Cell death induced by the *T. violacea* aqueous extract was characterized by cell shrinkage, translocation of phosphatidylserine, activation of caspase-3 and fragmentation of genomic DNA, indicating the occurrence of apoptosis (Lyantagaye, 2013a).

Some of the main phytochemicals identified in *T. violacea* include volatile sulfur-containing flavor compounds, which are responsible for the characteristic smell and taste in *Tulbaghia* species (Aremu and van Staden, 2013). Activity-guided fractionation of an aqueous extract of *T. violacea* led to the purification and chemical structure characterization of three apoptosis-inducing phytochemicals. These were identified as methyl-α-D-glucopyranoside, D-fructofuranose-β(26)-methyl-α-D-glucoyranoside and β-D-fructofuranosyl-(26)-α-D-glucopyranoside (Fig. 9) (Lyantagaye, 2013a; Lyantagaye, 2013b). Treatment of CHO cells with any one of these phytochemicals resulted in apoptosis, characterized by cell shrinkage, translocation of membrane phospholipids phosphatidylserine, loss of mitochondrial membrane potential, caspase-3 activity and genomic DNA fragmentation (Lyantagaye, 2014).

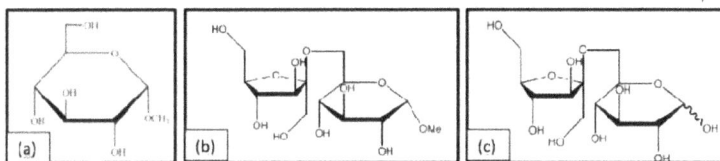

Fig. 9 Chemical structures of (a) methyl-α-D-glucopyranoside; (b) D-fructofuranose-β(26)-methyl-α-D-glucoyranoside and (c) β-D-fructofuranosyl-(26)-α-D-glucopyranoside.

Kedrostis foetidissima

Fig. 10 *K. foetidissima* plant (from bihrmann.com, no date)

K. foetidissima is called "umafuthasimba" in Zulu and is distributed throughout South Africa (South African National Biodiversity Institute (SANBI)) (Fig. 10). Extracts of *K. foetidissima* have exerted pro-apoptotic activity against breast cancer (MCF-7 and YMB-1) cells (Choene, 2012a). As a member of the Curcubitaceae plant family, *K. foetidissima* contains tetracyclic triterpenoids, called cucurbitacins. Cucurbitacins B, D, E and I have been identified in *K. foetidissima* (Fig. 11)

(Choene, 2012a). All of these compounds have documented anti-cancer activity.

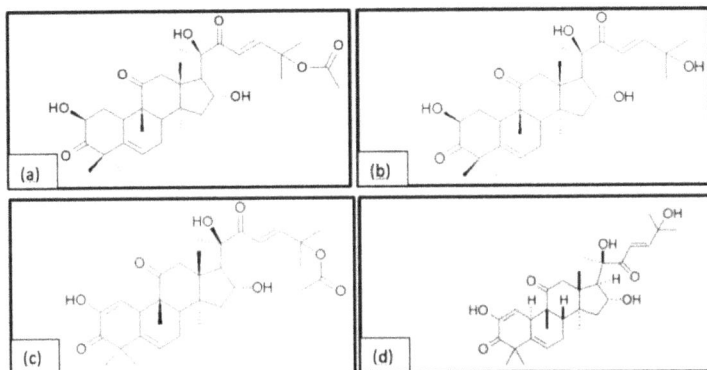

Fig. 11 Chemical structures of (a) Cucurbitacin B; (b) Cucurbitacin D; (c) Cucurbitacin E and (d) Cucurbitacin I

Cucurbitacin B has demonstrated cytotoxic activity against a variety of breast cancer (MCF-7, MDA-MB-231, MDA-MB-453, T47D, BT474, SK-BR-3 and ZR-75-1) (Wakimoto et al., 2008) and pancreatic cancer (MiaPaCa-2, PL45, and Panc-1, SU86.86 and AsPC-1, Panc-03.27 and Panc-10.05) (Thoennissen et al., 2009) cell lines. It was found to induce intrinsic apoptosis involving the inhibition of the JAK/STAT pathway (Thoennissen et al., 2009). Furthermore, cucurbitacin B was shown to inhibit the growth of MDA-MB-231 tumors in nude mice (Wakimoto et al., 2008). In another study, cucurbitacin D inhibited the growth of endometrial cancer (HHUA, HEC59 and Ishikawa) and ovarian cancer (SK-OV-3, OVCAR- 3, and TOV-112D) cells and triggered the intrinsic apoptotic pathway (Ishii et al., 2013). Additionally, cucurbitacin E was found to be cytotoxic to breast cancer (MCF-7, TNBC, MDA-MB-468), prostate cancer (PC-3), gastric cancer (NCI-N87) (Kong et al., 2014) and oral squamous cell carcinoma (SAS) cells but not to normal lung (MRC-5) or normal fibroblast (HS68) cells (Hung et al., 2013). It was further shown that cucurbitacin E induced intrinsic caspase-dependent apoptosis (Hung et al., 2013; Kong et al., 2014). Lastly, cucurbitacin I exhibited cytotoxicity against a lung cancer (A549, Calu-1) and breast cancer (MDA-MB-468) cell line. Cucurbitacin I was only able to induce apoptosis in cells that expressed constitutively activated tyrosine-phosphorylated STAT3 (Blaskovich *et al.*, 2003).

Artemisia afra

A. afra is commonly referred to as the African wormwood and is one of the most popular and commonly used herbal medicines in southern Africa (Fig. 12). It is widely distributed in Africa and occurs in Kenya, Tanzania, Uganda, Ethiopia, Namibia, Zimbabwe and South Africa (Liu *et al.*, 2009).

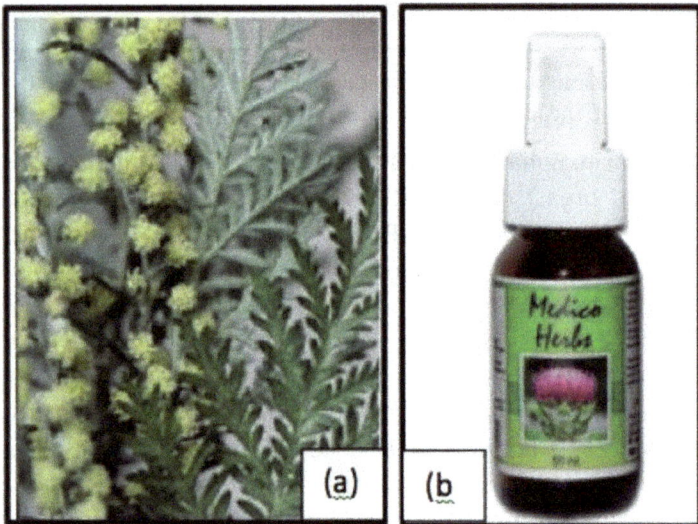

Fig. 12 (a) Flowering *A. afra* plant (from van Wyk, 2011); (b) tincture of *A. afra* from Medico Herbs

A. afra has been used in the treatment of a wide variety of ailments such as colds, influenza, cough, headaches, chills, stomach ailments gout, asthma, malaria, diabetes, bladder and kidney disorders, heart inflammation, rheumatism and is also used as a purgative (van Wyk, 2008; Liu et al., 2009). A tincture of A. afra is commercially available (Fig. 12 b).

Organic extracts of A. afra were shown to be cytotoxic to a number of cancer cell lines including non-small cell lung cancer (NCI-H522), melanoma (SK-MEL-5 and UACC62), colon (HT29), renal (TK10), breast (MCF7) (Fouche et al., 2008), cervical (HeLa) and leukemia (U937) cell lines (Spies et al., 2013). An ethanolic extract of A. afra was found to cause caspase-dependent apoptosis involving both the extrinsic and intrinsic pathways in HeLa and U937 cells (Spies et al., 2013).

Myrothamnus flabellifolius

M. flabellifolius is referred to as the resurrection plant (Fig. 13). It occurs in the northern part of South Africa and can extend as far north as Kenya (van Wyk *et al.*, 1997).

Fig. 13 *M. flabellifolius* plant (from SANBI, 2013)

Infusions and decoctions of *M. flabellifolius* are used to treat colds, respiratory ailments, influenza, kidney disorders, backache, hemorrhoids and gingivitis. Chest pains and asthma are treated with smoke from the burning leaves (van Wyk *et al.*, 1997; Dhillon *et al.*, 2014).

A methanolic extract of *M. flabellifolius* was found to be more cytotoxic to human leukemic cells (HL-60) compared to non-leukemic lymphocytes (TK6). *M. flabellifolius* induced apoptosis in a caspase-dependent manner (Dhillon *et al.*, 2014).

Commelina benghalensis

C. benghalensis is also known as Benghal dayflower, tropical spiderwort or wandering Jew (Fig. 14). It can be found in tropical and subtropical Asia and Africa. It is widely distributed in the eastern and northern parts of South Africa (Mbazima *et al.*, 2008).

Fig. 14 Flowers and leaves of *C. benghalensis* (from CABI.org)

C. benghalensis may be used to treat burns, skin growths, leprosy and opthalmia. The plant is also believed to have diuretic, anti-inflammatory and laxative properties (Mbazima *et al.*, 2008; Ibrahim *et al.*, 2010).

Extracts of *C. benghalensis* have been found to induce apoptosis in Jurkat T- cells. This was associated with increased Bax mRNA and protein levels and decreased Bcl-2 mRNA and protein levels. Increased p53 protein levels were also observed (Mbazima *et al.*, 2008; Lebogo *et al.*, 2014)

Euphorbia mauritanica

E. mauritanica is locally referred to as yellow milkbush or jackal's food (Fig. 15). It can be found in the Northern Cape, Eastern Cape, Western Cape, Free State and KwaZulu-Natal provinces of South Africa (SANBI).

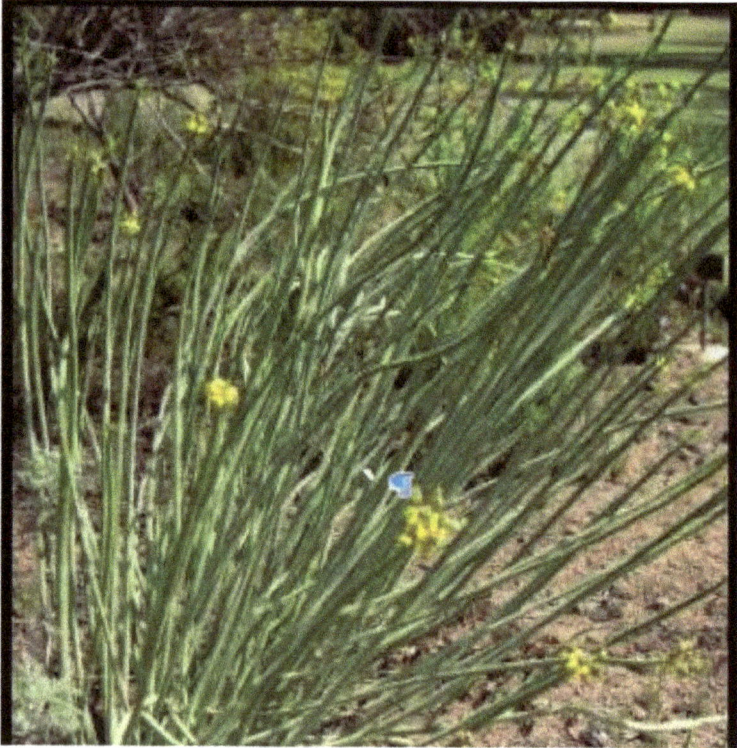

Fig. 15 Flowering *E. mauritanica* (from Bergh, 2006)

E. mauritanica exerted cytotoxic effects against breast (MCF-7) and cervical (HeLa) cancer cells as well as normal hamster ovary (CHO) cells. *E. mauritanica* extract was found to cause phosphatidylserine translocation, caspase-3 activation and DNA fragmentation in CHO cells indicating the induction of apoptosis (Essack, 2006; Choene, 2012b).

Cynanchum africanum

Fig. 16 Flowering *C. africanum* (from operationwildflower)

C. africanum or "bobbejaantou" in Afrikaans (directly translated as 'baboon rope') (Fig. 16) occurs in the Northern, Eastern and Western Cape of South Africa (SANBI). An aqueous extract of *C. africanum* was found to be cytotoxic to breast (MCF-7), cervical (HeLa) and hamster ovary (CHO) cells due to the induction of apoptosis (Essack, 2006).

Elytropappus rhinocerotis

Fig. 17 *E. rhinocerotis* plant (from SANBI)

E. rhinocerotis, or rhinoceros Bush, is found in the Northern, Eastern and Western Cape provinces of South Africa and into Namibia (SANBI) (Fig. 17). The young branches are typically infused and used to treat indigestion, dyspepsia, ulcers and stomach cancers (van Wyk *et al.*, 1997). An aqueous extract of *E. rhinocerotis* showed pro-apoptotic effects against breast (MCF-7), cervical (HeLa) and hamster ovary (CHO) cells (Essack, 2006).

Summary

Medicinal plants are a valuable source of new drugs, drug leads and chemical entities (Balunas and Kinghorn, 2005). Many of the cancer chemotherapeutic drugs currently in clinical use have been derived from plants (Cragg and Newman, 2005; Khazir *et al.*, 2014). Southern Africa is home to more than 30 000 species of higher plants (van Wyk *et al.*, 1997) and therefore represents a vast potential source of new anti-cancer drugs. The medicinal plants reviewed here show promising apoptosis-inducing anti-cancer activity. However, much research is needed before the full potential of these plants can be realized.

The cytotoxicity of these plants on normal primary cells needs to be established as much of the research to date has focused on the effects of the medicinal plants and/or their phytochemicals on cancer cell lines. Future studies should include further elucidation of the mechanisms of action of these plants. Although many of the medicinal plants mentioned in this review are commercially available for consumption, their safety and efficacy has not officially been established and their use is largely based on anecdotal evidence gained from traditional uses. Potential toxic or other side effects should be evaluated by means of rigorous clinical trials. In order to identify potential new anti-cancer compounds from these medicinal plants phytochemical analysis needs to be conducted. Furthermore, research into the potential synergy of the phytochemicals is needed as it may reveal combinations of bioactive compounds that maximize anti-cancer activity while minimizing side effects. However, several aspects of this research remain challenging in a socio-economically burdened country such as South Africa that prioritizes research on communicable diseases such as HIV-AIDS and tuberculosis (Schaay and Sanders, 2008).

References

African Traditional Medicine pharmacopoeia and resources for research (no date). A monograph with pharmacopoeia information on Centella asiatica. African Traditional Medicine, Pharmacopoeia and resources for research. Accessed on 20 October 2014 from http://www.tm.ukzn.ac.za/content/centella-asiatica

Albrecht C. F.; Theron E. J. and Kruger P. B. (1995) Morphological characterisation of the cell-growth inhibitory activity of rooperol and pharmacokinetic aspects of hypoxoside as an oral prod rug for cancer therapy. South African Medical Journal **85**: 853-860

Aremu A. O. and van Staden J. (2013) The genus Tulbaghia (Alliaceae)--a review of its ethnobotany, pharmacology, phytochemistry and conservation needs. Journal of Ethnopharmacology **149**: 387-400

Awad A.B.; Roy R. and Fink C. S. (2003) Beta-sitosterol, a plant sterol, induces apoptosis and activates key caspases in MDA-MB-231 human breast cancer cells. Oncology Reports **10**: 497-500

Awad A.B.; Chinnam M.; Fink C. S. and Bradford P. G. (2007) β-Sitosterol activates Fas signaling in human breast cancer cells. Phytomedicine **14**: 747–754

Babykutty S.; Padikkal J.; Sathiadevan P. P.; Vijayakurup V.; Karedath T.; Azis A.; Srinivas P. and Gopala S. (2009) Apoptosis induction of *Centella asciatica* on human breast cancer cells. African Journal of Traditional, Complementary, and Alternative Medicine **6**: 9 – 16

Balunas M. J. and Kinghorn A. D. (2005) Drug discovery from medicinal plants. Life Sciences **78**: 431 – 441

Bergh N. (2006) Elytropappus rhinocerotis (L.f.) Less. Plantzafrica.com. Retrieved on 20 October 2014 from http://www.plantzafrica.com/plantefg/elytrorhino.htm

Bihrmann.com (no date) Kedrostis foetidissima. Retrieved on 20 October 2014 from http://www.bihrmann.com/caudiciforms/subs/ked-foe-sub.asp

Blaskovich M. A.; Sun J.; Cantor A.; Turkson J.; Jove R and Sebti S. M. (2003) Discovery of JSI-124 (Cucurbitacin I), a Selective Janus Kinase/Signal Transducer and Activator of Transcription 3 Signaling Pathway Inhibitor with Potent Antitumor Activity against Human and Murine Cancer Cells in Mice. Cancer Research **63**: 1270–1279

Boukes G. J.; van de Venter M. and Oosthuizen V. (2008) Quantitative and qualitative analysis of sterols/sterolins and hypoxoside contents of three Hypoxis (African potato) spp. African Journal of Biotechnology **7**: 1624-1629

Boukes G. J.; Daniels B. B.; Albrecht C. F. and van de Venter M. (2009) Cell Survival or Apoptosis: Rooperol's Role as Anticancer Agent. Oncology Research Featuring Preclinical and Clinical Cancer Therapeutics, **18**: 365-376

Boukes G. J. and van de Venter M. (2011) Cytotoxicity and mechanism(s) of action of Hypoxis spp. (African potato) against HeLa, HT-29 and MCF-7 cancer cell lines. Journal of Medicinal Plants Research **5**: 2766-2774

Brinkhaus B.; Lindner M.; Schuppan D. and Hahn E. G. (2000) Chemical, pharmacological and clinical profile of the East Asian medical plant Centella asiatica. Phytomedicine **7**: 427-448

Bungu L.; Frost C. L.; Brauns S. C. and van de Venter M. (2006) Tulbaghia violacea inhibits growth and induces apoptosis in cancer cells in vitro. African Journal of Biotechnology **5**: 1936-1943

CABI.org (no date) Commelina benghalensis datasheet. Retrieved on 20 October 2014 from http://www.cabi.org/isc/datasheet/14977#toPictures

Chinkwo K. A. (2005) Sutherlandia frutescens extracts can induce apoptosis in cultured carcinoma cells. Journal of Ethnopharmacology **98**: 163–170

Cho C.W.; Choi D.S.; Cardone M.H.; Kim C.W .; Sinskey A.J. and Rha C. (2006) Glioblastoma cell death induced by asiatic acid. Cell Biology and Toxicology **22**: 393–408

Choene M. and Motadi L. R. (2012a) Anti-Proliferative Effects of the Methanolic Extract of Kedrostis Foetidissima in Breast Cancer Cell Lines. Molecular Biology **1**: 1-5

Choene M. S. (2012b) Evaluating the effect of South African herbal extracts on breast cancer cells. A dissertation submitted in fulfilment of the requirements for the degree of Masters in Science in the School of Molecular and Cell Biology, University of the Witwatersrand, South Africa

Cragg G. M. and Newman D. J. (2005) Plants as a source of anti-cancer agents. Journal of Ethnopharmacology **100**: 72–79

Dhillon J.; Miller V. A.; Carter J.; Badiab A.; Tang C. N.; Huynh A. and Peethambaran B. Apoptosis-inducing potential of Myrothamnus flabellifolius, an edible medicinal plant, on human myeloid leukemia HL-60 cells. International Journal of Applied Research in Natural Products **7**: 28-32

Ding Y.; Matsukawa Y.; Ohtani-Fujita N.; Kato D.; Dao S.; Fujii T.; Naito Y.; Yoshikawa T.; Sakai T. and Rosenthal G. A. (1999) Growth Inhibition of A549 Human Lung Adenocarcinoma Cells by L-Canavanine Is Associated with p21/WAF1 Induction. Japanese Journal of Cancer Research **90**: 69–74

Drewes S. and Liebenberg R. W. (1987) Patent number 4, 644, 085

Edkins A. (no date) Cancer: not an African problem? Retrieved from http://www.scienceinafrica.com/biotechnology/health/cancer-not-african-problem on 4 Nov 2014

Essack M. (2006) Screening extracts of indigenous South African plants for the presence of anti-cancer compounds. A thesis submitted in fulfillment of the requirements for the degree of Masters in Science in the Department of Biotechnology, Faculty of Science, University of the Western Cape, South Africa

Fennell C.W.; Lindsey K.L.; McGawb L.J.; Sparg S.G.; Stafford G.I.; Elgorashi E.E.; Grace O.M. and van Staden J. (2004) Assessing African medicinal plants for efficacy and safety: pharmacological screening and toxicology. Journal of Ethnopharmacology 94: 205–217

Fouche G.; Cragg G. M.; Pillay P.; Kolesnikova N.; Maharaj V. J. and Senabe J. (2008) In vitro anticancer screening of South African plants. Journal of Ethnopharmacology 119: 455–461

Harris S. (2004) Tulgaghia violacea Harv. Plantzafrica.com. Retrieved on 20 October 2014 from http://www.plantzafrica.com/planttuv/tulbaghviol.htm

Hölscher B. (2009) Hypoxis hemerocallidea Fisch., C.A.Mey. and Avé-Lall. South African National Biodiversity Institute. Retrieved on 20 October 2014 from http://www.plantzafrica.com/planthij/hypoxishemero.htm

Hsu Y.-L.; Kuo P.-L.; Lin L.-T. and Lin C.C. (2005) Asiatic Acid, a Triterpene, Induces Apoptosis and Cell Cycle Arrest through Activation of Extracellular Signal-Regulated Kinase and p38 Mitogen-Activated Protein Kinase Pathways in Human Breast Cancer Cells. The Journal of Pharmacology and Experimental Therapeutics 313: 333–344

Hung C.-M.; Chang C.-C.; Lin C.-W.; Ko S.-Y. and Hsu Y.-C. (2013) Cucurbitacin E as Inducer of Cell Death and Apoptosis in Human Oral Squamous Cell Carcinoma Cell Line SAS. International Journal of Molecular Sciences 14: 17147-17156

Hussin F.; Eshkoor S. A.; Rahmat A.; Othman F. and Akim A. (2014) The centella asiatica juice effects on DNA damage, apoptosis and gene expression in hepatocellular carcinoma (HCC). BMC Complementary and Alternative Medicine 14:32-38

Ishii T.; Kira N.; Yoshida T. and Narahara H. (2013) Cucurbitacin D induces growth inhibition, cell cycle arrest, and apoptosis in human endometrial and ovarian cancer cells. Tumor Biology 34: 285–291

Jang M. H.; Jun D. Y.; Rue S. W.; Han K. H.; Park W. and Kim Y. H. (2002) Arginine antimetabolite L-canavanine induces apoptotic cell death in human Jurkat T cells via caspase-3 activation regulated by Bcl-2 or Bcl-XL. Biochemical and Biophysical Research Communications 295: 283–288

Jemal A., Bray F., Center M. M., Ferlay J., Ward E. and Forman D. (2011) Global Cancer Statistics. A Cancer Journal for Clinicians **61**: 69–90

Khazir J.; Mir B. A.; Pilcher L. and Riley D. L. (2014) Role of plants in anticancer drug discovery. Phytochemistry Letters **7**: 173–181

Kim R.; Tanabe K.; Uchida Y.; Emi M.; Inoue H. and Toge T. (2002) Current status of the molecular mechanisms of anticancer drug-induced apoptosis. Cancer Chemotherapy and Pharmacology **50**: 343–352

Kong Y.; Chen J.; Zhou Z.; Xia H.; Qiu M.-H. and Chen C. (2014) Cucurbitacin E Induces Cell Cycle G2/M Phase Arrest and Apoptosis in Triple Negative Breast Cancer. PLOS ONE **9**: e103760

Korb V. C.; Moodley D. and Chuturgoon A. A. (2010) Apoptosis-promoting effects of Sutherlandia frutescens extracts on normal human lymphocytes in vitro. South African Journal of Science **106**: 64 – 69

Kruse P. F.; White P. B. and Carter H. A. (1959) Incorporation of Canavanine into Protein of Walker Carcinosarcoma 256 Cells Cultured in Vitro. Cancer Research **19**: 122-125

Lebogo K. W.; Mokgotho M. P.; Bagla V. P.; Matsebatlela T. M.; Mbazima V.; Shai L. J. and Mampuru L. (2014) Semi-purified extracts of Commelina benghalensis (Commelinaceae) induce apoptosis and cell cycle arrest in Jurkat-T cells. BMC Complementary and Alternative Medicine **14**: 65 -76

Lee Y. S.; Jin D.-Q.; Kwon E. J.; Park S. H.; Lee E.-S.; Jeong T. C.; Nam D. H.; Huh K. and Kim J.-A. (2002) Asiatic acid, a triterpene, induces apoptosis through intracellular Ca^{2+} release and enhanced expression of p53 in HepG2 human hepatoma cells. Cancer Letters **186**: 83–91

Liu N. Q.; van der Kooy F. and Verpoorte R. (2009) Artemisia afra: A potential flagship for African medicinal plants? South African Journal of Botany **75**: 185–195

Lyantagaye S. L. (2013a) Methyl-α-D-glucopyranoside from Tulbghia violacea extract induces apoptosis in vitro in cancer cells. Bangladesh Journal of Pharmacology **8**: 93-101

Lyantagaye S. L. (2013b) Two new pro-apoptotic glucopyranosides from Tulbaghia violacea. Journal of Medicinal Plants Research **7**: 2214-2220

Lyantagaye S. L. (2014) Characterization of the Biochemical Pathway of Apoptosis Induced by D-glucopyranoside Derivatives from Tulbaghia violacea. Annual Research & Review in Biology **4**: 962-977

Mahomoodally M. F. (2013) Traditional Medicines in Africa: An Appraisal of Ten Potent African Medicinal Plants. Evidence-Based Complementary and Alternative Medicine Volume 2013, Article ID 617459, DOI: 10.1155/2013/617459

Mbazima V.G.; Mokgotho M. P.; February F.; Rees D. J. G. and Mampuru L. J. (2008) Alteration of Bax-to-Bcl-2 ratio modulates the anticancer activity of methanolic extract of Commelina benghalensis (Commelinaceae) in Jurkat T cells. African Journal of Biotechnology **7**: 3569-3576

Millimouno F. M.; Dong J.; Yang L.; Li J. and Li X. (2014) Targeting Apoptosis Pathways in Cancer and Perspectives with Natural Compounds from Mother Nature. Cancer Prevention Research **7** (11): 1081-107

Moshe D. (1998) A Biosystematic Study of the Genus *Sutherlandia* Br. R. (Fabaceae, Galegeae). A dissertation presented in fulfilment of the requirements for the degree Magister Scientiae in Botany at the Faculty of Natural Sciences of the Rand Afrikaans University, South Africa

Ncube B.; Ndhlala A. R.; Okem A. and van Staden J. (2013) Hypoxis (Hypoxidaceae) in African traditional medicine. Journal of Ethnopharmacology **150**: 818–827

Operation Wildflower (no date) Cynanchum africanum. Retrieved on 20 October 2014 from http://www.operationwildflower.org.za/index.php/component/joomgallery/cl imbers/cynanchum-africanum-bobbejaantou-il-2-3063

Park B. C.; Bosire K. O.; Lee E.-S.; Lee Y. S. and Kim J.A. (2005) Asiatic acid induces apoptosis in SK-MEL-2 human melanoma cells. Cancer Letters **218**: 81–90

Pittella F.; Dutra R. C.; Junior D. D.; Lopes M. T. P. and Barbosa N. R. (2009) Antioxidant and Cytotoxic Activities of Centella asiatica (L) Urb. International Journal of Molecular Sciences **10**: 3713-3721

Phulukdaree A.; Moodley D. and Chuturgoon A. A. (2010) The effects of Sutherlandia frutescens extracts in cultured renal proximal and distal tubule epithelial cells. South African Journal of Science **160**: 54–58

Schaay N; Sanders D. (2008) International Perspective on Primary Health Care Over the Past 30 Years. In: Barron P, Roma-Reardon J, editors. South African Health Review 2008. Durban: Health Systems Trust. Pp 212, 217

Skerman N.B.; Joubert A. M. and Cronjé M. J. (2011) The apoptosis inducing effects of Sutherlandia spp. extracts on an oesophageal cancer cell line. Journal of Ethnopharmacology **137**: 1250–1260

Song J.; Dai Y.; Bian D.; Zhang H.; Xu X.; Xia Y. and Gong Z. (2011) Madecassoside Induces Apoptosis of Keloid Fibroblasts Via a Mitochondrial-Dependent Pathway. Drug Development Research **72**: 315–322

South African National Biodiversity Institute (no date) Tulbaghia rhizoma. Plantzafrica.com. Retrieved on 20 October 2014 from http://www.plantzafrica.com/medmonographs/tulbaghali.pdf

South African National Biodiversity Institute (2013) Myrothamnus flabellifolius Welw. Plantzafrica.com. Retrieved on 20 October 2014 from http://www.plantzafrica.com/plantklm/myrothamflabell.htm

Spies L.; Koekemoer T. C.; Sowemimo A. A.; Goosen E. D. and van de Venter M. (2013) Caspase-dependent apoptosis is induced by Artemisia afra Jacq. ex Willd in a mitochondria-dependent manner after G2/M arrest. South African Journal of Botany **84**: 104–109

Stander B. A.; Marais S.; Steynberg T. J.; Theron D.; Joubert F.; Albrecht C. and Joubert A. M. (2007) Influence of Sutherlandia frutescens extracts on cell numbers, morphology and gene expression in MCF-7 cells. Journal of Ethnopharmacology **112**: 312–318

Stander A.; Marais S.; Stivaktas V.; Vorster C.; Albrecht C.; Lottering M.-L. and Joubert A. M. (2009) In vitro effects of Sutherlandia frutescens water extracts on cell numbers, morphology, cell cycle progression and cell death in a tumorigenic and a non-tumorigenic epithelial breast cell line. Journal of Ethnopharmacology **124**: 45–60

Steenkamp V. and Gouws M. C. (2006) Cytotoxicity of six South African medicinal plant extracts used in the treatment of cancer. South African Journal of Botany **72**: 630–633

Swaffar D. S.; Ang C. Y.; Desal P. B. and Rosenthal G. A. (1994) Inhibition of the Growth of Human Pancreatic Cancer Cells by the Arginine Antimetabolite L-Canavanine. Cancer Research **54**: 6045 -6048

Tai J.; Cheung S.; Chan E. and Hasman D. (2004) In vitro culture studies of Sutherlandia frutescens on human tumor cell lines. Journal of Ethnopharmacology **93**: 9–19

Tang X.-L.; Yang X.-Y.; Jung Y.-J.; Kim S.-Y.; Jung A.-Y,; Choi D.-Y.; Park W.-C. and Park H. (2009) Asiatic Acid Induces Colon Cancer Cell Growth Inhibition and Apoptosis through Mitochondrial Death Cascade. Biological and Pharmaceutical Bulletin **32**: 1399—1405

Thamburan S. (2009) An investigation into the medicinal properties of Tulbaghia alliacea phytotherapy. A thesis in fulfillment of the requirements for the degree of Philosophiae Doctor in the Department of Pharmaceutical Sciences, School of Pharmacy, Faculty of Natural Sciences, at the University of the Western Cape, South Africa

Thoennissen N. H.; Iwanski G. B.; Doan N. B.; Okamoto R.; Lin P.; Abbassi S.; Song J. H.; Yin D.; Toh M.; Xie W. D.; Said J. W. and Koeffler H. P. (2009) Cucurbitacin B Induces Apoptosis by Inhibition of the JAK/STAT Pathway and Potentiates Antiproliferative Effects of Gemcitabine on Pancreatic Cancer Cells. Cancer Research **69**: 5876-5884

Treasure J. (2005) Herbal medicine and cancer: an introductory overview. Seminars in Oncology Nursing **21**: 177-183

Van Wyk B.-E. (2008) A broad review of commercially important southern African medicinal plants. Journal of Ethnopharmacology **119**: 342–355

Van Wyk B-E. and Albrecht C. (2008) A review of the taxonomy, ethnobotany, chemistry and pharmacology of Sutherlandia frutescens (Fabaceae). Journal of Ethnopharmacology **119**: 620–629

Van Wyk B.-E. (2008) A broad review of commercially important southern African medicinal plants. Journal of Ethnopharmacology **119**: (2008) 342–355

Van Wyk B.-E.; Oudtshoorn B and Gericke N. (1997) Medicinal plants of South Africa. Briza Publications, Pretoria, South Africa. Pp 64, 118, 176

Vorster C.; Stander A. and Joubert A. (2012) Differential signaling involved in Sutherlandia frutescens-induced cell death in MCF-7 and MCF-12A cells. Journal of Ethnopharmacology **140**: 123– 130

Wakimoto N.; Yin D.; O'Kelly J.; Haritunians T.; Karlan B.; Said J.; Xing H. and Koeffler H. P. (2008) Cucurbitacin B has a potent antiproliferative effect on breast cancer cells in vitro and in vivo. Cancer Science **99**: 1793–1797

Worthen D. R.; Chien L.; Tsuboi C. P.; Mu X. Y.; Bartik M. M. and Crooks P. A. (1998) L-Canavanine modulates cellular growth, chemosensitivity and P-glycoprotein substrate accumulation in cultured human tumor cell lines. Cancer Letters **132**: 229–239

Zhang H.; Zhang M.; Tao Y.; Wang G. and Xia B. (2014) Madecassic acid inhibits the mouse colon cancer growth by inducing apoptosis and immunomodulation. Journal of Balkan Union of Oncology **19**: 372-376

Index